ACS SYMPOSIUM SERIES **852**

Utilization of Greenhouse Gases

Chang-jun Liu, Editor
Tianjin University

Richard G. Mallinson, Editor
University of Oklahoma

Michele Aresta, Editor
University of Bari

Sponsored by the
ACS Division of Fuel Chemistry

Quality Seconds

American Chemical Society, Washington, DC

Library of Congress Cataloging-in-Publication Data

Utilization of greenhouse gases / Chang-jun Liu, editor, Richard G. Mallinson, editor, Michele Areste, editor ; sponsored by the ACS Division of Fuel Chemistry.

p. cm.—(ACS symposium series ; 852)

Includes bibliographical references and index.

ISBN: 978-0-8412-3827-5

1. Greenhouse gases—Congresses. 2. Carbon dioxide—Congresses. 3. Methane—Congresses. 4. Greenhouse gases—Congresses.

I. Liu, Chang-jun, 1964- II. Mallinson, Richard G., 1954- III. Aresta, M. (Michele), 1940- IV. American Chemical Society. Division of Fuel Chemistry. V. American Chemical Society. Meeting (223rd : 2002 : Orlando, Fla.) VI. Series.

TP244.G73U85 2003
628.5'32—dc21 2003043737

The paper used in this publication meets the minimum requirements of American National Standard for Information Sciences—Permanence of Paper for Printed Library Materials, ANSI Z39.48–1984.

Distributed by Oxford University Press

PRINTED IN THE UNITED STATES OF AMERICA

Foreword

The ACS Symposium Series was first published in 1974 to provide a mechanism for publishing symposia quickly in book form. The purpose of the series is to publish timely, comprehensive books developed from ACS sponsored symposia based on current scientific research. Occasionally, books are developed from symposia sponsored by other organizations when the topic is of keen interest to the chemistry audience.

Before agreeing to publish a book, the proposed table of contents is reviewed for appropriate and comprehensive coverage and for interest to the audience. Some papers may be excluded to better focus the book; others may be added to provide comprehensiveness. When appropriate, overview or introductory chapters are added. Drafts of chapters are peer-reviewed prior to final acceptance or rejection, and manuscripts are prepared in camera-ready format.

As a rule, only original research papers and original review papers are included in the volumes. Verbatim reproductions of previously publishcd papcrs arc not acccptcd.

ACS Books Department

Contents

Organic Synthesis with CO_2 as a Reactant

Biochemical Fixation of CO_2

Methane Valorization

Conversion of Methane and Other Greenhouse Gases via Plasmas or Microwave Heating

Application of Supercritical CO_2

Methane Combustion

Indexes

Preface

This book is based on the papers presented at the Symposium with the same title in the 223rd American Chemical Society (ACS) National Meeting, at Orlando, Florida, in April 2002. The Symposium focused on the utilization of greenhouse gases, principally methane and carbon dioxide, for the synthesis of more valuable chemicals or for the catalytic combustion of methane as an ultraclean fuel. As is widely discussed, a very serious issue of increasing importance to the world is the control of emissions of the major greenhouse gases, predominantly but not exclusively, carbon dioxide. With regard to the concept of emission control, the primary emphasis has been that carbon dioxide and other greenhouse gases are considered as a new class of pollutants. On this important issue, much negotiation and many arguments have been made. However, effective *and economic* methods to achieve major reductions of the net emissions of greenhouse gases, especially carbon dioxide, because it is the primary manmade contributor to the greenhouse effect, have not been found. On the other hand, it is recognized that nature uses carbon dioxide as a primary building block to provide all of the energy we consume every day. How exciting it is! Although replication of the perfection of nature can only be an aspiration, at least we can try our best to mimic the philosophy. In this regard, a significant worldwide effort has been made to find a feasible way to use the greenhouse gases. In this book, recent progress in the development of methods of utilization of greenhouse gases is presented. Focuses on the simultaneous utilization of greenhouse gases, innovative utilization technologies, and processes for the conversion of greenhouse gases are included.

The book's first section therefore begins with an overview of these concepts by covering the sources, separation, and utilization of carbon dioxide. The next section moves into simultaneous utilization of greenhouse gases (CH_4, CO_2, and NO_x). The simultaneous conversion of two greenhouse gases is to remove, eliminate, and convert one greenhouse gas by using another one as a co-reactant. Six papers are included in this book on this topic, including NO_x-catalyzed partial oxidation of CH_4 to syngas, drying reforming of methane to produce syngas, reaction of CH_4–CO_2 cofeed to produce liquid chemicals and partial reduction of SO_2 by CH_4. The present investigation will help us understand the reactive relationship between greenhouse gases and also has significant potential to afford new utilization technologies.

The next three to six sections are related to processes for conversion and innovative utilization technologies, which include biochemical fixation, plasma conversion, and microwave applications. The fundamental, economic, technical, and future development issues with these technologies are discussed. Reports of investigations on the use of the innovative technologies for the utilization of greenhouse gases are a typical feature of this book.

Supercritical carbon dioxide has been found to have many important applications. In the next section, one new application of supercritical CO_2 for the selective oxidation of alcohols over supported platinum catalysts is presented.

The book concludes with the next section that covers the utilization of methane for combustion. The catalytic combustion of methane has been considered to be one of the major techniques for reducing the emission of NO_x. The catalytic combustion of methane has also been investigated for the combustion of unburned methane effluents of natural gas fueled vehicles. A good catalyst is necessary for the cleanest methane combustion. The two contributions in this chapter show how the catalyst can be improved for better performance.

The editors acknowledge the ACS Division of Fuel Chemistry for the sponsorship of the symposium and also thank Dr. Robert Warzinski, the Program Chair of 2002 of the ACS Division of Fuel Chemistry for his instructions in making the symposium and therefore the book better. The instructions from Kelly Dennis, acquisition editor of the ACS Books Department, are also very appreciated. The efforts of Stacy VanDerWall, in acquisitions, in handling the correspondence with the authors and editors for completion are gratefully acknowledged. We also acknowledge Margaret Brown, Production Manager of the ACS Books Department for producing the book, lightly editing the titles, authors, affiliations, and the frontmatter of the book and making it ready for printing. We also appreciate the efforts of Jeanne Wynn in the Department of Chemistry of Tianjin University very much for her excellent job in correcting the use of English with the contributions from East Asia. Finally, we thank the efforts of the reviewers and the authors. With their great effort, we hope to have ensured the technical clarity and quality of papers.

We trust you will enjoy reading this book and find the papers informative and useful in dealing with the problems of greenhouse gases and developing a perspective for possible solutions for the tremendous challenge facing the world in this century

Chang-jun Liu
ABB Plasma Greenhouse Gas Chemistry Laboratory
State Key Laboratory of C_1 Chemistry and Technology
Tianjin University
Tianjin 300072, China
changliu@public.tpt.tj.cn

Richard G. Mallinson
Institute for Gas Utilization Technologies
School of Chemical Engineering and Materials Science
The University of Oklahoma
Norman, OK 73019
mallinson@ou.edu

Michele Aresta
Department of Chemistry and METEA Research Centre
University of Bari
70126 Bari, Italy
aresta@metea.uniba.it

General Overview

Chapter 1

Carbon Dioxide Utilization: Greening Both the Energy and Chemical Industry: An Overview

Michele Aresta

Department of Chemistry and METEA Research Centre, University of Bari, Via Celso Ulpiani 27, 70126 Bari, Italy (phone: 39 080 544 2430; fax 39 080 544 2429; email aresta@metea.uniba.it)

Carbon dioxide is the most abundant waste produced by human activities. Its further accumulation in the atmosphere must be avoided in order to prevent risky situations derived by the reinforcement of the greenhouse effect, to which CO_2 is estimated to be the major contributor. The energy production short- and medium-term scenarios, demonstrate that fossil fuels will be the major energy source for the coming 30-40 years. Therefore, our Society has to cope with the increase of the carbon dioxide emission, if more energy has to be used for ameliorating the standard of life of emergent and developing country populations. New technologies for carbon dioxide control are necessary, that imply its recovery from various sources. Effective separation technologies are available, and their cost is known to depend upon the concentration and quality of the source gas, and the required efficacy of separation. Also, the end use of recovered carbon dioxide depends on its quality. For example, use for the food (drink) sector, demands carbon dioxide characterized by a high degree of purity, without toxic pollutants. Only selected, pure sources can, thus, be used to this end. Supercritical carbon dioxide is now finding larger and larger utilization in several industrial sectors, as solvent and reagent. The use of CO_2 for methane reforming is an interesting application that would avoid CO_2 separation after extraction of LNG. In fact, should it be found a suitable catalyst or a technology for low temperature, effective conversion of CH_4 and CO_2 into syn-gas, the direct conversion of LNG at the extraction site into liquid fuels as methanol or gasoline would reduce the distribution costs with respect to methane separation and pumping. The utilization of carbon dioxide by other sectors of the chemical industry may contribute to implement greener synthetic technologies with respect to existing ones, and to reduce the carbon dioxide emission in terms of waste reduction at source, solvent shift, more direct synthetic procedures, energy saving. Mimicking natural processes for synthetic purposes can also address the development of innovative syntheses with CO_2 emission reduction. The fixation of CO_2 by algae (micro and macro) is a promising area that is worth to investigate for assessing the potential of aquatic biomass as soure of fuels. The assessment of the real amount of avoided carbon dioxide is not an easy and simple exercise. It is not only represented by the amount of recycled/fixed CO_2: the fraction of carbon dioxide not produced by implementing an innovative synthetic technology must also be taken into account. The Life Cycle Assessment methodology is the proper tool for such evaluation. Incentives are needed for running appropriate integrated programmes aimed at assessing the potential of the utilization option.

Carbon dioxide, CO_2 is produced in essentially all human activities, including the basic life process, respiration. An average person emits 900 g_{CO2}/d, or 23 t during an average life of 70 years. The world population (6 billion persons), thus, produces 2 Gt/y of CO_2 at zero activity. Carbon dioxide produced by human activities (26 Gt/y, or 7.1 Gt/y as carbon) represents 1% of the estimated amount resident in the atmosphere (720 Gt as C), or 3.5 % of the amount of carbon involved in the natural carbon cycle (203 GtC/y), as shown in Figure 1.

Although the anthropogenic CO_2 is only a minor fraction of the total amount of carbon in the carbon cycle, Nature cannot balance it out as demonstrated by the fact that the atmospheric CO_2 concentration is grown from 270 ppm of the pre-industrial era to the actual 380 ppm. The source of anthropogenic CO_2 is the combustion of carbon based fuels, essentially fossil fuels as coal, oil, LNG. Fresh biomass (wood) used for energy purposes represents less than 2% of the total. Fuels are burned for different purposes, the production of electric energy representing the major use (Fig. 2).

The four major sources of CO_2 emission can be categorized as dispersed (transportation) or point (electric energy production, industries, heating of civil or industrial sites). This is relevant to the question of carbon dioxide capture, that can conveniently be done from point sources. As fossil fuels will represent for the coming 30-40 years the bulk of energy source (60-70%), there is a lot of concern about the increasing amount of CO_2 that will be emitted. As a matter of fact, a lot of security issues are still associated to the use of nuclear energy, while solar-wind-water power cannot respond to the request of production of any important fraction of bulk energy, also because the availability of the source depends on the geographical position of a country.

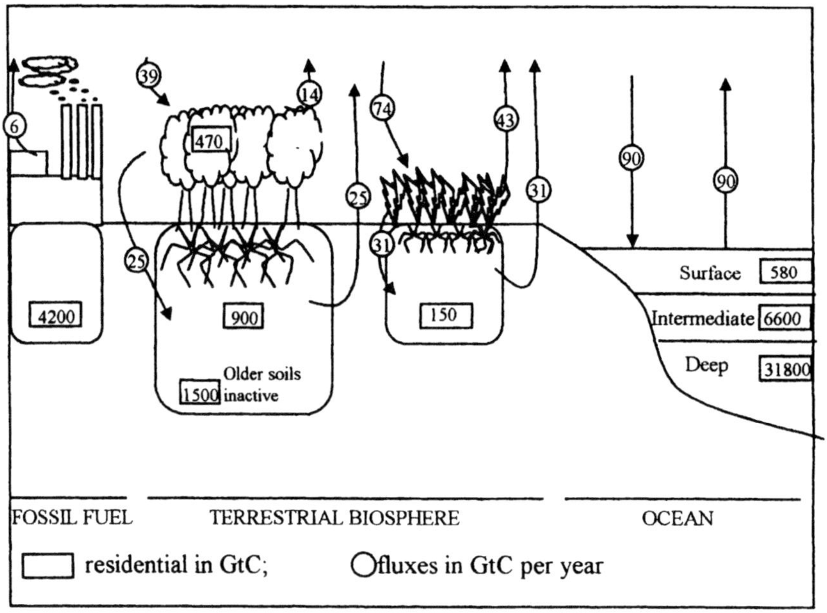

Figure 1. The carbon cycle

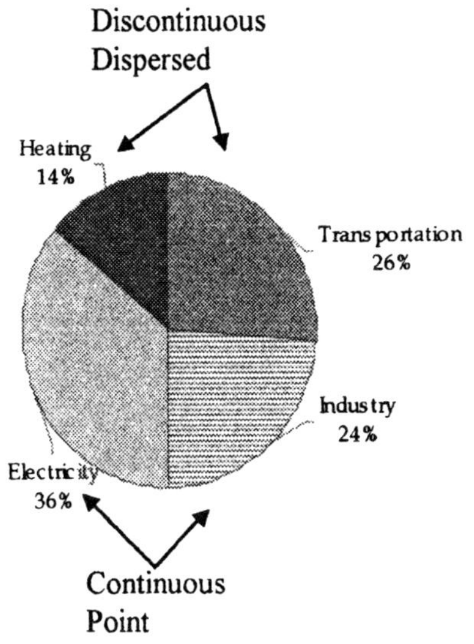

Figure 2. Sources of carbon dioxide (ca 27 Gt/y)

The use of biomass leaves open the question of the correct technology to be implemented if heavy energy has to be produced. Its use for responding to the demand at the family size is better suited, but does not contribute to solve big problems. Hydrogen will be a source of energy in the future. Unless water scission by solar energy, either thermal or photovoltaic or photochemical, will be developed at a completely different level than today, hydrogen production will mean CO_2 emission, as hydrogen will be produced through decarbonisation of fossil fuels. With respect to the existing situation, the use of hydrogen as fuel will represent a shift of the emission relevant to transportation from the dispersed- to the point-emission category. In fact, hydrogen will be produced by large facilities where fuels will be treated. This will not eliminate the problem of CO_2 production, but will make possible its capture. It is obvious that if hydrogen will be generated from water, without any important use of carbon-based fossil fuels, then the emission of carbon dioxide will be cut-off and a new scenario will open. But this stage is still too far away.

However, as for now, the reduction of carbon dioxide emission can be achieved by implementing efficiency strategies in the production of electric energy and in the use of any form of energy. Innovative technologies for electricity production may increase the efficiency of conversion of the chemical energy from 28-32% of the recent past, to 52% with IGCC. Fuel shift can also contribute to the reduction of the emission, shifting from coal to oil to LNG the amount of carbon dioxide emitted per kWh goes down from 1 kg to 0.75 to 0.5. A recent study commissioned by the European Union *(1)* has given the statistics of energy use (Table I) within EU-15 with data for the energy use by industrial sector (Table II) and the baseline emission scenario to 2030 (Table III) considering the tendency to fuel shift.

By sector	3077.7	3049.4	3144.3	3297.4	3314.9	3235.4	3224.3	3122.1	3058.0	0.6	0.5	-0.3	-0.5
Industry	571.6	521.1	500.4	460.9	452.6	431.1	413.0	381.1	368.9	-0.8	-1.0	-0.9	-1.4
Tertiary	202.8	212.8	205.6	216.8	217.8	202.7	184.6	175.5	167.8	-0.7	0.6	-1.6	-1.0
Household	436.5	417.8	424.5	454.9	478.0	476.9	467.8	457.6	448.5	0.3	1.2	-0.2	-0.4
Transports	738.6	803.6	912.9	1016.6	1024.7	987.9	976.3	946.9	920.3	2.6	1.2	-0.5	-0.6
Electricity-steam production	961.0	925.6	942.3	1015.4	1013.2	1021.1	1080.1	1067.3	1075.8	0.4	0.7	0.6	0.0
District heating	36.7	23.6	9.4	7.2	7.2	10.5	7.4	5.8	6.0	-16.8	-2.6	0.2	-2.1
New fuels (hydrogen) production	0.0	0.0	0.0	0.2	0.2	0.3	0.6	1.0	0.9			14.2	4.4
Energy branch	130.5	145.0	149.2	125.3	121.3	104.8	9.4	86.8	79.8	0.6	-2.0	-2.6	-1.7
CO2 emissions index (1990=100)	100.0	99.1	102.2	107.1	107.7	105.1	104.8	101.4	99.4				
Industry	100.0	91.2	87.5	80.6	79.2	75.4	72.3	66.7	62.8				
Tertiary	100.0	105.0	101.4	106.9	107.4	100.0	91.1	86.6	82.7				
Household	100.0	95.7	97.3	104.2	109.5	109.3	107.2	104.8	102.7				
Tansports	100.0	108.8	123.6	137.6	138.7	133.8	132.2	128.2	124.6				
Electricity-steam production	100.0	96.3	98.1	105.7	105.4	106.2	112.4	11.1	111.9				
District heating	100.0	64.1	25.6	19.7	19.7	28.5	20.1	15.9	16.2				
New fuels (hydrogen etc.) production	-	-	-	-	-	-	-	-	-				
Energy branch	100.0	111.1	114.3	96.0	92.9	80.3	72.3	66.5	61.2				

It has shown the potential of such shift with an average annual reduction of the emission ranging from 0.9 to 1.4% over three decades for the EU-15 countries.

Obviously, it is not possible to use only LNG for electric energy production, but the application of technologies known for years (coal gasification) may be helpful for reducing the emission of carbon dioxide. Moreover, the introduction of the coal gasification combined cycle (CGCC) may make available flue gases more concentrated in carbon dioxide making less expensive the separation *(2)*.

Despite the increase of efficiency will generate a foreseeable reduction of the carbon dioxide emission, the expansion of the worldwide demand of energy will produce a net increase of the emission. In order to comply with the recent international agreements, and have a net reduction of the immission into the atmosphere of the produced carbon dioxide, the capture technology can be the winning strategy. Methodologies for carbon dioxide separation from flue gases are well established and their limit of application and costs are well known. Table IV presents a summary of the existing technologies. All of them are or can be exploited at the industrial level.

Table IV. The separation technologies

Absorption
Cryogenic distillation
Adsorption
Membranes

Separation of Carbon Dioxide from Industrial Sources

The separation of carbon dioxide from flue gases can be achieved by using any of the technologies reported in Table IV. The cost of the separation depends on: i) the concentration of carbon dioxide in the source gas; ii) the required efficiency of recovery; iii) the presence in the gas mixture of contaminants that may reduce the efficiency of the separation phase. Table V summarizes the cost per each technique, highlighting major drawbacks.*(1)* The purity of the separated gas is a most important characteristic for its end use.

Table V. Cost of the separation technologies

Technology	*Drawbacks*	*Cost \$/tCO_2*	*Purity of CO_2*
Absorption (MEA)	Foam Presence of H_2S/SO_2 Particulate (max level 100-150 mg/m^3) Oxygen (5% V at 0.1 MPa)	35-40	>98%
Membranes	Cost increases with the purity of separated CO_2	45-48 71	50% 90%
Adsorption	High energy consumption Deactivation	84-264	75%
Cryogenic	Not yet mature	40	99%

It is obvious that the quality and cost of the separated gas depend on the nature of the source and the purification process. Scheme 1 gives a classification of sources. For some specific applications, like food industry, carbon dioxide should not contain any toxic species. Therefore, in order to avoid high costs for the separation of species present at a low concentration, the selection of the source is very important.

CONTINUOUS	DISCONTINUOS
POINT	DISPERSED
CONCENTRATED	DILUTED

The D sources are not economically viable for recovery

Good sources	**Poor sources**
CEMENT	ELECTRIC ENERGY
LIME	
AMMONIA	
ETHENE OXIDE	
FERMENTATION	
ALUMINIUM	

Scheme 1. Categorization of carbon dioxide emission sources

The work done within the RUCADI Project *(1)* has categorized sources as "good" (purity of CO_2 higher than 98%) and "poor", according to the CO_2 concentration. Moreover, the quaality of the source has also been defined according to the presence or not of pollutants that may be dangerous to human health. Therefore, the electric power sector, despite being a point and continuous source, because of the low carbon dioxide content and the potential presence in flue gases of toxic species, is not that indicated for the supply of carbon dixide for the food sector. Other sectors that are large producers of carbon dioxide are categorized in Table VI.

Table VI. Industrial sectors that are continuous or point sources of carbon dioxide

i) Sugar industry and breweries	Most large breweries now start recovering the CO_2 produced during their fermentation process. The production is not continuous.
ii) H_2 plants	Not enough data (at least in Europe) to assess the importance.
iii) Landfills	Too scattered to be of economic relevance
iv)Natural gas on/off-shore extraction	Not enough data
v) Chemical industries	
vi) Steel & coke	
vii) Refineries	
viii) District heating	
ix) Food-drink-tobacco industry	
x) Paper industry	

Sectors vi-x in Table VI are not well defined for their production. Indeed, sectors such as refineries are in constant evolution: more and more plants are built to produce light hydrocarbons by reaction of heavy hydrocarbons with dihydrogen. Consequently it is difficult to have a reliable set of data and define their importance for the production of carbon dioxide.

The existing technologies may, thus, make available a large amount of carbon dioxide, with a different purity degree, suitable either for disposal (low purity), or for chemical and technological use (average purity) or for direct utilization in the food sector (high purity). In principle, the cost of recovery of carbon dioxide should not be higher than, and its purity comparable with, that of extracted (from natural reservoirs) CO_2 in order the former to find an application. Therefore, sources that require an extended treatment for carbon dioxide purification are not to be considered for carbon dioxide supply, but CO_2 should be better considered for disposal. Let us now consider the possible utilization of carbon dioxide.

Utilization of Carbon Dioxide

The utilization of carbon dioxide is receiving great attention in these days for its relevance to carbon dioxide immission control *(2)*. Carbon dioxide finds an application in several sectors. Through this paper, the uses will be categorized as:

i) Technological use (carbon dioxide is not converted into other species)
ii) Chemical use (carbon dioxide is used as source of carbon for the synthesis of chemicals)
iii) Biological use (carbon dioxide is fixed in terrestrial or aquatic biomass).

A specific paragraph will be devoted to supercritical carbon dioxide.

Noterworthy, the utilization of carbon dioxide integrated into the recovery-disposal strategy may make economically viable the disposal technology that otherwise has today a very high cost.

i) Technological Use

The properties of carbon dioxide that make this gas particularly useful for many applications and for which it finds a large use (Table VII) are: inertness, acid character, critic temperature and pressure, high density. Moreover, carbon dioxide is non conducting, good solvent in dense phase, bacteriostatic, fungicide and pesticide leaving no residues, and can be used to enhance the photosynthetic activity of plants in greenhouses. Also, the solid at room temperature sublimates without liquid formation. (Table VII).

Table VII. Commercial use of CO_2 *(1)*

	Value %	*Mass* %	*Consumption/ capita (*)*	*Cost: $/t*
Total:	2 700 M$	13.5 Mt/y	Average 2.4 kg/y	
Americas	37	57	9.6-25	130
Asia	31	21	0.8-10	326
Europe	26	19	5-10	248
Others	6	3	0.15-6	400

(*) High consumption per capita is noticed in countries with: high degree of industrialization, high standard of living, availability of low cost CO_2, R&D dedication, high innovation.

Table VIII presents the USA merchant market of CO_2, while the EU market is given in Table IX.

The merchant market of carbon dioxide is expected to enlarge more and more, also due to the expansion of the use of supercritical CO_2 as solvent. Curiously enough, up to now a limitation to the enlargement has been in the availability of good quality CO_2. The utilization in the food sector may expand, especially for the demand by the refrigeration and packaging sectors, with a consequent larger

Table VIII. Merchant market of CO_2 in the USA *(1)*

Total Market, 1996: 5.3 Mt at 130 $/t.Total value = 689 M$.

Approximate split:

	Market:	*kt/y*	*%*
1	Food processing	3 100	58.0
2	Carbonated beverages	930	17.0
3	Chemical processing	400	7.5
4	Metal fabrication (welding)	200	3.8
5	Agriculture	50	0.9
6	Plastic & Rubber (Blowing Agent)	20	0.4
7	Others	600	11.3
	Total:	**5 300**	**100.0**

Table IX. Consumption of CO_2 in EU-15 Countries *(1)*

Application	*Northern Europe: (kg/person y)*	*Southern Europe: (kg/person y)*
Soft drinks & mineral waters	1.4	1.3
Breweries	1.5	1.1
Chemical Ind. (inerting, reactive)	1.0	0.12
Refrigeration	1.0	0.20
Freezing	0.4	0.1
Nuclear power stations	0.4	N/A
Greenhouses	0.2	N/A
Water neutralization	0.2	0.1
Welding + foundries	0.15	0.37
Slaughterhouses (pigs, poultry)	0.1	0.05
Blow molding + foaming	0.06	0.08
Fire extinguishers	0.05	0.07
Supercritical extraction	0.05	0

exploitation of good sources. A map of such sources and their exploitation level can be helpful for the industries active on the market.
As reported in Table VII, the total market of carbon dioxide used for technological uses, ranges around 13.5 Mt/y. The amount used for enhanced oil recovery, EOR has not been considered. Curiously, the carbon dioxide used in this sector up to now has been extracted from natural wells. To use recovered carbon dioxide would be double beneficial: the extraction from natural wells would be avoided, and part (*ca.* 50%) of the injected fluid would remain in the well, adsorbed by the walls. The EOR can, thus, operate also as a disposal technology of CO_2. This sector may considerably contribute to increase the market of merchant CO_2. In fact, a single well could require up to 50 kt/y of carbon dioxide.

ii) Chemical Utilization

The amount of carbon dioxide used by the chemical industry ranges around 120 Mt/y, much less than the emitted amount (Table X). Including the technological sector roughly 120 Mt/y of CO_2 are used, plus EOR.

Table X. Amount of CO_2 used by the Chemical Industry

Technological use	13.5 Mt/y
Synthesis of urea	60 Mt/y
Salycilic acid	20 kt/y
Propylene carbonate	
Polycarbonate	few kt/y
Methanol	10 Mt/y
Inorganic Carbonates	$\approx$ 30 Mt/y

As Table XI shows, the utilization of carbon dioxide in the chemical industry is a practice more than one century old.

Table XI. Utilization of carbon dioxide for synthetic purposes

Organic compounds

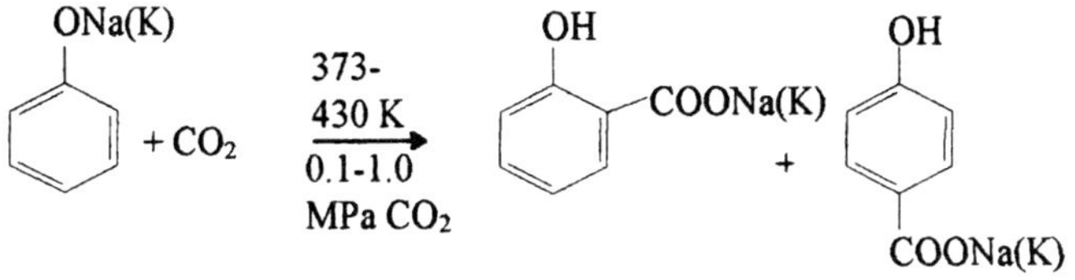

1869

$2\ NH_3 + CO_2 = H_2NCONH_2 + H_2O$ **1870**

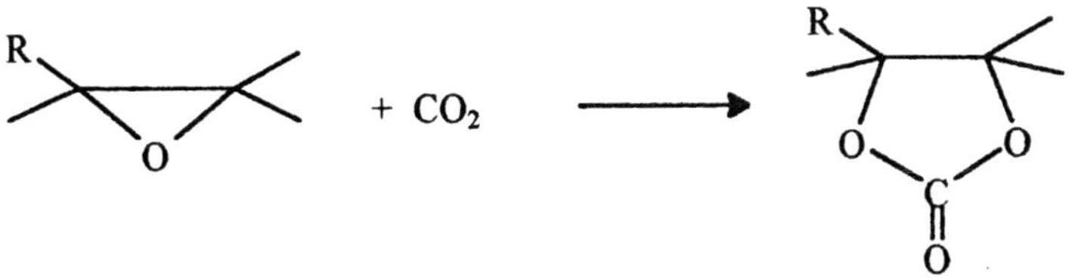

1973

$CO_2 + 3H_2 \longrightarrow CH_3OH + H_2O$ **1970**

Inorganic compounds

$CO_2 + M_2O, MO \longrightarrow M_xCO_3$

$x = 1, M = Ba$
$x = 2, M = Na, K$

It is worth to note that both the synthesis of urea and salycilic acid do not require any catalyst: they are pure thermal reactions. Also, the synthesis of urea represents the largest single process for the utilization of carbon dioxide. Additionally, urea can be considered as an active form of carbon dioxide: in fact it can react, for example, with alcohols to afford organic carbonates (Eq. 1), a reaction not exploited so far. Ammonia can be separated and recycled. This reaction represents an interesting route to organic carbonates, avoiding the use of phosgene (See below).

$$H_2NCONH_2 + 2\ ROH = (RO)_2CO + 2\ NH_3 \quad (1)$$

It is interesting to note that, although the synthesis of salycilic acid is known since very long, still the reaction mechanism has not been completely elucidated and the process is characterized by low yield and selectivity. The two other industrial processes in which carbon dioxide is used demand a catalysts. Fig.3, gives a perspective on the possible reactions based on carbon dioxide *(3, 4)*. Each reaction requires a catalyst.

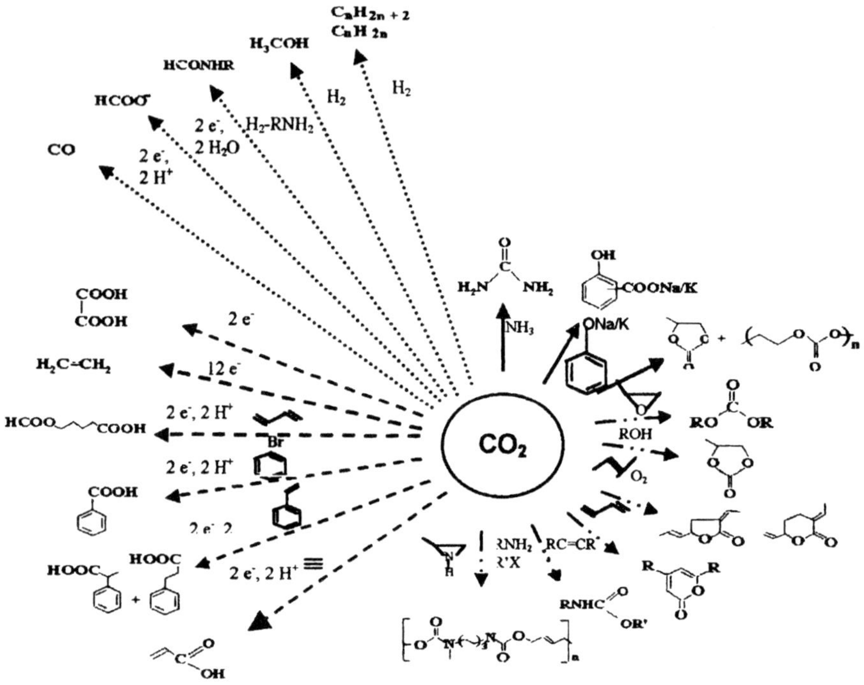

Figure 3. Known reactions for the synthesis of chemicals from carbon dioxide

Figure 3 also gives at glance the potential of carbon dioxide in synthetic applications and demonstrates its versatility as source of carbon. This is not surprising if one thinks that carbon dioxide and its hydrated form HCO_3^- are the source of carbon for a myriad of compounds, both organic and inorganic, in Nature. Despite man has had both the knowledge of that since ever, and the need of developing syntheses of compounds containing the carboxylic, –COO, functionality, the number of processes developed to this end remains very limited. The existing barriers to the enlargement of the use of carbon dioxide are of various nature. For sure thermodynamic and kinetic factors play an important role. The fact that carbon dioxide lays at the bottom of a potential well (in fact it is, with water, the ultimate product of combustion processes and metabolic pathways that provide energy) has discouraged chemists to develop syntheses based on its conversion. Table XII shows the free energy of formation of C1 molecules: it is clear that going from carbon dioxide to any other species energy is needed.

Table XII. Free energy of formation of C1 molecules

C_1 species	*C formal oxidation state*	*Free Energy/kJ mol^{-1}*
CH_4	-4	-50.75
CH_3OH	-2	-166.1
CH_2O	0	-109.93
C	0	0
HCN	+2	+119.07
CO	+2	-137.15
HCOOH	+2	-345.09
CO_2	+4	-394.01

The common belief has been so far that too much energy is necessary for any conversion of carbon dioxide to occur. Consequently, other routes, apparently based on more reactive species, have been exploited for the synthesis of compounds, that could be made using CO_2. A careful analysis of some well established processes shows that they imply a much larger number of steps, with respect to those eventually based on CO_2, and that one of these steps may require a large energy consumption. Their apparent convenience stands if one looks only at the final reaction for the synthesis of a chemical, and does not take in consideration how reagents were obtained. Later on in this paper the use of carbon dioxide as alternative to phosgene, $COCl_2$ will be discussed. Phosgene is much more reactive that carbon dioxide towards alcohols, for example. But in a correct energy and mass balance perspective, one has to consider that $COCl_2$ is formed from CO and chlorine, two reagents that require quite energy intensive synthetic routes: syn-gas production and electrochemical oxidation of chlorides, respectively. Two new facts may produce a rethinking of many established synthetic technologies. The reduction of the production of waste is now an urgent problem. Consequently, the attitude to make energy and mass balance for a synthetic process has grown very recently with the introduction of the Life Cycle Assessment, LCA a tool used for the evaluation of the environmental, energetic, and economic convenience of products and processes *(5, 6)*. The use of such methodology is the only way to demonstrate wether or not an exploited route to a given product implies a lower use of energy and reagents/solvents with respect to an innovative path based on carbon dioxide. It is not trivial to comparate energy and mass balance of existing and innovative processes and it is quite clear that many of the "easy to go routes" are not characterised by a high degree of sophistication. Some of them are just thermal processes (emitting, thus, a large amount of carbon dioxide!) and do not require engineerized catalysts. The question arises if the barriers to the utilisation of carbon dioxide have a kinetic more than a thermodynamic nature. As a matter of fact, a number of processes that have a positive free energy are exploited for synthetic purposes right now, supposed that the appropriate conditions (catalyst, temperature) are used. Carbon dioxide requires very specific catalysts, performing like enzymes in Nature. The bad thermodynamics of the carbon dioxide conversion reaction has been so far a justification for avoiding the important investment (human resources, time and budget) necessary for developing efficient technologies for its utilization. On the other hand, there was no urgency to develop such technologies because chemicals were prepared by other ways. Today we are facing a different reality: there may be a convenience in developing synthetic strategies based on carbon dioxide, for avoiding taxes and for saving resources and also energy. Therefore, the search for new catalysts that may drive reactions with high yield and selectivity is becoming more and more popular. But, as it has been mentioned above, the development of a new synthetic process industrially viable requires means, time, and human resources. The chemistry of carbon dioxide, in this respect, has received much less attention than that of carbon monoxide in the last century. This is the third barrier, besides thermodynamics and kinetics: CO_2 has been considered so far as a waste, CO as a resource. An innocuous waste, already abundant in Nature and Nature was expected to get us free of it. But it appears now that carbon dioxide is not innocent and Nature needs our cooperation for the elimination of CO_2.

Thermodynamics and kinetics of carbon dioxide conversion

The utilization of carbon dioxide occurs in Nature essentially according to two routes, namely the carboxylation reactions and reduction of carbon dioxide to other C1 or Cn molecules. A similar utilization may occur in the chemical industry, as specified in Scheme 2 *(7-9)*.

Fixation of the entire CO_2 moiety	*CO_2 reducion to other C_1 molecules*
Acids (RCOOH)	$CO_2 \rightarrow CO$
Esters (RCOOR)	$CO_2 \rightarrow H_2CO$
Carbamates RHNCOOR'	$CO_2 \rightarrow CH_3OH$
Organic Carbonates $(RO)_2CO$	$CO_2 \rightarrow HCOOH$
Ureas R'HNCONHR"	$CO_2 \rightarrow CH_4$
Amides RHNCOR'	

Scheme 2. The utilization of carbon dioxide

These two classes of reaction demand a different source and amount of energy *(10)*. (Scheme 3)

The substrate may be the energy vector	Various forms of energy (thermal, vibrational, UV /IR) and electrons are used
- Electron rich species - HO^-, RR'R"N, H^-	$CO_2 \xrightarrow{h\nu} CO + \text{"O"}$ $CO_2 + 2e^- + 2\,H^+ \longrightarrow CO + H_2O$

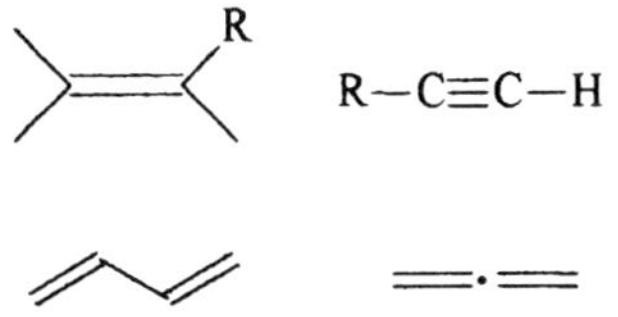

Scheme 3. Source of energy for carboxylation and reduction reactions

Carboxylation reactions occurring in Nature imply a nucleophilic attack to the carbon atom of carbon dioxide by an electron rich species *(11)*. Amines or amides, hydroxo-groups, carbanions and hydrides are the elected species in Nature, affording carbamate, hydrogencarbonate, carboxylate, and formate anions, respectively. (Scheme 4)

O=C=O	+	OH^-	=	HCO_3^-
		RNH^-	=	$RNHCOO^-$
		$RR'HC^-$	=	$RR'HC\text{-}COO^-$
		H^-	=	$HCOO^-$

Scheme 4. Reaction of nucleophiles with carbon dioxide and the reaction products

The reaction with carbon dioxide takes place at room temperature, or lower. Some of these reactions are quite common, like the reaction of the hydroxo-group, other are less. The formation of carbanions and hydrides requires energy: in Nature such energy finds its origin in solar energy and is transferred through a complex electron transfer chain. In artificial systems, chemicals are used for producing such species that may also be generated electrochemically. An example is the formation of Grignard reagents, RMgX, where R is an organic group, that promptly react with carbon dioxide to afford carboxylates (Eq. 2)

$$RX + Mg = RMgX; \qquad RMgX + CO_2 = RCOOMgX \quad (2)$$

Similarly, amides can be produced by reactions of amines with elemental first Group metals, Eq.3

$$RR'NH + Na = RR'NNa + \tfrac{1}{2} H_2;$$
$$RR'NNa + CO_2 = RR'NCOONa \quad (3)$$

Such reactions do not have any practical importance from the synthetic point of view, as they use elemental metals for the fixation of carbon dioxide. In fact, after reaction with water of the carboxylate, or carbamate, (Eq. 4) the metal remains in its oxidized form and cannot repeat the carboxylation process depicted in Eq. 2-3.

$$RCOOMgX + H_2O = RCOOH + MgX(OH) \quad (4)$$

Therefore, such reactions are not used for the carboxylation of organic substrates, although there is a need for direct processes that produce carboxylates, especially long chain aliphatic carboxylic acids or aromatic acids. Other processes have been studied and developed that may afford such species with an economic and energetic convenience, using transition metal systems as catalysts. In the conversion of the carbon dioxide molecule a key question is if it requires a base or acid-base catalysis. The molecular orbital energy diagram of the linear molecule in the ground state and of its bent form *(12)* is reported in Fig. 4.

The reactive site of the CO_2 molecule is the electrophilic carbon atom. The attack at carbon displaces a negative charge on the oxygen atoms, so that the coordinated carbon dioxide is stabilized by an interaction with an electrophile. (Fig. 5)

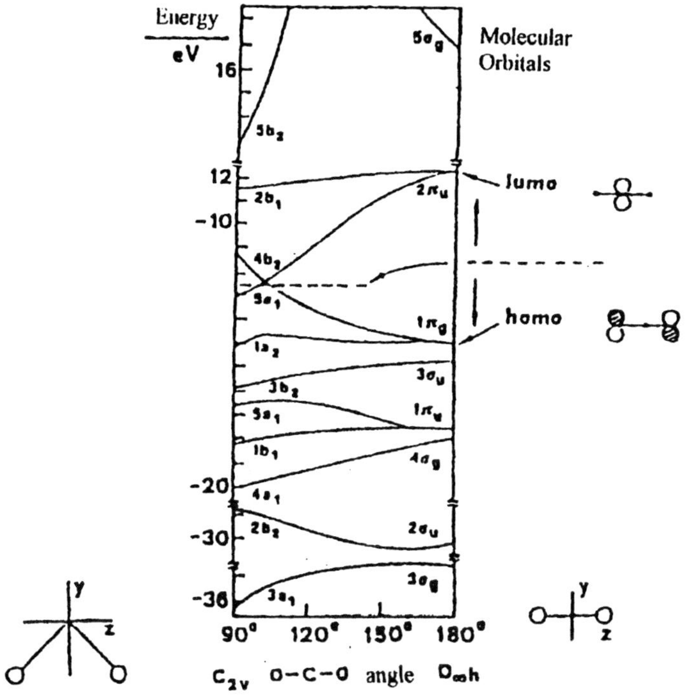

Figure 4. Walsh molecular orbital diagram for linear and bent carbon dioxide

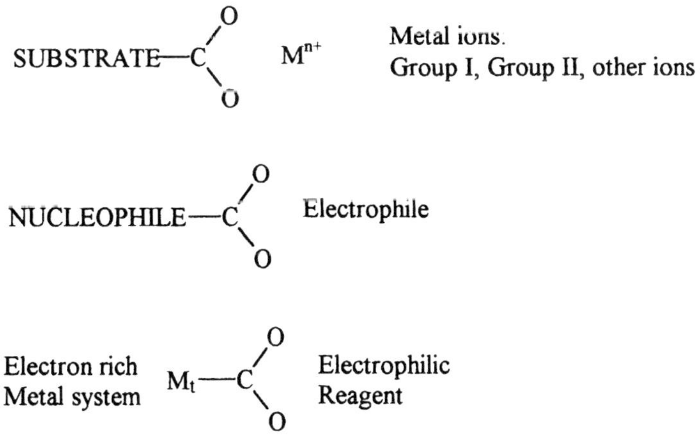

Figure 5. Role of the metal in CO_2 fixation

How Much CO_2 is Avoided by Developing Carbon Dioxide based Syntheses: the Assessment Methodology

When carbon dioxide is used for the synthesis of a chemical, the evaluation of the amount of avoided carbon dioxide must take into consideration several parameters. In fact, the quantity of avoided CO_2 is not simply represented by the amount of CO_2 incorporated into the product. The comparison of two synthetic procedures must take into account: reagents used, yield and selectivity of the reaction, nature of the by-products and their potential use as secondary raw materials, amount and quality of waste produced, amount and quality of solvent used, technologies for waste solvents/residues treatment, energy used during the various phases of the reaction and for the separation/purification of the products *(18)*. This means that the exact evaluation of the advantages produced by a synthetic procedure based on carbon dioxide can be quantified only after the Life Cycle Assessment methodology has been applied to the old and innovative technology and their environmental, energetic, and economic aspects are evaluated *(19)*. However, to reject or accept a carbon dioxide based synthesis is not trivial. The fact that carbon dioxide is eventually fixed is not per se an advantage, the fact that carbon dioxide lies in a potentially energy well is not a good reason for an *a priori* exclusion of the approach. In the following paragraphs some specific syntheses will be discussed.

Synthesis of Carboxylates

Key issues for the carboxylation of organic substrates by direct CO_2 incorporation has been already reviewed within a recent ACS Symposium Series book *(20)*. In this paragraph a different set of considerations will be developed, namely the advantages hidden by such synthetic technologies. In particular, the contribution of innovative syntheses to waste reduction at source will be highlighted. It is worth to recall that, as it has been already specified above, using more direct synthetic procedures, increasing the selectivity of a process, lowering the reaction temperature, using less organic solvents, means to reduce the specific carbon dioxide emission relevant to the final product. These aspects will be discussed below.

Long chain carboxylic acids, $CH_3(CH_2)_nCOOH$ and $HOOC(CH_2)_nCOOH$

Such compounds are essentially used as additives and in the production of surfactants. Their global market ranges around 10 Mt/y. They are prepared by oxidation of linear alcohols (Eq. 5) or of cyclic olefins (Eq. 6). The oxidation process has a low atom efficiency and is not selective.

$$HOH_2C(CH_2)_nCH_2OH \xrightarrow{ox} HOOC(CH_2)_nCOOH \qquad (5)$$

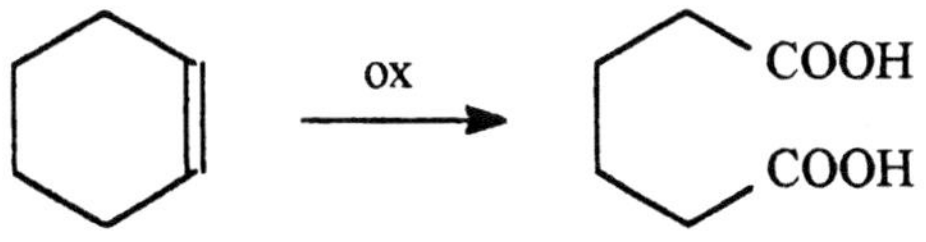

(6)

A synthetic technology based on the use of olefins and carbon dioxide, (Eq. 7) would greatly reduce the production of waste.

$$CH_3(CH_2)_nCH{=}CH_2 + CO_2 + H_2 \rightarrow CH_3(CH_2)_{n+2}COOH \qquad (7)$$

The direct carboxylation of olefins has not yet been developed in a catalytic way for several reasons *(20)*. Nevertheless, the above reaction can be carried out by electrocatalytic carboxylation of terminal olefins with high yield *(21)*.

Alternatively, the reaction of dienes with carbon dioxide *(1, 22-24)* can be used. Butadiene reacts with carbon dioxide under mild conditions and the nature of the products is driven by the catalysts used. Cyclic or linear carboxylic compounds are obtained, that can be either used as such or converted into linear saturated long chain acids by hydrogenation and hydrolysis (Scheme 5) *(24)*.

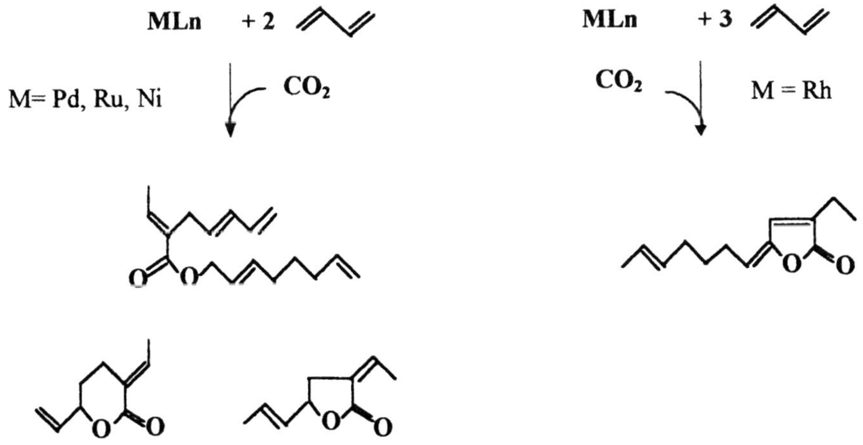

Scheme 5. Products of coupling of butadiene and carbon dioxide

This synthetic strategy may be of great interest for reducing the waste production and, then, the CO_2 emission.

Formic acid, HCOOH

Formic acid is today produced through a complex sequence of reactions. An example is given by Eqs. 8.

$$CO + 2H_2 \rightarrow CH_3OH \quad (8a)$$

$$CH_3OH + CO \rightarrow HC(O)OCH_3 \quad (8b)$$

$$HC(O)CH_3 + H_2O \rightarrow HCOOH + CH_3OH \quad (8c)$$

A more direct approach is represented by Eq. 9, in which carbon dioxide and dihydrogen are reacted to afford the desired product with high atom economy .

$$H_2 + CO_2 \rightarrow HCOOH \quad (9)$$

Recently, this synthesis has been achieved in $scCO_2$, avoiding the use of solvents *(25,26)*. As it will be discussed later on, this approach represents a great advantage. The exclusion of organic solvents in particular and of a solvent in general, means a reduction of CO_2 emission. Formic acid, besides having an industrial importance per se, could also find an utilization as hydrogen vector in a hydrogen-based economy.

Acetic acid, CH_3COOH

Acetic acid, CH_3COOH has a large market (> 8 Mt/y) and several uses as reagent. It is currently made by reaction of methanol with CO (homologation of methanol, Monsanto process), but could be prepared from carbon dioxide and methane (Eq. 10)

$$CH_4 + CO_2 \rightarrow CH_3COOH \quad (10)$$

through a formal insertion into the C-H bond. Such reaction has been demonstrated to occur in CH_4-CO_2 plasmas *(27)*, also if the reaction mechanism has not been elucidated. In principle, either CH_3 radicals, CO_2 and H radicals or CH_3, CO and OH radicals could be implied.

The formation of acetic acid from methane and carbon dioxide has also been achieved by using heterogeneous catalysts *(28)*. The fact that the reaction occurs, despite it is endoergonic, represents an interesting result. In both cases, yield and selectivity are too low for the reactions could have any practical application right now.

Carbonates (cyclic and linear) and polycarbonates, ROC(O)OR

Organic carbonates find many application as solvents and selective reagents, or monomers for polymers. More recently, their use as additive to gasoline for improving the combustion has been sought. The market is around 80 kt/y for the monomers and 1.7 Mt/y for the polymeric materials. The linear (Scheme 6) and cyclic (Scheme 7) monomers are mainly synthetized by using phosgene, although alternative processes have been developed based either on CO *(29-38)* or CO_2 *(39)*.

Processes based on carbon dioxide are of great interest, as they are more direct and can help to save energy and avoid pollution with respect to the

Phosgene route

$$ROH + COCl_2 \longrightarrow ROC(O)Cl + HCl \qquad i$$

$$ROC(O)Cl + ROH \longrightarrow (RO)_2CO + HCl \qquad ii$$

Carbon monoxide route

$$2\ CH_3OH + CO \xrightarrow{CuCl} (CH_3O)_2CO + H_2O \qquad iii$$

ENIChem

$$2\ CH_3OH + CO \xrightarrow[NO]{Pd(II)} (CH_3O)_2CO + H_2O \qquad iv$$

Ube

Carbon dioxide route

$$(H_2N)_2CO + 2\ ROH \longrightarrow (RO)_2CO + 2\ NH_3 \qquad v$$

$$2\ ROH + CO_2 \longrightarrow (RO)_2CO + H_2O \qquad vi$$

Scheme 6. Synthesis of linear carbonates

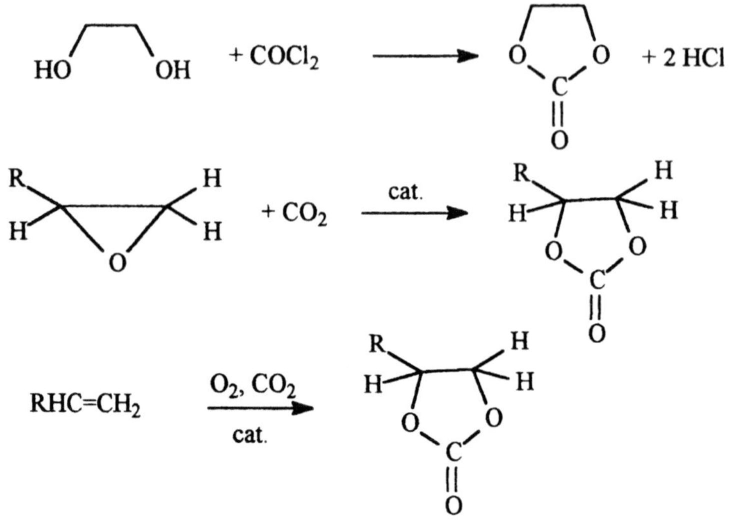

Scheme 7. Synthetic pathways for cyclic carbonates

phosgene route *(40)*. As reported above, phosgene requires the synthesis of CO and chlorine, characterized by a high energy intensity. Moreover, the reaction of phosgene with alcohols is carried out in chlorinated solvents, that rises the question of waste treatment. However, reacting phosgene or carbon dioxide with an alcohol requires different conditions: phosgene reacts at room temperature without any catalyst, carbon dioxide requires a catalyst and high temperature. All parameters considered, the route based on carbon dioxide is more convenient from the environmental, energetic and economic point of view. A few catalytic systems have been discovered able to drive the synthesis of carbonates from methanol and carbon dioxide *(41, 42)*. A major problem to be solved is the formation of water that may destroy the catalyst or reverse the reaction (Scheme 6, Eq.vi).

The same considerations are valid for cyclic carbonates (Scheme 7) The process on stream as alternative to the use of phosgene is the carboxylation of epoxides. Also the oxidative carboxylation of olefins *(43)* may find an application if selective catalysts are developed. Due to actual and future uses of such chemicals, and to their expanding market there is a need to develop alternative synthetic routes to phosgene. In fact, due to the existing limitation to build large phosgene plants, the diversification of the raw material is highly wished.

Carbamates and isocyanates, RHNCOOR', RNCO

Carbamates are used for several purposes in agriculture and in the chemical and pharmaceutical industry *(44)*. Their market ranges around several Mt/y. The synthesis is currently based on the use of phosgene, but new synthetic methodologies have been developed based on the reaction of amines with carbon dioxide and alkylating agents *(45-47)*. An interesting role of carbon dioxide as catalyst has been discovered in the reaction of aliphatic amines with organic carbonates *(48)*. Carbamates of primary amines can be used as source of isocyanates, that are used as monomers for polymers. (Scheme 8)

$$2\,RNH_2 + CO_2 \rightarrow RNH_3^{+}\,{}^{-}O_2CNHR$$

$$RNH_3^{+}\,{}^{-}O_2CNHR + OC(OMe)_2 \rightarrow RNHC(O)OMe + CO_2 + RNH_2 + MeOH$$

$$RNHC(O)OMe \xrightarrow{\Delta} RNCO$$

Scheme 8. Synthesis of carbamates and isocyanates

Again, there is an opportunity for a long term sequestration of carbon dioxide into polymers. The interrelation among carbonates, carbamates and isocyanates is quite interesting, as these species can be interconverted as shown in Scheme 9. However, does not matter which of the compounds is more easily synthetised from carbon dioxide, the two others can be obtained very easily. As mentioned above, urea could be used as an active form of carbon dioxide for the synthesis of carbonates or urethanes (Scheme 9, upper right part). The recycling of ammonia allows to operate in a closed cycle. Catalysts are needed for running the reaction to completion, avoiding the intermediate formation of urethanes of formula H_2NCOOR that may decompose into ROH and trimeric HNCO with reduction of both yield and selectivity. Alcoholysis of urea has a great potential as a convenient route to carbonates, that will convert urea from agrochemical into raw material for the chemical industry.

Other nitrogen derivatives

Formamides and methylamines have been obtained by reacting carbon dioxide with ammonia and dihydrogen. (Scheme 10)

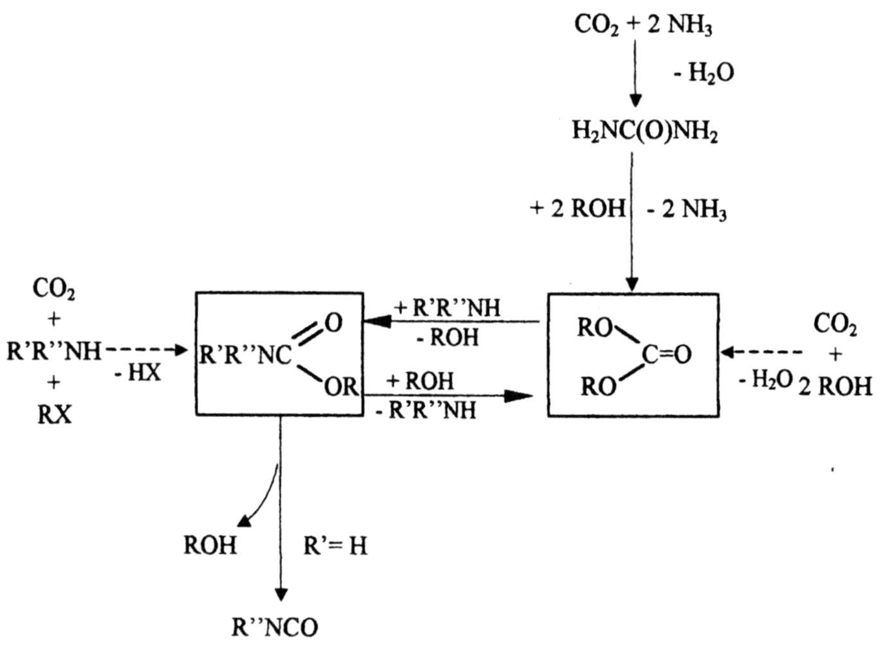

Scheme 9. Trans-esterification reactions

$$CO_2 + NH_3 + 2\ H_2 \rightarrow CH_3NH_2 + 2\ H_2O$$

$$CO_2 + NH_3 + H_2 \rightarrow NH_2CHO + H_2O$$

Scheme 10. Carbon dioxide conversion into methylamines and formamides

This methodology has some potential interest as it may supplant more complex reaction paths, opening the route to organic amines.

Syn-gas and methanol from carbon dioxide

Each of the chemicals discussed above may have a market ranging from Mt/y to tens Mt/y. As already said, the contribution to the reduction of carbon dioxide emission is much more important than the amount of CO_2 fixed. A large interest has the chemical utilization of carbon dioxide for the synthesis of products of the energy industry, like syn-gas, used for the synthesis of methanol and gasoline that have a market of hundreds Mt/y. The reduction of carbon dioxide to methanol (Eq. 11) has encountered a great interest as methanol can be used both as intermediate in the chemical industry and as fuel.

$$CO_2 + 3\,H_2 \rightarrow CH_3OH + H_2O \qquad (11)$$

Eq. 11 says that with respect to the synthesis from carbon monoxide (Eq. 12), one extra mol of hydrogen is required when carbon dioxide is used.

$$CO + 2\,H_2 \rightarrow CH_3OH \qquad (12)$$

This fact has opened the question if it is worth or not to use CO_2, considered that hydrogen is now produced from hydrocarbons or coal. An analysis of the conditions under which the synthesis of methanol from CO_2 can meet the environmental, energetic and economic convenience has recently been made *(19)*. An important point that deserves attention is if carbon dioxide is reduced to methanol in a two step reaction implying the formation of CO (Eq. 13) or if a

$$CO_2 + H_2 \rightarrow CO + H_2O \qquad (13)$$
$$CO + 2\,H_2 \rightarrow CH_3OH \qquad (12)$$

more direct route, through the formate formation (Eq. 14) is operative.

$$M + H_2 \rightarrow M(H)_2 \qquad (14a)$$
$$M(H)_2 + CO_2 \rightarrow MH(OCHO) \qquad (14b)$$

In the former case, the first step represents the reverse water gas reaction that produces CO from CO_2 and then the formation of methanol goes through the same path observed for syn-gas. In the latter case, a different reaction mechanism could be operative with different energy barriers, and possible higher yield and selectivity in the production of methanol, that would justify the use of carbon dioxide instead of CO. Besides the synthesis of methanol, a process that has a great potential for the synthesis of liquid fuels is the combination of the methane selective oxidation with carbon dioxide reduction, a reaction (Eq. 15) known as dry reforming of methane. It gives CO and hydrogen in equimolar amount that is not enough for the synthesis of methanol
(Eq 13b).

$$CH_4 + CO_2 \rightarrow 2CO + 2\,H_2 \qquad (15)$$
$$CH_4 + H_2O \rightarrow CO + 3\,H_2 \qquad (16)$$

Nevertheless, if the dry reforming is coupled with wet reforming (Eq. 16) that gives an excess of hydrogen with respect to the ratio required by Eq. 12, then it is possible to produce a syn-gas with the correct ratio CO/H_2 (Eq. 17) to be used in the synthesis of methanol or gasoline. The thermodynamics of the reactions is presented in Table XIII.

$$2\,CH_4 + CO_2 + H_2O \rightarrow 3CO + 5H_2 \quad (17)$$

Table XIII. $\Delta H°_r$ for some reactions relevant to the synthesis of methanol

Reaction	$\Delta H°_r$ (kcal/mol)
$CO + 2\,H_2 = CH_3OH$	-30.58
$C + H_2O = CO + H_2$	31.38
$CH_4 + H_2O = CO + 3\,H_2$	49.21
$CO + H_2O = CO_2 + H_2$	-9.83
$CH_4 + CO_2 = 2\,CO + 2\,H_2$	59.1
$CO_2 + 3\,H_2 = CH_3OH_{(g)} + H_2O_{(g)}$	-11.83
$CO_2 + 3\,H_2 = CH_3OH_{(l)} + H_2O_{(l)}$	-31.29

It is evident that reactions 15 and 16 are endoergonic as hydrogen is produced . In order to support the process, a partial oxidation of methane with dioxygen may be performed , as suggested by the "tri-reforming" process *(49)*. The major convenience of coprocessing methane and carbon dioxide in the fact that it may allow the conversion of methane into a liquid fuel at the LNG extraction well. Noteworthy, LNG contains methane and CO_2 in a 70-65/30-35 molar ratio, that is quite close to the 2:1 ratio required by reaction 17 for the synthesis of syn-gas. The direct conversion of LNG into liquid fuels would have a practical interest as it would eliminate the necessary separation of methane from CO_2 that is currently carried out before pumping methane in pipelines. If such process is implemented, it would make eventually possible to distribute liquid fuels instead of gas. The major barrier to the implementation of this technology is represented by the high temperature required for reaction 15 to occur. Usually, it requires 750-900 °C as at lower temperature the coke that is formed during methane dehydrogenation disactivates the catalyst and reduces the efficiency of the process *(50)*. Synthesis of more active catalysts that would allow to operate at lower temperature is the aim of several research programmes today *(51)*. Also, the plasma technology has been used for the conversion of carbon dioxide and methane, trying to avoid the high temperature required by the thermal catalytic processes *(52)*. In this case, too, yields and selectivity need to be improved for the process could find any practical application.

Supercritical carbon dioxide as solvent and reagent

The use of $scCO_2$ is becoming more and more popular. It is used as solvent since long (Table XIV).

Its utilization takes the advantages of the inertness of carbon dioxide with respect to other reaction media and of its volatility. In fact, once the reaction has gone to completion, it is possible to separate and recover carbon dioxide by a simple expansion, that makes also quite easy the isolation of the reaction products.
A new and more interesting use of $scCO_2$ is its use as solvent and reagent at the same time. This application gathers in one many positive aspects: solvents can be avoided, direct syntheses can be carried out, and the products are easily recovered. Example of such approach are the direct synthesis of formic acid from dihydrogen in $scCO_2$ *(26)* see Eq. 9, or the synthesis of carbonates from ketals (Eqs. 18-20) *(53-55)*.

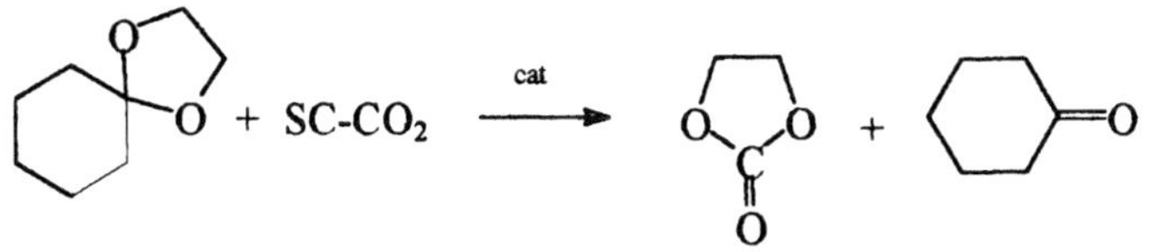

(18)

(19)

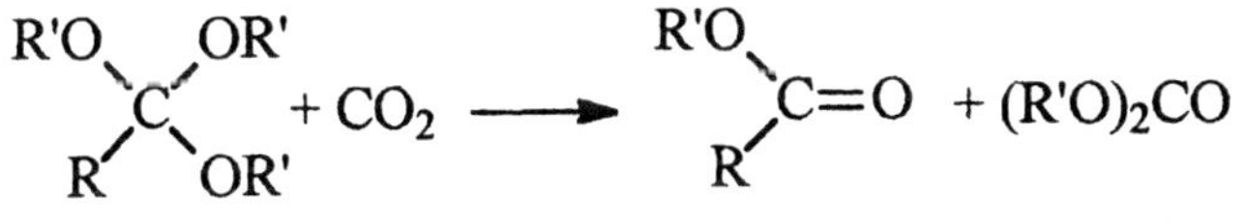

(20)

iii) Bioenergy production form micro- and macro-algae

The use of biomass is one of the technologies under assessment for its contribution to carbon dioxide mitigation. In fact, biomass may be produced either from atmospheric carbon dioxide or from CO_2 recovered from power stations or industrial activities. The energetic use of biomass allows both to reduce the utilization of natural fossil fuels and to set up an ideal carbon-recycle process, or a "quasi zero-emission" production of electric energy. The interest of the scientific community has focused so far on the following technologies for the utilisation of biomass:

Table XIV. Uses of SC-CO_2

Micro-encapsulation
Dyeing processes
Reactive modifiers, complexing agents
Waste stream detoxification:
    Removal of plutonium and uranium waste in the nuclear industry
Supercritical fluid in analytical chemistry
Reaction medium in catalysis, both heterogeneous and homogeneous
Oligomer processing
Polymerization in sc CO_2:
    Free radical polymerizations
    Metal complex promoted polymerization reactions
    Cationic polymerizations
    Step-growth polymerizations
    Modification of polymers
    Impregnation and extraction of residual impurities by swelling of the polymer matrix
Dendritic micelles stabilization

- Utilisation of fast growing crops for co-firing energy power plants.
- Thermal processes (pirolysis, liquefaction, gasification) or extraction processes for the conversion of biomass into utilizable fuels.
- Fermentation processes (methanation of waste, production of ethanol)

The maximum advantage from biomass can be obtained by converting it into an energy vector such as fuel (gas, liquid or solid) or into electric power. In Table XV the most common biomass conversion processes are reported.

Table XV. Bioenergy conversion options

Process	*Product*
Biological Processes	
Anaerobic Digestion	Biogas (mixture methane-CO_2 60/40)
Fermentation - Bioconversion	Ethanol; Hydrogen; Volatile fatty acids
Chemical Process	
Transesterification	Biodiesel
Thermochemical Processes	
Combustion	Production of electric power by steam cycles
Gasification	Syn-gas and fuel gas for fueling gas turbines
Pyrolysis	Liquid hydrocarbons and syn-gas

The biodiesel production has attracted the interest of the scientific community and efforts have been made for evaluating the real potential and profitability of this technology. Biodiesel is an ester that can be used in normal diesel engine without any major drawbacks. As a matter of facts, biodiesel-compatible tractors exist on the market. Consequently, major industrial-scale "biodiesel" refineries (7-10 ML/y production) are on stream in several EU countries and USA.
A limit to such application has been put so far by the price of crude-oil, which was as low as *ca* 20 \$/barrel until two years ago. The recent increase of the price of oil (which ranges around 25-27 \$/barrel) has renewed the interest in biofuel. The strategy of using biomass for energetic purposes is one of the key objectives for the future, and the forecast is that in ten years the share of biomass on the energy market will increase by an order of magnitude.
Among possible biomass source (fast growing crops, plants, plant seeds, residual biomass, algae) the utilisation of algae for energy production seems to be the most promising because of their: higher efficiency in solar energy conversion with respect to terrestrial plants, high adaptability to environmental conditions, possibility to grow in fresh- and sea-water avoiding the more expensive use of land.
A major interest laid so far in fresh-water grown microalgae, which have a high photosynthetic activity and ensure a fast grow with high carbon-fixation potential. The use of fresh-water microalgae has an intrinsic limitation in water availability. A noticeable interest was deserved so far in assessing the potential of utilizing cooling-water discharged by power-stations as the growing medium. Studies have been performed, especially in USA (SERI, DOE) and Japan (RITE, NEDO) for assessing the potential of bubbling in water carbon dioxide recovered from flue gases for producing algae. A complete economic study is now available on this issue. The utilisation of biodiesel as a substitute for conventional fuels is an environmental advantageous option as it is a renewable source of energy and has the potential of reducing the level of pollutants and carcinogens emission *(56)*. Biofuel has no intrinsic toxicity and is biologically degradable. Several authors report that biofuel has a content of polycyclic aromatic hydrocarbons in exhaust gases 3-40 fold less than fossil diesel fuels *(57)* and reduced fume production. In addition to that its combustion gases contain less solid particles *(58)*, have a total weaker carcinogenic and mutagenic potential *(59)* if compared to fossil diesel fuels and a negligible sulphur content (SO_2 emissions are avoided). Tests of biodegradability *(60)* have shown that biofuel biodegradability is 99 % within 21 days compared to 72 % for traditional fuels *(61)*.

Conclusions

If the capture of CO_2 from flue gases and industrial processes will be expanded, large amounts of CO_2 will be made available, either for utilization or for disposal. Supposed that good quality sources were used, the cost of recovery will be such to make the utilization of recovered CO_2 economically viable, while poor quality CO_2 could be disposed.

CO_2 could be used for the synthesis of chemicals of various molecular complexity, under reaction conditions milder than those used today for their synthesis or through more selective pathways that can drive the reactions with waste reduction at source. Table XVI gives a perspective on the use of carbon dioxide.

Table XVI.Estimate of the potential of CO_2 utilization

	Actual Mt/y	*Perspective Mt/y*
Technological uses	13.5	several tens (EOR)
Synthesis of urea	60	> 100
Propylene carbonate	0.04	
Polycarbonates	-	several
Cyclic carbonates	-	tens
Linear carbonates	-	tens
Carbamates		several
Acids		~ 10
Other chemical		> 100
Methanol	10	tens

It is foreseeable that in a different scenario, it would be possible to expand the utilization of carbon dioxide by a factor of ten with respect to the the actual limit of 120 Mt/y. Should fuels be synthesized form carbon dioxide, or should algae find a large use as source of fuels, the amount of recycled carbon could easily go beyond the estimated actual limit of 7-10% of the emitted carbon dioxide.
New catalysts must be developed for an efficient conversion of carbon dioxide into fuels.
Synthetic methodologies based on CO_2 are twice beneficial as they recycle carbon, reducing the exploitation of natural resources, and avoid carbon dioxide production by several ways. Major benefits of CO_2 utilization as reagent and/or solvent are:
* Direct synthesis with raw materials diversification and energy saving
* Atom economy with waste reduction at source
* Solvent shift with waste solvent reduction
All this may give a much more substantial contribution to CO_2 mitigation than right now foreseen considering the amount of CO_2 fixed in chemicals.
The international cooperation is essential for setting short- medium-term programmes aimed at the assessment of the potential of the utilization option.

References

1. RUCADI, Recovery and Utilization of Carbon Dioxide, BRITE Project BRRT-CT98-5089, Final Report edited by Aresta M.
2. *CO_2 Conversion and Utilization*, Song, C., Gaffney, A.M. and Fujimoto, K. Eds; ACS Symp Series 809, 2002.
3. Aresta, M. *Energy Convers. Mgmt.* **1993**, *34 (9-11)*, 745-752.
4. Marks, T., Aresta, M. et al. *Chem. Rev.* **2001**, *101*, 953-996.
5. Aresta, M., Galatola, M. *J. Cleaner Production* **1999**, 7, 181.
6. Aresta, M., Caroppo, A., Dibenedetto, A. *221 ACS Natl Meeting*, San Diego April 1-5 **2001**, Fuel Division, *Abs 61*.
7. Aresta, M., Tommasi, I. *Energy Convers Mgmt.* **1997**, *38*, S373-S378.
8. *Carbon Dioxide as a Source of Carbon*, Aresta, M. and Forti, G. Eds., NATO-ASI Series C, **1987**, 206, 331.
9. *Enzymatic and Model Reaction for Carbon Dioxide Carboxylation and Reduction Reactions* Aresta, M. and Schloss, J.V. Eds., Elsevier Publ.,1990, pp. 239–258.
10. Aresta, M. *Advances in Chem. Conv. For Mitigating Carbon Dioxide*, T. Inui, M. Anpo, K. Izui, S. Yanagida, T. Yamaguchi (Eds), 1998, vol. 114, 65-76.
11. Aresta, M., Ciccarese, A., Giannoccaro, P., Quaranta, E., Tommasi, I. *Gazz. Chim. Ital.*, **1995**, *125*, 509.
12. Rabalais, J.W., McDonald, J.M., Scherr, V., McGlynn, S.P. *Chem. Rev.* **1971**, *71*, 73.
13. Gibson, D.H. *Chem. Rev.* **1996**, *96(6)*, 2063-2095.
14. Aresta, M., Quaranta, E., Tommasi, I. *J. Chem. Soc. Chem. Commun.*, **1988**, 450.
15. Aresta, M., Nobile C.F., Albano, V.G., Forni, E., Manassero, M. *J. Chem. Soc. Chem. Commun.*, **1975**, 675.
16. M. Aresta et al. Inorg. Chem. 24, 1977, L49.
17. Tsuda, T., Sanada, S.I., Saegusa T., *J. Organometal. Chem.* **1976**, *116*, C10.
18. Aresta, M., Dibenedetto, A., Tommasi, I., *Energy&Fuels*, **2001**, *15*, 2, 269-273.
19. Aresta, M., Caroppo, A., Dibenedetto A. and Narracci, M. *Environmental challenges and greenhouse gas control for fossil fuel utilization in the 21st century*, Maroto Valer Ed., ACS Symposium Series, 2002 in press.
20. Aresta, M., Dibenedetto, A. *CO_2 conversion and Utilization*, C. Song, A.M. Gaffney, and K. Fujimoto Eds, ACS Symp Series 809, 2002, p.55-70.
21. Dunach, E., Dinjus E., Tascedda P. *ICCDU IV*, September, 7-11, **1997**, 38.
22. Behr, A. *Chem. Ber.*, **1984**, *272*, 29.
23. Aresta, M., Quaranta, E., Tommasi, I. *New J. Chem.*, **1994**, *18*, 133
24. Dinjus, E. et al. *ICCDU IV*, September, 7-11, **1997.**
25. Leitner, J. *J. Chem. Soc., Chem. Comm.*, **1993**, 1465-1466.
26. Jessop, P. *Nature*, **1994**, *368*, 231-233.

27. Liu, C.J., He, M.X., Zo, J.J., Eliasson, B., Am. Chem. Soc. Div. Fuel Chem. Prepr., **2002**, 47, 1, 328.
28. Taniguchi, Y.; Kitamura, T., Fujiwara, Y., 4th International Conference on Carbon Dioxide Utilization, Kyoto, Japan, September 1997, P-030
29. Romano, U., Vimercate, F., Rivetti, S., Di Muzio N.*US Patent* 4 318 862, *Chem. Abstr.* **1982**, volume 95, 80141w (ENIChem).
30. Romano, U. *EP Patent* 365 083, 1989, *Chem. Abstr.* **1990**, volume 113, 97039g (ENIChem).
31. Di Muzio, N., Fusi, C., Rivetti, S., Sasselli G.*EP Patent* 460 732, 1991, *Chem. Abstr.* **1992**, volume 116, 62045r (ENIChem).
32. Paret, G., Donati, G., Ghirardini M.*EP Patent* 460 735, 1991, *Chem. Abstr.* **1985**, volume 116, 62046s (ENIChem).
33. Dreni, D., Rivetti, S., Delledonne, D. *US Patent* 5 322 958, *Chem. Abstr.* **1994**, volume 120, 76901m (ENIChem).
34. Rivetti, S., Romano, U., Garone, G., Ghirardini M.*EP Patent* 634 390, *Chem. Abstr.* **1995**, volume 122, 136770w (ENIChem).
35. Nishihira, K., Mizutare, K., Tanaka S. *EP Patent* 425 197, 1991, *Chem. Abstr.* **1991**, volume 115, 8100w (Ube Ind. Ltd.)
36. Matsuzaki, T., Shimamura, T., Fujitsu, S., Toriyahara Y.*US Patent* 5 292 916, *Chem. Abstr.* **1994**, volume 117, 47917d (Ube Ind. Ltd.)
37. Nishihira, S., Tanaka, K. Kodama, T. Kaneko, T. Kawashita, Y. Nishida, T. Matsuzaki, K. Abe *US Patent* 5 380 906, *Chem. Abstr.* **1995**, volume 117, 194021k (Ube Ind. Ltd.)
38. Yoshida, S., Tanaka S. *EP Patent* 655 433, *Chem. Abstr.* **1995**, volume 123, 86588r (Ube Ind. Ltd.).
39. Darensbourg, D.J., Holtcamp, M.W. *Coord. Chem. Rew.*, **1996**, 153-155.
40. Aresta, M., Dibenedetto, A., Caroppo, A., Magarelli, A. *Greenhouse gas control technologies*, D. Williamd, B. Durie, P. McMullan, C. Paulson, A. Smith Eds, CSIRO Publ., **2001**, 923-928.
41. Ikeda, Y., Furusawa, Y., Tomishige, K., Fujimoto, K. *CO_2 Conversion and Utilization*, C. Song, A.M. Gaffney, and K. Fujimoto Eds, ACS Symp Series 809, **2002**, 71.
42. Ballivet-Tkatchenko, D., Douteau, O., Stutzmann, S. *Organometallics* **2000**, *19*, 4563-4567.
43. Aresta, M., Dibenedetto, A. *J. Mol. Cat.*, **2002**, in press.
44. Aresta, M., Quaranta, E. **Chem. Tech**. **1997**, *27*, 32-40.
45. Aresta, M., Dibenedetto, A., Quaranta, E. *Tetrahedron* **1998**, *54*, 14145-14156.
46. Aresta, M., Dibenedetto, A., Quaranta, E. *Green Chemistry* **1999**, 237-242.
47. Aresta, M., Bosetti, A., Quaranta, E. *Ital. Pat. Appl.*, 002202, **1996**.
48. Aresta, M., Dibenedetto, A. *Chemistry-A Eur. J.* **2002**, *8*, 685-690.
49. Pan, W., Zheng, J., Song, C. Am. Chem. Soc. Div. Fuel Chem. Prepr., **2002**, 47, 1, 262.

50. Kroll, V. C., Swaan, H. M., Mirodatos, C., J Catal. , 1996, 161, 409.
51. Pan, W., Song, C. *CO_2 Conversion and Utilization*, C. Song, A.M. Gaffney, and K. Fujimoto Eds, ACS Symp Series 809, **2002**, 316.
52. Mallinson, R. G.,Supat, K., Chavdej, S., Lobban, L., Larkin, D. Am. Chem. Soc. Div. Fuel Chem. Prepr., **2002**, 47, 1, 269.
53. Aresta, M., Dibenedetto, A., Dileo, C., Tommasi, I., Amodio E. *J. Sup. Fluid* **2002**, in press.
54. Sakakura, T., Choi, J.-C., Saito, Y., Sako, T. *Polyhedron*, **2000**, 19, 573.
55. Isaacs, N. S., O'Sullivan, B., *Tetrahedron*, **1999**, 55, 11949.
56. Krawczyk T., *INFORM*, **1996**, *7*, 801-829.
57. Worgetter M., *Biokraftstoffe fur Dieselmotoren*, Symposium N° 13982/68.236, June **1991**.
58. Scharf H., Proc. Journées Techniques - Business Workoshop on "*Rational Use of Biodiesel in Urban Fleets*", Ademe, Amiens, 24-29 September **1993**, p 67-77.
59. Ruiz-Altisent M., (1994) *Biofuels*, European Commission, Directorate General XII, EUR 16547 EN, Paris.
60. Cvengros J., Povazanec F., *Bioresource Technology*, **1996**, *55*, 145-152.
61. Korbiz M., Proc. Proc. Journées Techniques – Business Workoshop on "*Rational Use of Biodiesel in Urban Fleets*", Ademe, Amiens, 24-29 September **1993**, p. 91-105.

Simultaneous Utilization

Chapter 2

Catalysis in the Simultaneous Utilization of Greenhouse Gases

Zhen Yan, Hongshan Shang, Chaoxian Xiao, and Yuan Kou*

College of Chemistry and Molecular Engineering, Peking University, Beijing 100871, China

CH_4, NO_x, and CO_2 are greenhouse gases. Catalytic conversion of these harmful gases to environmentally benign products or efficient feedstocks represents one of the outstanding challenges of modern science and technology. Simultaneous conversion of two greenhouse gases, i.e., removal/elimination/conversion of one greenhouse gas by using another one as a reactant has been investigated, but an integrated strategy has not been established. Preliminary work currently underway in our laboratory includes NO_x-catalyzed partial oxidation of CH_4 to syngas, reaction of CH_4-CO_2 co-feed to produce transportable liquid materials and partial reduction of SO_2 by CH_4 and waste biomass inside a coal-fired boiler. Studying these new reaction pathways will significantly improve our understanding of the relationship between greenhouse gases and has great potential to afford a new basis for greenhouse gas control.

NO_x-catalyzed homogeneous partial oxidation of CH_4

Catalytic conversion and catalytic combustion of methane have received extensive attention because of their potential to synthesize useful chemicals and to reduce pollutant emissions. Both processes require activation of methane, which has been mostly achieved over solid catalysts to date.

Although the partial oxidation of methane and ethane promoted by small amounts of NO_2 was reported by Soukup *et al.* as early as 1928 (*1*), the promotional effects of NO_x on the oxidation of methane or other alkanes have only received significant attention in the past few years (*2-8*). Bañares *et al.* reported the beneficial effects of NO addition in the selective oxidation of methane to MeOH and HCHO over V_2O_5/SiO_2 catalysts (*5*). Otsuka *et al.* obtained >6% yield for oxygenates (HCHO + MeOH) in the gas phase oxidation of methane catalyzed by NO (*6, 7*). It is noteworthy that all these reactions were carried out with an excess of methane (O_2/CH_4 = 0.1-0.5). Recently, Kašpar *et al.* reported that the catalytic combustion of methane over ZrO_2-, Al_2O_3- and CeO_2-based oxide catalysts could be promoted by the addition of catalytic amounts of nitric oxide to the feed under strong oxidizing conditions, leading to a remarkable increase in CO formation (*8*). Here we show that under either weak or strong oxidizing conditions (O_2/CH_4 = 1.5-5), NO_x can act as a good catalyst for the partial oxidation of methane to CO without the use of any solid catalyst.

The experiments were carried out in a conventional gas flow system with an empty quartz tube reactor (inner diameter 5 mm). The gases were controlled by mass flow controllers and premixed before entering the reactor. The temperatures indicated in this paper are the temperatures in the reaction zone of the reactor as measured with a thermocouple. O_2, CH_4 and CO_2 in the products were analyzed by on-line gas chromatography. NO, NO_2, CO, CH_4, HCHO, and C_2H_4 were analyzed by *in situ* or on-line FT-IR using a Bruker Vector 22 spectrometer with a gas cell of 10 cm length. The concentrations were determined automatically using the OPUS Quantitative Method with appropriate calibration curves.

In the absence of NO, the oxidation of CH_4 requires temperatures higher than 850 °C. In the presence of NO, the reaction starts at a lower temperature, ca. 550 °C, and the selectivity for CO is about 40% at a temperature of 650 °C. At the same time, formaldehyde and ethylene are observed in the products formed under weak oxidizing conditions (O_2/CH_4 = 1.5). The infrared spectrum of the product mixture shows the absorption bands of CO (2200-2100 cm^{-1}), NO (1930-1830 cm^{-1}), HCHO (1745 cm^{-1}), NO_2 (1628, 1600 cm^{-1}), and C_2H_4 (949 cm^{-1}). The optimum temperature for the formation of HCHO is in the range 600-800 °C whereas that for the formation of C_2H_4 is 650-900 °C.

Under strong oxidizing conditions (O_2/CH_4 = 5), NO also exhibits considerable catalytic activities and selectivity (~68%) for the conversion of methane to CO. The results are summarized in Table I. It can be seen that, in the absence of NO, the temperatures required for the activation of CH_4 are higher than 800 °C. In the presence of NO however, the activation temperature is reduced to below 570 °C. The lowest temperature for the activation of CH_4 is only 520 °C. The $T_{50\%}$ (defined as the temperature at which there is 50% CH_4 conversion) is higher than 866 °C in the absence of NO, but this is reduced to around 600 °C when NO is present. Parallel experiments show that no reaction is observed for the CH_4-NO system (without free-oxygen) under the reaction conditions, indicating that the formation of NO_2 is a prerequisite for the activation of CH_4. In all the experiment whose results are shown in Table I, no decrease of NO/NO_2 concentrations (based on the area of infrared absorption) is observed up to 950 °C, indicating that NO_2 and other possible nitrogen oxides cannot be reduced by CH_4 in the absence of a solid catalyst. It can be seen that the reactivity in the partial oxidation of methane (without NO_x catalyst) depends significantly on the flow rate. When the flow rate increases from 20 to 40 to 80 mL/min, the $T_{50\%}$ rises from 866 to 911 to 950 °C. The selectivity for CO improves slightly with increasing temperature. The influence of the flow rate on the NO-catalyzed reactions seems less significant however. When the flow rate increases, the $T_{50\%}$ increases in steps of about 20 °C and the selectivity for CO remains unchanged. Under strong oxidizing conditions (O_2/CH_4 = 5), formation of C_2H_4 is observed at only the ppm level. No HCHO was detected. This indicates that the NO-catalyzed partial oxidation of methane is highly selective for production of CO (68%).

Table I. CH_4 conversion and CO selectivity in the partial oxidation of methane [a]

NO / ppm	*Flow rate/ mL/min*	*Temp. range of CO prodn. /°C*	*CO selec. (max.)[b] /%*	*CH_4 conv. at CO max. /%*	*Temp. for max. CO prodn.. /°C*	*$T_{50\%}$ / °C*	*TOF[d] s^{-1}*
-	20	800-880	16	79	868	866	/
-	40	840-920	25	42	910	911	/
-	80	875-	30[c]	54[c]	>950	950	/
200	40	540-760	66	88	630	595	28
600	20	520-760	68	83	610	590	5
600	40	560-780	69	90	645	612	9
600	80	570-840	67	94	680	635	13

[a] Reaction conditions: CH_4 : 0.4 mol%; O_2: 2 mol%; temperature: 200-950 °C.

[b] CO selectivity: mol CO/(mol CO + mol CO_2 + mol CH_4)

[c] Values measured at 950 °C

[d] Values measured at 650 °C

Figure 1 shows the influence of NO concentration in the feed on the partial oxidation of methane at 650 °C. Although the conversion of CH_4 and the selectivity for CO reaches a maximum when the concentration of NO is 400-600 ppm, the highest turnover frequency (TOF, methane molecules converted per second per NO) of 40 is obtained with an NO concentration of 100 ppm. The formation of NO_2 clearly implies that either NO or NO_2 itself is actually the catalyst precursor rather than the actual catalyst. Even so, the TOF is still higher than the average value obtained for conventional homogeneous catalysis in the liquid-phase, suggesting that the reactions described here are a gas-phase homogeneous catalytic reaction.

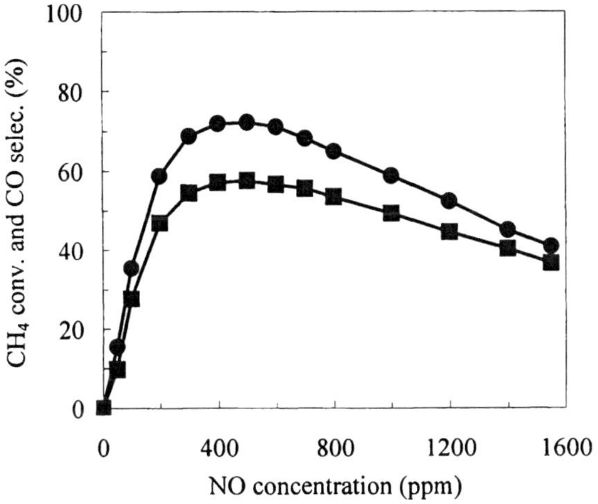

Figure 1. Effect of NO concentration on CH_4 conversion (•) and CO selectivity (■). (CH_4 0.4%, O_2 2%, flow rate 80 ml/min, temperature 650 °C)

Furthermore, in comparison with the results of Kašpar *et al.*, who observed a maximum CO selectivity of 59% at 730 °C over pure a ZrO_2 catalyst with CH_4 conversion of 85%, our results show that higher CO selectivity (>65%) can be observed at much lower temperature (by about 100 °C) in the absence of any solid catalyst. The selectivity for oxygenates or for CO may change with the conditions. Because the catalytic efficiency is very high and the system is simple, it provides an excellent opportunity for in situ investigation of the mechanism and the factors controlling the activity and selectivity in the catalytic partial oxidation of methane. In addition, the catalytic activity of NO_x demonstrated in this work may have practical application in the catalytic conversion and catalytic combustion of methane in view of the advantages of gas-phase reactions over heterogeneous reactions.

A CO_2 neutral process for the removal of SO_2 by biomass

Studies on recyclable solid sorbents (*9-13*) in dry desulfurization processes have been the subject of much recent attention. For example, in the NOXSO and Powder-Particle Fluidized Bed (PPFB) processes (*9-11*), $Na_2CO_3/\gamma\text{-}Al_2O_3$ has been employed as a recyclable sorbent for SO_2 in order to realize the subsequent transformation of SO_2 at high operating temperatures. The outlook for the actual application of the processes in currently operating power stations however, is not very promising due mainly to the complex processes involved, the high cost of the sorbents and the high energy consumption. Therefore, a cheap, easily obtainable and disposable sorbent would seem to be required for the development of a highly efficient, industrially acceptable desulfurization system. Utilization of biomaterials or abandoned biomaterials (BIOM), of which the major component is cellulose $(C_6H_{10}O_5)_n$, is a creative new approach to the problem (*14, 15*). The BIOM includes straw, rice husks and stalks, dried branches and leaves, waste paper and similar materials. The BIOM are generally buried or burnt and finally converted to CO_2, indicating that not only are they useless but also CO_2-producers.

The wheat straw employed as the BIOM in this work had a composition of 40.50% C, 5.58% H and 45.26% O (Elementar Vario EL, Germany). The typical preparation procedure for an 80 wt% Na_2CO_3/straw mixture has been described elsewhere (*14, 15*). The reference sorbent $Na_2CO_3/\gamma\text{-}Al_2O_3$ was prepared according to the literature procedure (*10*).

The SO_2 removal experiment was carried out under atmosphere pressure at 70-300 °C, using a fixed-bed quartz reactor in which the sorbent (0.33 g) was supported on a quartz frit of medium porosity. The flow rate was 40 mL/min. The simulated flue gas contained 1960 ppm SO_2 with N_2 as the balance gas. An FT-IR spectrometer (Vector 22, Bruker) with an on-line cell was used to monitor the SO_2 concentrations before and after the reactor. The adsorption efficiency η (breakthrough/stoichiometric (*16, 17*)/saturation) of the sorbent is given by $\eta = f_s / f_c \times 100$ %, where f_c is the amount in moles of Na_2CO_3 in the sorbent and f_s is the amount in moles of SO_2 adsorbed by the sorbent. The Na/S ratio is given by $r = 2 f_c / f_s$, and in theory is equal to 2.

Table II shows a comparison of Na_2CO_3/straw, $Na_2CO_3/\gamma\text{-}Al_2O_3$ and pure Na_2CO_3 for removal of SO_2 at a temperature of 80 °C. The breakthrough and stoichiometric adsorption efficiencies of 80 wt% Na_2CO_3/straw are 48.9% and 80.6% respectively, whereas those of 80 wt% $Na_2CO_3/\gamma\text{-}Al_2O_3$ are 7.6 % and 24.0 %, respectively. The activity of the 80 wt% Na_2CO_3/straw is also significantly higher than 20 wt% $Na_2CO_3/\gamma\text{-}Al_2O_3$, which is a typical sorbent employed in PPFB processes (*10*). As shown in Table II, the breakthrough adsorption efficiency of 80 wt % Na_2CO_3/straw is 11.7 percent higher than that for 20 wt % $Na_2CO_3/\gamma\text{-}Al_2O_3$, whereas the stoichiometric efficiency is 5.7 percent higher. In addition, with respect to saturation of the sorbents, the efficiency of 80 wt% Na_2CO_3/straw is significantly higher than that of the 80 wt% $Na_2CO_3/\gamma\text{-}Al_2O_3$, and is almost equal to 20 wt% $Na_2CO_3/\gamma\text{-}Al_2O_3$. It should be noted that the higher saturation efficiency of 20 wt% $Na_2CO_3/\gamma\text{-}Al_2O_3$ compared with that of 80 wt% Na_2CO_3/straw is due mainly to the adsorption properties of the $\gamma\text{-}Al_2O_3$ support (*10, 18, 19*). In brief, the data show that the adsorption efficiency and Na/S ratio for breakthrough and stoichiometric adsorption of the 80 wt% Na_2CO_3/straw are the best of all the samples investigated. XPS analysis demonstrated that the main product in each case is Na_2SO_3.

Table II. Comparison of different sorbents for removal of SO_2 at a temperature of 80 °C

	Breakthrough			*Saturation*		*Stoichiometric*	
	Efficiency / mol%	*Time / hours*	*Na/S*	*Efficiency / mol%*	*Na/S*	*Efficiency / mol%*	*Na/S*
a	0	0	-	8.6	23.2	8.6	23.2
b	7.6	0.90	26.6	24.2	8.26	24.0	8.34
c	37.2	1.10	5.38	92.1	2.17	74.9	2.67
d	48.9	5.80	4.09	85.1	2.35	80.6	2.48

NOTE: a, Na_2CO_3; b, 80 wt% $Na_2CO_3/\gamma\text{-}Al_2O_3$; c, 20 wt% $Na_2CO_3/\gamma\text{-}Al_2O_3$; d, 80 wt% Na_2CO_3/straw.

It should be noted that the reducing properties of cellulose-based materials have been reported previously. Examples include the reduction of NO, bivalent copper and hexavalent chromium by cellulose-based BIOM (*20-22*). Based on both the previous work and this study, a more efficient and complete flue gas desulfurization (FGD) process can be proposed, as shown in Figure 2.

The key steps in the proposed process are the enrichment of SO_2 and the subsequent reduction of SO_2 to S. The reactions can be described schematically as follows:

$$Na_2CO_3/BIOM + SO_2 \rightarrow Na_2SO_3/BIOM + CO_2$$

$$C_6H_{10}O_5 + 6SO_2 \rightarrow 6S + 6CO_2 + 5H_2O$$

It can be seen that in the above reactions SO_2 is first enriched by the BIOM system and then selectively reduced to elemental S by BIOM under oxygen-free conditions. This new approach combines energy efficiency, waste treatment and flue gas desulfurization in one process, and is a CO_2 neutral process. It is thus very economical in terms of cost and space and has great prospects in terms of actual industrial application.

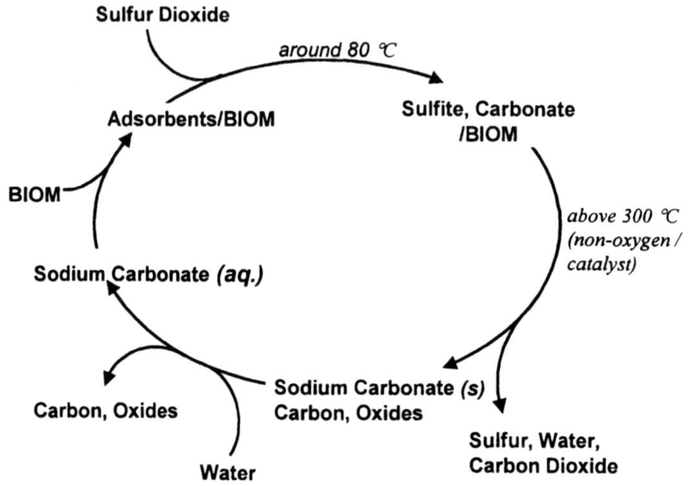

Figure 2. New strategy for the removal of SO_2 from flue gas by BIOM.

CO_2-related aromatization of lower alkanes including methane

Partial oxidation of lower alkanes, especially methane, to afford useful liquid hydrocarbons represents one of the outstanding challenges of the chemical industry. Conversion of methane through the addition of CO_2 in the feed has received significant attention in recent years. The use of CO_2 in oxidative coupling of methane was found to inhibit the further oxidation of C_2H_4, causing an increase in C_2H_4 selectivity (*23*). In the partial oxidation of methane to syngas or aromatics, it was reported that the presence of CO_2 could inhibit coke deposition (*24*). It is still a matter of dispute however, especially in the case of aromatization, whether the presence of CO_2 results in a loss of activity (*25*, *26*) or a promotional effect (*27*) towards the formation of benzene. Although conversion of C_1-C_3 alkanes in the presence of CO_2 has been investigated for many years, the role of the CO_2 under the reaction conditions and exactly what products may be obtained from the reaction are both still unclear.

Two recent studies have expanded our knowledge of methane activation in the presence of CO_2. First was the partial oxidation of methane by carbon dioxide (*28, 29*):

$$2CH_4 + 2CO_2 \rightarrow C_2H_4 + 2CO + 2H_2O$$

Second was our initial observation of Simultaneous Activation of Methane and Carbon Dioxide to Aromatics (SAMCA) (*30*, *31*):

$4CH_4 + 3CO_2 \rightarrow C_6H_6 + CO + 5H_2O$

The former reaction was estimated to have a theoretical methane conversion of 15-25%, but in fact only 6% was obtained. The reasons for this we will discuss below. The unique feature of the latter reaction, as revealed by isotopic tracer experiments using $^{13}CO_2$ in the CH_4-CO_2 co-feed, was the participation of the carbon from carbon dioxide in building up the benzene ring.

Wesendrup and Schwarz proposed in early 1995 that it would be possible to obtain coupling products if a transition metal cation were able to react simultaneously with both methane and carbon dioxide (*32*). In 1997 we proposed a bi-cationic system for SAMCA (*33*), which could hydrogenate CO_2 first to CO and subsequently to CH_x fragments. Since CH_x fragments would be produced simultaneously by oxidation of CH_4, combination of CH_x would afford C_2 and aromatic hydrocarbons, i.e.

$CO_x \rightarrow CH_x \leftarrow CH_4$

$\downarrow$

AROMATICS

This possibility has been finally realized (*27*, *31*).

Combination of the exothermic partial oxidation of CH_4 with the endothermic CO_2-reforming reaction has previously been realized in the partial oxidation of methane to syngas (*34*). Here we perform a thermodynamic analysis of SAMCA as well as Simultaneous Conversion of Alkane and Carbon Dioxide to Aromatics (SCACA). The results show that both SAMCA and SCACA are more thermodynamically favorable than other reactions occurring in the CH_4 (C_2H_6/C_3H_8)-CO_2 co-feed.

CO_2 as oxidant: n→n reaction

Direct conversion of methane at 1073 K using carbon dioxide as an oxidant is fairly endothermic when the products are C_2 hydrocarbons:

$$2CH_4 + 2CO_2 \rightarrow C_2H_4 + 2CO + 2H_2O \quad (\Delta H_{1073} = 285.8 \text{ kJ/mol } C_2) \qquad (1)$$

$$2CH_4 + CO_2 \rightarrow C_2H_6 + CO + H_2O \quad (\Delta H_{1073} = 107.8 \text{ kJ/mol } C_2) \qquad (2)$$

It is easy to see from the stoichiometry of the reactions that CO_2 merely acts as an oxidant i.e. it is an n→n ($nCO_2 \rightarrow nCO$) reaction. The calculated conversions of CH_4 at equilibrium for reactions (1) and (2) at 1073 K are 21% and 25% respectively. As shown in Table III, when $CH_4/CO_2 = 1$ the following reaction, an atom economy reaction, has a much higher conversion of ca. 45%,

$$6CH_4 + 6CO_2 \rightarrow C_6H_6 + 6H_2O + 6CO + 3H_2 \quad (\Delta H_{1073} = 805.0 \text{ kJ/mol}C_6H_6)$$

The reason that aromatics cannot be obtained via this reaction is due to its extremely high endothermic character. It is by no means clear, however, that aromatics cannot be obtained via other routes.

It is also worthy of note that aromatization of methane without free-oxygen is also a highly endothermic reaction (*25, 35*),

$$6CH_4 \rightarrow C_6H_6 + 9H_2 \qquad (\Delta H_{1073} = 602.6 \text{ kJ/mol } C_6H_6)$$

Furthermore, the addition of CO_2 to the reactant gas may make the reaction more endothermic and therefore even less thermodynamically feasible, as shown in Table III.

Several reactions may occur competitively to different degrees under methane activation conditions, as shown in Table IV. With regard to reactions (1) and (2), simultaneous occurrence of the water-gas shift reaction may make both reactions less endothermic and give higher methane conversion. For example, the reaction

$$2CH_4 + CO_2 \rightarrow C_2H_4 + H_2O + CO + H_2 \quad (\Delta H_{1073} = 252.1 \text{ kJ/mol } C_2H_4)$$

has a methane conversion of ca. 34.2%.

SAMCA(1): n→n-1 reaction

It is easy to see that the stoichiometry of SAMCA, in which the carbon dioxide is used not only to oxidize methane but also to take part in the construction of the hydrocarbon chain, is an nCO_2 to (n-x)CO reaction, where x is the number of the carbon dioxide molecules whose carbon atoms participate in the hydrocarbon construction. In SAMCA, the carbon dioxide acts as both the oxidant and a carbon source.

A series of SAMCA reactions with this stoichiometry have been recognized. Calculations of thermodynamic equilibria predict that a nCO_2 to (n-3)CO SAMCA reaction (i.e., x = 3) is impossible. The thermodynamic barrier is dramatically reduced however when x decreases from 3 to 2 and then to 1. In the reaction where ethylene is the product, an n → n-1 reaction would be

$$3CH_4 + CO_2 \rightarrow 2C_2H_4 + 2H_2O$$

Conversion of methane in this system at equilibrium at 1073 K is ca. 0.6%, and it is thus a thermodynamically disfavored reaction. However, when the product is benzene,

$$5CH_4 + CO_2 \rightarrow C_6H_6 + 2H_2O + 5H_2 \quad (\Delta H_{1073} = 410.6 \text{ kJ/mol} C_6H_6) \qquad (3)$$

the conversion of methane is ca. 13%. This reaction seems to be highly selective to aromatics without C_2 hydrocarbon by-products. In comparison with the reaction ($6CH_4 \rightarrow C_6H_6 + 9H_2$, ΔH_{1073} = 602.6 kJ/molC_6H_6), the value of ΔH_{1073} is reduced by about 200 kJ per mole of benzene, indicating that reaction

(3) is a much more feasible reaction than the aromatization of methane without free- oxygen. This suggests that the addition of CO_2 to the reaction mixture may result in a higher CH_4 conversion or a higher C_6H_6 yield, i.e., CO_2 has a "promotional effect" on the formation of aromatics.

Table III. Free Energy change (ΔG) and heat of reaction (ΔH) for the methane aromatization reactions with CO_2 as oxidant (per mole of aromatic product)

Reaction	*1073K* ΔH/ kJ/mol	ΔG/ kJ/mol	c/% [a]
$6CH_4 \rightarrow C_6H_6 + 9H_2$	602.6	109.6	22.1
$6CH_4 + CO_2 \rightarrow C_6H_6 + H_2O + CO + 8H_2$	636.3	110.0	32.2
$6CH_4 + 2CO_2 \rightarrow C_6H_6 + 2H_2O + 2CO + 7H_2$	670.1	110.3	38.4
$6CH_4 + 3CO_2 \rightarrow C_6H_6 + 3H_2O + 3CO + 6H_2$	703.8	110.7	42.3
$6CH_4 + 4CO_2 \rightarrow C_6H_6 + 4H_2O + 4CO + 5H_2$	737.5	111.1	44.5
$6CH_4 + 5CO_2 \rightarrow C_6H_6 + 5H_2O + 5CO + 4H_2$	771.3	111.4	45.5
$6CH_4 + 6CO_2 \rightarrow C_6H_6 + 6H_2O + 6CO + 3H_2$	805.0	111.8	45.5
$6CH_4 + 7CO_2 \rightarrow C_6H_6 + 7H_2O + 7CO + 2H_2$	838.8	112.2	44.6
$6CH_4 + 8CO_2 \rightarrow C_6H_6 + 8H_2O + 8CO + H_2$	872.5	112.5	42.8
$6CH_4 + 9CO_2 \rightarrow C_6H_6 + 9H_2O + 9CO$	906.2	112.9	39.0

a: The conversion is calculated for CH_4:CO_2 ratios corresponding to the reaction stoichiometry at atmosphere pressure.

Table IV. Free Energy change (ΔG) and heat of reaction (ΔH) for several reactions which may occur under methane activation conditions

Name	*Reaction*	*1073K* ΔH/ kJ/mol	ΔG/ kJ/mol	c/% [a]
Steam-Reforming	$CH_4 + H_2O \rightarrow CO + 3H_2$	225.8	-46.5	91.6
Partial Oxidation	$2CH_4 + O_2 \rightarrow 2CO + 4H_2$	-45.2	-469.8	~100
CO_2-Reforming	$CH_4 + CO_2 \rightarrow 2CO + 2H_2$	259.5	-46.1	93.2
Boudouard Reaction	$2CO \rightarrow C + CO_2$	-169.4	18.4	18.6
Water-Gas Shift	$CO + H_2O \rightarrow CO_2 + H_2$	-33.7	-0.37	50.5
Oxidative Coupling	$2CH_4 + O_2 \rightarrow C_2H_4 + 2H_2O$	-278.4	-307.7	~100

a: The conversion is calculated for reactant ratios corresponding to the reaction stoichiometry at atmosphere pressure.

Similarly, the following reaction is also viable, having an equilibrium conversion of CH_4 of ca. 11.0%,

$$6CH_4 + CO_2 \rightarrow C_7H_8 + 2H_2O + 6H_2 \quad (\Delta H_{1073} = 464.3 \text{ kJ/mol } C_7H_8) \quad (4)$$

Excessive addition of CO_2 may inhibit the water-gas shift reaction, and may make reactions (3) and (4) more endothermic, as shown in Table V.

Table V. Free Energy change (ΔG) and heat of reaction (ΔH) for n→n-1 SAMCA reactions (per mole of aromatic product)

Reaction	*1073K*		
	ΔH/ kJ/mol	*ΔG/ kJ/mol*	*c/%**
$5CH_4 + CO_2 \rightarrow C_6H_6 + 2H_2O + 5H_2$	410.6	156.5	13.3
$5CH_4 + 2CO_2 \rightarrow C_6H_6 + 3H_2O + CO + 4H_2$	444.3	156.8	19.9
$5CH_4 + 3CO_2 \rightarrow C_6H_6 + 4H_2O + 2CO + 3H_2$	478.0	157.2	23.7
$5CH_4 + 4CO_2 \rightarrow C_6H_6 + 5H_2O + 3CO + 2H_2$	511.8	157.6	25.7
$5CH_4 + 5CO_2 \rightarrow C_6H_6 + 6H_2O + 4CO + H_2$	545.5	157.9	25.9
$5CH_4 + 6CO_2 \rightarrow C_6H_6 + 7H_2O + 5CO$	579.3	158.3	23.7
$6CH_4 + CO_2 \rightarrow C_7H_8 + 2H_2O + 6H_2$	464.3	193.0	11.0
$6CH_4 + 2CO_2 \rightarrow C_7H_8 + 3H_2O + CO + 5H_2$	498.0	193.3	16.4
$6CH_4 + 3CO_2 \rightarrow C_7H_8 + 4H_2O + 2CO + 4H_2$	531.7	193.7	20.1
$6CH_4 + 4CO_2 \rightarrow C_7H_8 + 5H_2O + 3CO + 3H_2$	565.5	194.1	22.4
$6CH_4 + 5CO_2 \rightarrow C_7H_8 + 6H_2O + 4CO + 2H_2$	599.2	194.4	23.6
$6CH_4 + 6CO_2 \rightarrow C_7H_8 + 7H_2O + 5CO + H_2$	633.0	194.8	23.6
$6CH_4 + 7CO_2 \rightarrow C_7H_8 + 8H_2O + 6CO$	666.7	195.2	21.7

*The conversion is calculated for CH_4:CO_2 ratios corresponding to the reaction stoichiometry at atmosphere pressure.

SAMCA(2): n→n-2 reaction

The other type of thermodynamically feasible SAMCA reaction is an n→n-2 reaction in which nCO_2 gives (n-2)CO. In this case, ethylene as a product is also impossible and only aromatics are thermodynamically allowed:

$$4CH_4 + 2CO_2 \rightarrow C_6H_6 + 4H_2O + H_2 \quad (\Delta H_{1073} = 218.5 \text{ kJ/mol } C_6)$$

$$5CH_4 + 2CO_2 \rightarrow C_7H_8 + 4H_2O + 2H_2 \quad (\Delta H_{1073} = 272.2 \text{ kJ/mol } C_7)$$

The reactions are both endothermic reactions with theoretical CH_4 conversions of ca. 2.9% and ca. 4.2%, respectively. In comparison with reaction of aromatization of CH_4 without free-oxygen and reaction of oxidation of CH_4 by CO_2 to produce C_2H_4, these reactions are more thermodynamically feasible, suggesting that all studies of conversion of CH_4 in the presence of CO_2 as either an additive or a by-product should carefully check whether some aromatics are formed in the product mixture. If aromatics are formed, the calculated methane conversion based on the assumption that the carbon in methane is converted only to C_2 and CO (*28*) will be too low.

The water-gas shift reaction may, in principle, occur incompletely and the reactions may then be formulated as (5) and (6):

$$4CH_4 + 3CO_2 \rightarrow C_6H_6 + CO + 5H_2O \quad (\Delta H_{1073} = 252.3 \text{ kJ/mol } C_6) \quad (5)$$

$$5CH_4 + 3CO_2 \rightarrow C_7H_8 + 5H_2O + CO + H_2 \quad (\Delta H_{1073} = 306.0 \text{ kJ/mol } C_7) \quad (6)$$

The participation of the incomplete water-gas shift reaction increases the conversion of CH_4, which is ca. 4.2% for reaction (5) and ca. 6.5% for reaction (6). In principle, there should be no ethylene, but significant amounts of CO and H_2 in the products. In our previous studies (0.4 MPa, CH_4/CO_2 = 7/5), the products obtained over a Cr(VI)-Cr(III)/SiO_2 catalyst were aromatics (48.7%, benzene/toluene=2), CO (46.5%), H_2 (CO/H_2 = 3). The results are consistent with the stoichiometry of 2/3 × Eq. (2) + 1/3 × Eq. (12), leading to the total reaction:

$$13/3CH_4 + 3CO_2 = (2/3C_6H_6 + 1/3C_7H_8) + 5H_2O + CO + 1/3H_2 \quad (7)$$

The CH_4/CO_2 molar ratio required by reaction (7) is calculated to be 1.43, which is in full agreement with the ratio in the feed (7/5 = 1.40). By comparison of the calculated conversion of CH_4 (4.9%) with the result obtained (3.8%), it can be concluded that equilibrium has been reached. It is likely that the C_2 hydrocarbons observed (4.8%) are by-products obtained via reaction (1).

SCACA

Thermodynamic analysis of the SAMCA reaction suggests that, in principle, formation of aromatics from other lower alkanes, especially C_2/C_3 alkanes (*36-38*), instead of methane is feasible. A new type of catalytic reaction, simultaneous conversion of alkane and carbon dioxide to aromatics (SCACA) is therefore proposed. Some simple examples of such reactions are compared with dehydrogenation and aromatization without free-oxygen in Tables VI and VII.

It can be seen from the tables that, in terms of thermodynamics, SCACA is competitive with the dehydrogenation of lower alkanes to produce olefins, and is more feasible than the aromatization of lower alkanes without free-oxygen. This

Table VI. Free Energy change (ΔG) and heat of reaction (ΔH) for SCACA (Alkane = C_2H_6) reactions in comparison with other possible reactions (per mole of olefin or aromatic product)

Reaction	1073K		
	ΔH/ kJ/mol	*ΔG/ kJ/mol*	*c/ %**
$2C_2H_6 \rightarrow C_2H_4 + 2CH_4$	80.1	-23.6	98.8
$C_2H_6 \rightarrow C_2H_4 + H_2$	144.1	19.3	27.3
$C_2H_6 + CO_2 \rightarrow C_2H_4 + H_2O + CO$	179.7	24.4	34.6
$3C_2H_6 \rightarrow C_6H_6 + 6H_2$	377.4	-32.9	83.8
$3C_2H_6 \rightarrow C_6H_6 + 6H_2O + 6CO$	590.7	-2.3	62.3
$3C_2H_6 + CO_2 \rightarrow C_7H_8 + 2H_2O + 3H_2$ (SCACA)	240.1	18.9	54.0
$6C_2H_6 + 2CO_2 \rightarrow 7C_2H_4 + 4H_2O$ (SCACA)	100.5	33.1	5.7

*The conversion is calculated for reactant ratios corresponding to the reaction stoichiometry at atmosphere pressure.

Table VII. Free Energy change (ΔG) and heat of reaction (ΔH) for SCACA (Alkane = C_3H_8) reactions in comparison with other possible reactions (per mole of olefin or aromatic product).

Reaction	1073K		
	ΔH/ kJ/mol	*ΔG/ kJ/mol*	*c/%**
$C_3H_8 \rightarrow C_3H_6 + H_2$	129.2	22.1	17.6
$C_3H_8 + CO \rightarrow C_3H_6 + H_2O + CO$	166.4	32.3	20.9
$2C_3H_8 \rightarrow C_6H_6 + 5H_2$	316.3	-21.4	88.2
$2C_3H_8 + 5CO_2 \rightarrow C_6H_6 + 5H_2O + 5CO$	502.2	29.5	50.9
$2C_3H_8 + CO_2 \rightarrow C_7H_8 + 2H_2O + 2H_2$ (SCACA)	180.9	-0.4	68.8

*The conversion is calculated for reactant ratios corresponding to the reaction stoichiometry at atmosphere pressure.

suggests that all studies of conversion of lower alkanes in the presence of CO_2 as either an additive or a by-product (*36-38*) should also carefully check whether some aromatics are present in the product mixture.

In conclusion, three types of reactions of CO_2 potentially important in the activation of lower alkanes, namely, n→n, n→n-1, and n→n-2, have been identified on the basis of thermodynamic analysis. We believe that further study of the dynamics of these reactions is warranted and that the principles established here provide a basis for such a study.

Acknowledgement

This project was supported by the Ministry of Science and Technology, China (G1999022202).

References

1. Layng, T. E.; Soukup R. *Ind. Eng. Chem.* **1928,** *20*, 1052-1055.
2. McConkey, B. H.; Wilkinson, P. R. *I&EC Proc. Des. Dev.* **1967,** *6*, 436-440.
3. Irusta, S.; Lombardo, E. A.; Miró, E. E. *Catal. Lett.* **1994,** *29*, 339-348.
4. Han, L. B.; Tsubota, S.; Haruta, M. *Chem. Lett.* **1995,** 931-932.
5. Bañares, A. M.; Cardoso, J. H.; Hutchings, G. J.; Bueno, J. M. C.; Fierro, J. L. G. *Catal. Lett.* **1998,** *56*, 149-153.
6. Otsuka, K.; Takahashi, R.; Amakawa, K.; Yamanaka, I. *Catal. Today* **1998,** *45*, 23-28.
7. Otsuka, K.; Takahashi, R.; Yamanaka, I. *J. Catal.* **1999,** *185*, 182-191.
8. Fornasiero, P.; Kašpar, J.; Fagotto, S.; Graziani, M. *J. Catal.* **2000,** *189*, 463-466.
9. Ma, W. T.; Haslbeck, J. L. *Environ. Prog.* **1993,** *12*, 163-168.
10. Xu, G.; Luo, G.; Akamatsu, H.; Kato, K. *Ind. Eng. Chem. Res.* **2000,** *39*, 2190-2198.
11. Snip, O. C.; Wood, M.; Korbee, R. Schouten, J. C. *Chem. Eng. Sci.* **1996,** *51*, 2021-2029.
12. Deng, S. G.; Lin, Y. S. *Ind. Eng. Chem. Res.* **1996,** *35*, 1429-1437.
13. Kiel, J. H. A.; Prins, W. *Appl. Catal. B: Environ.* **1992,** *1*, 13-39.
14. Kou, Y.; Shang, H. S. China Patent application number 01115505, 2001.
15. Kou, Y.; Shang, H. S. China Patent application number 01115506.X, 2001.
16. Stejskalova, K.; Bastl, Z.; Mocek, K. *Collect. Czech. Chem. Commun.* **1996,** *61*, 1141-1157.
17. Mocek, K.; Stejskalova, K.; Bach, P. *Collect. Czech. Chem. Commun.* **1996,** *61*, 825-836.
18. Pieplu, A.; Saur, O; Lavalley, J. C. *J. Catal.* **1996,** *159*, 394-400.
19. Yoo, K. S.; Kim, S. D.; Park, S. B. *Ind. Eng. Chem. Res.* **1994,** *39*, 1786-1791.
20. Bhalla, S.; Sarma, V. *Proc. - Annu. Int. Pittsburgh Coal Conf. 16th*, **1999,** 404-418.

21. Andreas, R. M.; Dirk, S. R.; Hans, H.; Gerard, A. V.; Soebhaas, R. I. European Patent EP908525, 1999.
22. Jorge, F. S.; Santos, T. M.; de Jesus. J. P. *Wood Sci. Technol.* **1999,** *33*, 487-501.
23. Lunsford, J. H. *Angew. Chem. Int. Ed. Engl.* **1995,** *34*, 970-980.
24. Erdöhelyi, A.; Cserényi, J.; Solymosi, F. *J. Catal.* **1993,** *141*, 287-299.
25. Wang, D.; Rosynek, M.; Lunsford, J. H. *Topics Catal.* **1996,** *3*, 289-297.
26. Wang, D.; Lunsford, J. H.; Rosynek, M. *J. Catal.* **1997,** *169*, 347-358.
27. Liu, S.; Dong, Q.; Ohnishi, R.; Ichikawa, M. *Chem. Commun.* **1998,** 1217-1218.
28. Wang, Y.; Takahashi, Y.; Ohtsuka, Y. *J. Catal.* **1999,** *186*, 160-168.
29. Wang, Y.; Ohtsuka, Y. *J. Catal.* **2000,** *192*, 252-255.
30. Kou, Y.; Wang, H.-L.; Zhang, H.; Yang, X.-Y. *Catal. Today* **1999,** *51*, 47-57.
31. Zhang, H.; Kou, Y. *Chem. Commun.* **1999,** 1729-1730.
32. Wesendrup, R.; Schwarz, H. *Angew. Chem. Int. Ed. Engl.* **1995,** *34*, 2033-2035.
33. Kou, Y.; Wang, H.-L. 5th China-Japan Bilateral Symposium on Effective Utilization of Carbon Resources, Invited Lecture, Proceedings, Chengdu, China, 1997, pp 5-.
34. Ashcroft, A. T.; Cheetham, A. K.; Green, M. L. H.; Vernon, P. D. F. *Nature* **1991,** *352*, 225-226.
35. Liu, W.; Xu, Y. *J. Catal.* **1999,** *185*, 386-392, and references therein.
36. Krylov, O. V. *Catal. Today*, **1992,** *13*, 481-486.
37. Krylov, O. V.; Mamedov, A. K.; Mirzabekova, S. R. *Catal. Today* **1995,** *24*, 371-375.
38. Krylov, O. V.; Mamedov, A. K.; Mirzabekova, S. R. *Ind. Eng. Chem. Res.* **1995,** *34*, 474-482.

Chapter 3

Novel Highly Active Reforming Catalyst System and Production of Pure H_2 from CH_4 and CO_2 with Pd–Ag Membrane Reactor

Linsheng Wang*, Kazuhisa Murata, and Megumu Inaba

National Institute of Advanced Industrial Science and Technology, AIST, Tsukuba Central 5, Tsukuba, Ibaraki 305, Japan
***Corresponding author: email: linsheng.wang@aist.go.jp**

Re/HZSM-5 as a novel highly active reforming catalyst system for production of highly pure hydrogen and syngas from CO_2 and methane at mild reaction temperatures has been studied in the present paper. Compared to the traditional group VIII metal catalysts, Re/HZSM-5 is much more active for methane reforming with CO_2 at mild temperatures. The novel catalyst system exhibits quite high stability for the CO_2 reforming reaction of methane with time on stream. High activity is still maintained well after methane reforming for 80hrs. The high activity of the novel catalyst system at mild temperature makes it more suitable for production of highly pure hydrogen by integration of the CO_2 reforming reaction and hydrogen separation with a Pd type membrane reactor. In addition, the much higher conversions of methane and CO_2 beyond the thermodynamic equilibrium are attained when hydrogen is continuously separated from the product by hydrogen selective permeation through the membrane tuble in the reactor.

Pure hydrogen is a clean energy resource for use in fell cell apparatus and a basic petrochemical feedstock. Syngas ($CO+H_2$) is a valuable feedstock for production of gasoline and oxygenates (Fischer-Tropsch reaction). Recently CO_2 reforming reaction of methane on Group VIII metals and Mo or W carbide based catalysts at high temperatures has attracted a increasing attention [2-14] since the reaction was first studied by Fischer and Tropsch in 1928 [1]. SiO_2, Al_2O_3, TiO_2, MgO, ZrO_2, La_2O_3, NaY, NaZSM-5 were used as supporting materials of the catalysts for the reforming reaction. Among them the Rh based catalyst had been reported as one of the most active catalyst for the CO_2 reforming reaction of methane. Additionally, most of the reported reforming catalysts are exactly that which have been used for catalyzing the homologation of methane into higher hydrocarbons by two steps [15-17]. Methane is firstly dehydrogenated into CHx (x=0-3) at a higher temperature, which are hydrogenated into higher hydrocarbons such as ethane at a lower temperature in second step. Activation of methane into CHx intermediate specie on the catalysts is the first step for both reforming and homologation reaction of methane. Therefore, it can be assumed that the homologation and reforming reaction of methane take place on the same active phase of the catalysts. More recently, some improvement on the homologation reaction of methane has been achieved because the bifunctional Re based catalysts have been used for the activation of methane. Higher hydrocarbons such as aromatics and C_2 hydrocarbons can be produced from methane in one step on Re/HZSM-5 [18, 19]. However, the catalytic performance of Re based catalysts for the reforming reaction of methane and CO_2 has not been reported so far. In addition, more attention has recently been paid to the CO_2 reforming reaction because it produced the higher CO/H_2 ratio syngas than steam reforming or partial oxidation reaction. It also provides an elegant route for the utilization of CO_2. However, no promising catalyst has been reported for CO_2 reforming reaction of higher hydrocarbons such as ethane although steam reforming of ethane on nickel/alumina catalysts has been studied [20,21]. It has been an unsolved problem in the current catalysis research over the past years. In the present paper, a novel highly active catalyst system for CO_2 reforming reaction of methane and ethane is reported and highly pure hydrogen is directly produced from methane and ethane on the novel catalyst at mild temperatures by integration of the reforming reaction and H_2 separation with a Pd membrane reactor.

Experimental Section

Catalyst preparation: Re/HZSM-5 and Rh/HZSM-5 were prepared by impregnating NH_4ZSM-5 (SiO_2/Al_2O_3=30) with NH_4ReO_4 and $RhCl_3$ aqueous solution respectively, followed by drying at 383 K for 6 h and calcination at 773 K for 6 h. ReK/HZSM-5 and RhK/HZSM-5 were prepare by impregnating the

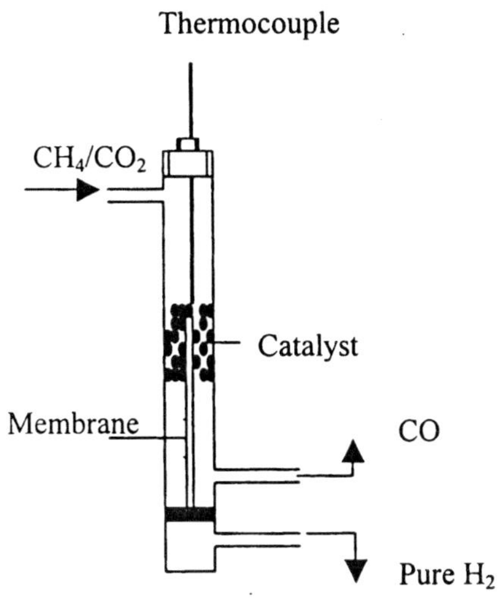

Fig 1. Scheme for the Pd-Ag Membrane reactor

Re/HZSM-5 and Rh/HZSM-5 with KNO_3 aqueous solution, followed by drying at 383 K for 6 h and calcination at 773 K for 6 h. The samples were finally pressed, crushed and sorted into 20-40 meshes.

Membrane reactor:

The membrane reactor is with 5/8" diameter and by 10" tall. It contains four membrane tubes with 1/8" diameter and by 7" tall. H_2 was permeated through the membrane into the tubes and removed by vacuum pump immediately after produced. The selectivity for hydrogen permeation through the membrane is 100% and ultra high pure H_2 was produced as a result. The membrane reactor is schemed as Fig 1.

Catalytic tests: Catalytic tests were carried out with about 200-2000 mg catalyst placed in a fixed bed continuous-flow membrane reactor. The catalysts were reduced by hydrogen at 623-673K for 2h before the reforming reaction. Conversion, selectivity and formation rate of products were calculated by an internal standard analyzing method. About 10% N_2 was added into methane feed and about 2% N_2 was added into ethane feed as internal standard.

Results and discussion

1. Bifunctional effect of the novel catalyst system

One of the reported most promising catalysts for CO_2 reforming reaction of methane is supported noble metal Rh based catalyst, which is active for homologation of methane into higher hydrocarbons by two steps. Recently, we have reported that supported bifunctional Re based catalysts are highly active for conversion of methane into hydrogen and higher hydrocarbons in one step. The catalytic performance of the Re based catalysts for CO_2 reforming reaction of methane has been studied as a novel catalyst system and compared with traditional Rh based catalyst in table 1.

Table 1. Activity of Re/HZSM-5 catalyst for reforming reaction of CH_4 with CO_2 at 973 K

Catalyst	Conversion (%) CH_4	Conversion (%) CO_2	Consumption rate ratio of CH_4/CO_2	H_2/CO ratio in product
Re/HZSM-5	83	80	1.07	0.81
Rh/HZSM-5	78	72	0.96	0.80
Rh/KZSM-5	78	78	0.89	0.80
Rh/Al_2O_3	77	69	0.97	0.76
Re/Al_2O_3	78	69	0.99	0.76
Re/KZSM-5	77	67	1.01	0.76

Methane SV=5000 ml/h/g
Feed ratio (CH_4/CO_2)=1.05
Temperature=973K

The novel Re based catalyst exhibits a quite high activity for the reforming reaction. The high methane and CO_2 conversions of 83% and 80% were attained on Re/HZSM-5 at 973K as shown in table1. It is more active than the Rh based catalyst system. The conversions of methane and carbon dioxide on Rh/HZSM-5 are 78% and 72% at the exactly same reaction conditions. In addition, H acid sites on Rh/HZSM-5 seem not necessary for the CO_2 reforming reaction because Rh/HZSM-5 and Rh/KZSM-5 exhibit the similar activity as shown in table 1. The high activity of the novel bifunctional Re/HZSM-5 for the reforming reaction is due to its high activity for one-step activation of methane. It is also indicated in table 1 that the Bronsted acid of ZSM-5 support are favourable for the high activity of Re/HZSM-5. The monofunctional Re/KZSM-5 (B acid sites of HZSM-5 support were poisoned by K) and bifunctional Re/Al_2O_3 give a little bit less activity for carbon dioxide reforming of methane. In addition to its high activity, the stability of the Re/HZSM-5 catalyst with time on stream is also quite high. The conversions of CH_4 and CO_2 and the selectivity to H_2 on the Re/HZSM-5 catalyst are still quite high even after 80hrs on the stream as shown in fig 2.

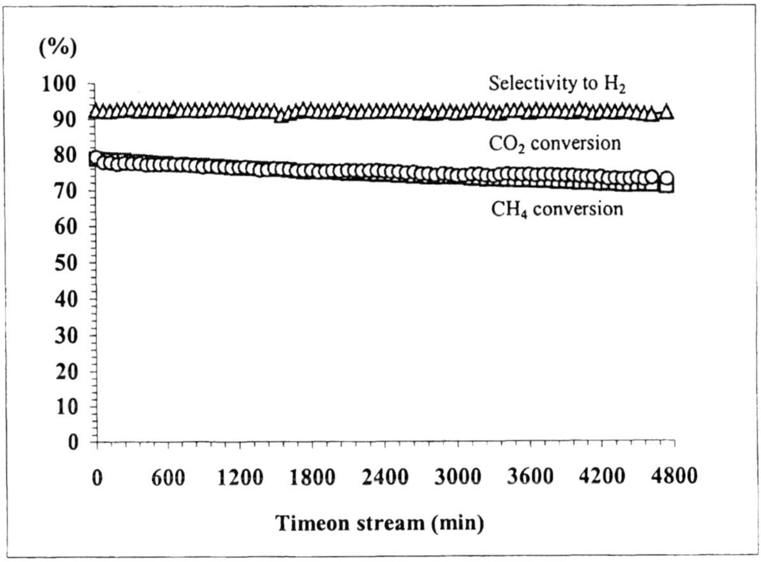

Fig 2. The conversions of CH_4 and CO_2 and the selectivity to H_2 with time on stream for the reforming reaction over Re/HZSM-5 at SV of 10000 ml/h/g and 973K

The high activity of Re based catalyst is especially desirable for hydrogen production from the greenhouses of methane and carbon dioxide with Pd based membrane reactor at mild temperatures because Pd based membrane materials exhibits better durability at mild temperatures (at 873K or lower). The low-temperature activity of Re/HZSM-5 catalyst for the dry reforming reaction is also evaluated and the result is listed in table 2.

Table 2. Activity of Re/HZSM-5 for the reforming reaction of CH_4 with CO_2 at lower temperatures

Temperature (K)	Conversion (%) CH_4	CO_2	Consumption rate ratio of CH_4/CO_2	Formation rate ratio of H_2/CO
773	15	26	0.61	0.15
818	30	44	0.70	0.60
858	41	49	0.85	0.66
973	83	80	1.07	0.81

Methane SV=5000 ml/h/g
Feed ratio (CH_4/CO_2)=1.05

It is indicated in table 2 that the Re/HZSM-5 exhibits a considerable activity even at as low as 773-858K and the conversions of methane and CO_2 are 41% and 49% at 858K. The excellent low temperature activity of Re/HZSM-5 makes it have many advantages for the integration of the catalytic reaction of the reforming and hydrogen separation because the Pd type membrane is limited to use only at relatively mild temperatures.

In the lower reaction temperatures, the better catalytic performance of Re based catalysts for CO_2 reforming of methane becomes more clear. The formation rates of CO and H_2 for CO_2 reforming of CH_4 at 858K on Re/HZSM-5 and Rh/HZSM-5 with time on stream are shown in fig 3. It is clearly indicated in fig 3 that the formation rates of CO and H_2 for CO_2 reforming of CH_4 on Re/HZSM-5 is higher than that on Rh/HZSM-5. Futhermore, the formation rates of CO and H_2 on Re/HZSM-5 are quite stable with time on stream. However, the formation rates of CO and H_2 on Rh/HZSM-5 decline with time on stream. Therefore, the Re based catalyst exhibits much better resistance to coke deposition than Rh based catalyst.

2 The activity of Re based catalysts for CO_2 reforming of ethane

Over the past years, CO_2 reforming reaction only focused on methane feed. CO_2 reforming reaction of higher hydrocarbon such as ethane has not been reported so far. Table 3 lists the results for CO_2 reforming reaction of ethane on Re and Rh based catalysts. The Re based catalysts exhibits a unique high activity for ethane reforming with CO_2. Ethane conversion on the Re based catalysts is about 100% at a reaction temperature of 973 K and ethane SV of 640ml/min as shown in table 3. However, ethane conversion on Rh based catalysts is only about 62% at the same reaction conditions and it is much lower than that on the Re based catalyst. Therefore, ZSM-5 supported Re catalyst is also a highly active novel catalyst system for CO_2 reforming of ethane. It is also indicated in table 3 the formation rate of CH_4 is only about 1/30 of the formation rate of CO. Therefore, the Re based catalysts exhibit quite high selectivity for syngas production from carbon dioxide and ethane.

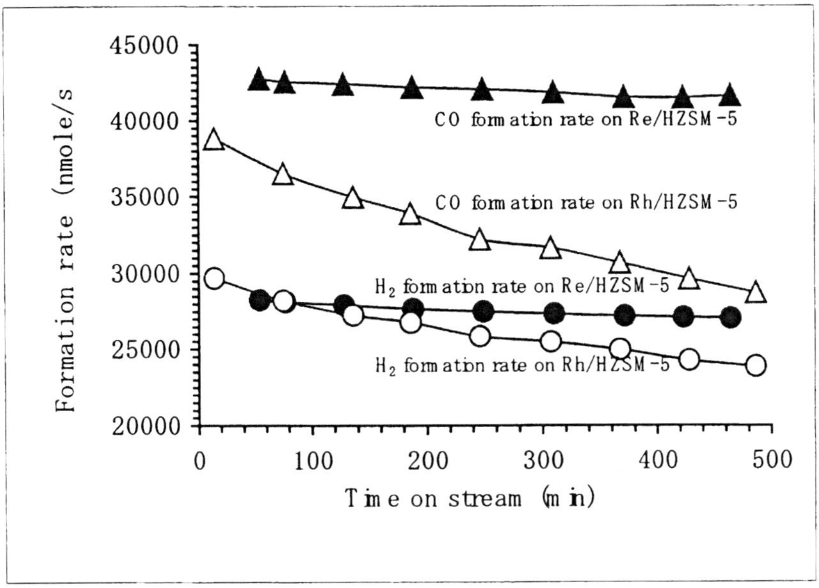

Methane SV=5000 ml/h/g
Feed ratio (CH_4/CO_2)=1.05
Reaction temperature=858K

Fig 3. Comparison of activity and stability of Re/HZSM-5 and Rh/HZSM-5 for CO_2 reforming of CH_4 into CO and H_2

Table 3. Activity of the Re based catalysts for CO_2 reforming of ethane and their comparison with Rh based catalysts

Catalyst	Conversion (%)		Formation rates (μ mole/s/g)			
	C_2H_6	CO_2	H_2	CO	CO/H_2	CH_4
Rh/HZSM-5	62.1	56.0	10.6	17.6	1.66	0.48
Rh/KZSM-5	47.3	49.6	8.1	13.3	1.64	0.21
Re/HZSM-5	100	75.4	17.4	30.3	1.74	1.10
Re/KZSM-5	100	78.7	17.8	30.1	1.69	0.99

Reaction temperature=973K
Ethane SV=640 ml/h/g
$CO_2/C_2H_6/N_2$=2.30/1/4

Table 4. Effect of CO_2/C_2H_6 ratio in feed on reforming reaction on Re/HZSM-5

$CO_2/C_2H_6/N_2$	Conversion(%)		Formation rates (μ mole/s/g)			
Ratio in feed	C_2H_6	CO_2	H_2	CO	CO/H_2	CH_4
1.1/1/4	100	94.5	12.1	20.3	1.68	4.9
1.5/1/4	100	81.5	15.4	25.1	1.63	2.8
2.3/1/4	100	75.4	17.4	30.3	1.74	1.1
2.7/1/4	100	71.3	17.3	31.5	1.81	0.85

Reaction temperature=973K
Ethane SV=640 ml/min

To optimize the reaction conditions for the reforming reaction of ethane with carbon dioxide, table 4 shows the effect of CO_2/C_2H_6 ratio in feed on formation rates of H_2, CO and CH_4. It is indicated in table 4 that the formation rates of H_2 and CO are remarkably increased and formation of CH_4 is effectively suppressed with increasing CO_2/C_2H_6 ratio. The maximum formation rates of H_2 and CO are attained when CO_2/C_2H_6 ratio is about 2.3. The over-amount of CO_2 in feed is more favorable for suppressing formation of methane by the secondary reaction of methane with CO_2 to produce H and CO. It also can be seen from table 4 that the CO_2 reforming of ethane produces the syngas with much high CO/H_2 ratios than that from the reforming of methane. The CO/H_2 ratio of the product for ethane reforming is about between 1.7 and 1.8 as shown in table 4. This is because the reforming reaction of methane with CO_2 proceeds according to the equation (1) and the reforming reaction of ethane with CO_2 proceeds according to the equation (2).

$$CO_2 + CH_4 = 2CO + 2H_2 \qquad (1)$$

$$C_2H_6 + 2CO_2 = 4CO + 3H_2 \qquad (2)$$

Therefore, the CO_2 reforming reactions of methane and ethane produce the syngas with different CO/H_2 ratio.

3 Production of highly pure H_2 by integration of the reaction and H_2 separation

Pure hydrogen is highly desired energy resource for fuel cell. Highly pure hydrogen is directly produced from CO_2 reforming of methane or ethane on the novel Re based catalyst by integration of the reforming reaction and hydrogen separation with a Pd based membrane reactor as shown in figure 1. In addition, the reforming reaction is greatly promoted by hydrogen separation in the membrane reactor as shown in table 5.

Table 5. Effect of H_2 separation on the conversions of CH_4 and CO_2 at different SV

SV of CH_4 (ml/h/g)	Permeability of H_2 (%)	Conversion(%) CH_4	CO_2	Ratio of consumption rate of CH_4/CO_2	H_2/CO ratio in product
325	92	76.9	60.5	1.15	*
	0	48.3	48.0	0.91	0.63
650	92	69.2	45.6	0.98	*
	0	47.9	34.4	0.90	0.71
1300	90	62.7	42.6	0.95	*
	0	47.1	33.3	0.92	0.71

Feed ratio (CH_4/CO_2)=0.69
Temperature=858K
* H_2 separated

It is clear in table 5 that methane conversion for the reforming reaction is about 47-48% and it changes little with varying methane SV from 1300 to 325 ml/h/g when hydrogen was not separated from product at the low temperature of 858K. However, the much higher methane conversion of about 77% has been achieved at the same temperature when hydrogen was separated from product by selective permeation through the membrane tubes in the reactor. CO_2 conversion is also enhanced by the hydrogen separation at the same time. This indicates that the conversions of CH_4 and CO_2 are not high at the low temperature when hydrogen is not separated because the reaction is constrained by thermodynamic equilibrium. However, the much higher conversions are attained when hydrogen is continuously removed from product with the membrane reactor because the thermodynamic equilibrium is disturbed by hydrogen separation. In addition, the ratio of feed consumption rates (CH_4/CO_2) is more near unity when hydrogen is separated from product. This is because some side reaction such as reduction of CO_2 by H_2 in to CO is effectively suppressed by the continuously hydrogen separation from product with the membrane.

4 Optimization of the reaction conditions for reforming reaction of methane with CO_2

The conversions of CH_4 and CO_2 and the H_2/CO ratio of product for the reforming of methane with CO_2 are also partially dependent on the CH_4/CO_2

ratio in feed. The effect of CH_4/CO_2 ratio in feed on the conversions of CH_4 and CO_2 is shown in table 6. The maximum methane conversion is obtained when the CH_4/CO_2 ratio is near 1. However, CO_2 conversion increases with increasing the ratio of CH_4/CO_2 in feed. The H_2/CO ratio is near unity is at the higher CH_4/CO_2 ratio. In addition, the ratio of the consumption rates of methane and carbon dioxide is about 1 even at the different feed ratio (CH_4/CO_2).

5 Advantages of the novel catalyst system for a commercialized process

Production of pure hydrogen or syngas from the reforming reaction of light hydrocarbons with CO_2 has been highly desired for the chemical industries because it is not only a fascinating rout for the utilization of naturals gas but also it is of significance for consumption of the greenhouse gases. Although Rh based catalyst have been reported as a most promising catalyst system for CO_2 reforming reaction of methane, the reforming reaction of ethane with CO_2 has been kept as an unsolved problem because no promising catalyst has been reported for the reaction so far. It can be assumed that the discovery of the novel highly active Re based catalysts for the CO_2 reforming reaction of light hydrocarbons will play a promoting effect on development of a practical process for production of pure hydrogen or syngas from light hydrocarbons and CO_2. The Advantages of the novel Re based catalyst system for use in a commercialized process are summarized as the following points: i) The novel catalyst system is not only highly active for CO_2 reforming of methane but also a promising catalyst for the ethane reforming reaction. The syngas with much

Table 6. Dependence of CH_4 and CO_2 conversions on the feed ratio (CH_4/CO_2)

Feed ratio CH_4/CO_2	Conversion (%) CH_4	CO_2	Ratio of consumption rates of CH_4 and CO_2	H_2/CO ratio in product
0.81	77	59	0.89	0.67
0.91	83	64	1.01	0.71
1.05	83	80	1.07	0.81
1.47	66	83	0.99	0.84
1.64	45	94	0.99	0.92

Eethane SV=5000 ml/h/g
Temperature=973K
Catalyst: Re/HZSM-5

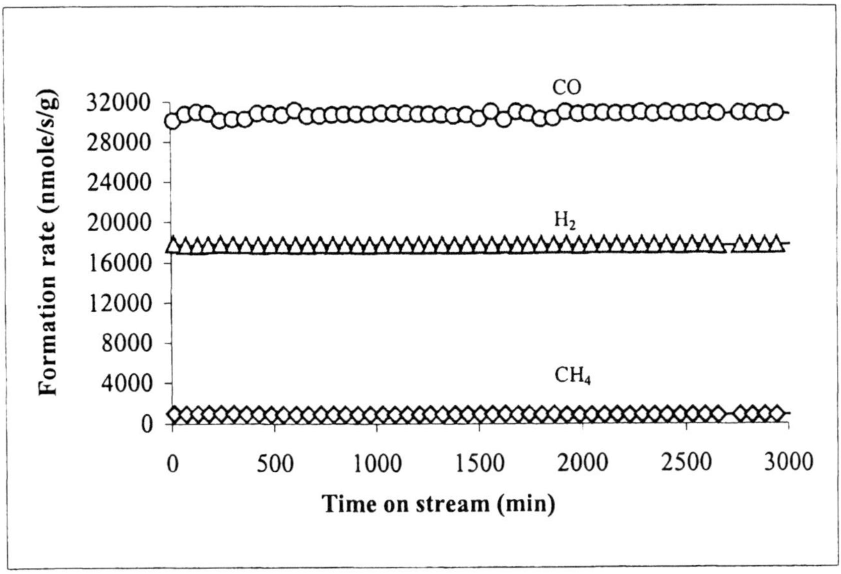

Fig 4. Formation rates of the products for CO_2 reforming of ethane with time on stream

higher CO/H_2 ratio can be produced from the reforming reaction of ethane with CO_2. Therefore, it opens a new route for production of high CO/H_2 syngas from ethane and CO_2. ii) The high activity, especially the high activity at lower temperatures of the novel catalyst system is highly desired for production of pure H_2 by integration of the reforming reaction and H_2 separation with the Pd based membrane reactor. The conversion of methane and CO_2 beyond the thermodynamic equilibrium has been achieved in the Pd type membrane reactor for production of hydrogen. iii) The novel catalyst system exhibits a high stability for the CO_2 reforming reaction of methane but also exhibits unique high stability for the CO_2 reforming reaction of ethane with time on stream as shown in figure 4.

Conclusion

Re/HZSM-5 is a novel highly active reforming catalyst system for production of highly pure hydrogen and syngas from CO_2 and light hydrocarbons such as methane and ethane at mild reaction temperatures. The novel catalyst system is much more active both for methane reforming and ethane reforming with CO_2 than the traditional group VIII metal catalysts at mild reaction temperatures. The syngas with higher CO/H_2 ratio are produced by ethane reforming with CO_2 on the novel catalyst system. The novel catalyst system is highly stable for CO_2 reforming of methane and ethane with time on stream. The high activity of the novel catalyst is maintained well even after ethane reforming for over 50 hrs and methane reforming for 80hrs. Highly pure hydrogen is produced as an alternative product by integration of the CO_2 reforming reaction and hydrogen separation with a Pd type membrane reactor because the high activity of the novel catalyst system at mild temperature. In addition, the much higher conversions of methane and CO_2 beyond the thermodynamic equilibrium are attained when hydrogen is continuously separated from the product by hydrogen selective permeation through the membrane.

References

1. F. Fischer and H. Tropsch, Brennst. Chem. 1928, 3, 39
2. F. Solymosi, Gy. Kutsan and A. Erdohelyi, Catalysis Letters, 2000, 11, 149
3. Z. L. Zhang, V. A. Tsipouriari, A. M. Efstathiou and X. E. Verykios, Journal of Catalysis, 1996, 158, 51
4. M. C. J. Bradford and M. A. Vannice, Journal of Catalysis, 1998, 173, 157
5. M. C. J. Bradford and M. A. Vannice, Journal of Catalysis, 1999, 183, 69
6. M. F. Mark and W. F. Maier, Journal of Catalysis, 1996, 164, 122
7. J. H. Bitter, K. Seshan and J. A. Lercher, Journal of Catalysis, 1998, 176, 93
8. R. N. Bhat and W. M. H. Sachtler, Applied Catalysis A: General, 1997, 150, 279
9. M. Tsuji, T. Miyao and S. Naito, Catalysis Letters, 2000, 69, 195
10. A. J. Brungs, A. P. E. York, J. B. Claridge, C. Marquez-Alvarez and M. L. H. Green, Catalysis Letters, 2000, 70, 117
11. M. E. S. Hegarty, A.M. O'Connor and J. R. H. Ross, Catalysis Today, 1998, 42, 225
12. S. M. Stagg-William, F. Noronha, G. Fendley, and D. E. Resasco, Journal of Catalysis, 2000, 194, 240
13. S. Murata and K. I. Aika, Applied Catalysis A, 1992, 82, 1
14. V. C. H. Kroll, H. M. Swaan, S. Lacombe, M. Mirodatos, Journal of Catalysis, 1997, 164, 38
15. L. Guczi, K. V. Sarma and L. Borko, Journal of Catalysis, 1997, 167, 495
16. M. Belgued, A. Amariglio, L. Lefort, P. Pareja and H. Amariglio, Journal of Catalysis, 1996, 161, 282
17. T. Koerts, M. J. A. G. Deelen and R. A. Van Santen, Journal of Catalysis, 1992, 138, 101
18. L. Wang, R. Ohnishi and M. Ichikawa, Journal of Catalysis, 2000, 190, 276
19. L. Wang, K. Murata, A. Sayari, B. Grandjean and M. Inaba, ChemComm., 2001, 19, 1952
20. B. Kneale, J. R. H. Ross, Faraday Discussion. Chem. Soc., 1982, 72, 157
21. J. R. Rostrup-Nielsen, Journal of Catalysis, 1973, 31, 173

Chapter 4

Dry Reforming of Ethane on Trimetallic Perovskites $LaCo_xFe_{1-x}O_3$: Characterizations and Reactivity

G. Rodriguez[1,2], L. Bedel[1], A. C. Roger[1], L. Udron[1], L. Carballo[2], and A. Kiennemann[1,*]

[1]LMSPC-ECPM, UMR CNRS 7515, 25, rue Becquerel 67087 Strasbourg Cedex 2, France

[2]Universidad Nacional Colombia, Santa Fé de Bogota, Colombia

Performances of $LaCo_xFe_{1-x}O_3$ perovskites as catalysts in dry reforming of ethane have been studied. Depending on x, the crystalline structure of the perovskite can be either orthorhombic or rhombohedral. The catalytic activity of the perovskite is tightly linked to its structure. If $x \leq 0.5$ the reducing catalytic atmosphere leads to the formation of Co^0 particles in strong interaction with a $LaCo_yFe_{1-y}$ ($y<x$) remaining oxide. Catalysts with higher Co loading ($x>0.5$) are more easily reduced during the reactivity tests and lead to the formation of Co^0 on a bimetallic LaFe perovskite. The catalytic activity strongly increases when the perovskites crystalline structure turns from orthorhombic to rhombohedral.

Introduction

Natural gas is abundant, clean and easy to drill. However natural gas field can contain large percentage of ethane and it is of interest to convert methane and ethane on the production sites into liquid fuels via singas (CO, CO_2, H_2) (*1*). Their transformation occurs via partial oxidation with oxygen, carbon dioxide or often water (respectively oxy, dry and vaporeforming) but most of the results have been obtained on methane. Due to a fast turnover rate, a low cost and a relatively long-term stability, nickel deposited on α-alumina is usually used in industry (*2*). Other metals such as cobalt, iron but also noble metals (Pt, Rh, Pd, Ru) have been tested more recently (*3-5*). The fast deactivation by excess of carbon deposits is the main drawback of metal supported systems (*6*). Basic supports like MgO, La_2O_3, CaO have been used to increase the catalyst lifetime (*7*). Recently, another way has been explored through formation of well-defined structures: pyrochlore, zircon and perovskite containing both active metals and rare earths (*8-11*). These defined structures are either partially or totally reduced in order to form active metal entities. However, for bimetallic perovskites such as $LaNiO_3$ or $LaCoO_3$, the total reduction temperature is much lower than the reaction temperature. As a consequence, these catalysts are quickly reduced and then act as supported metals (*12*).

Addition of a third element could be beneficial to increase the reduction temperature. In the case of ABO_3 perovskites, cations A or/and B can be partially substituted with formation of compounds such as $Ca_{0.8}Sr_{0.2}Ti_{0.8}Ni_{0.2}O_3$; $La_{0.66}Sr_{0.34}Ni_{0.3}Co_{0.7}O_3$ or $LaNi_xAl_{1-x}O_3$ (13-15). We have shown for example that the partial substitution of Ni by Fe in $LaNiO_3$ increases the nickel reduction temperature from 550 up to 850°C depending on the iron content (16, 17). Strong Ni-trimetallic LaNiFe perovskite interactions decrease the sintering of nickel particles and as a consequence increase the catalysts lifetime through the decrease of carbon formation. Due to the good ability of perovskites to be B-site substituted, Ni could be replaced by any other metal which has a metal-oxygen distance, when six-fold co-ordinated, so that Goldschmidt's tolerance factor t of the considered perovskite is between 0.8 and 1.05. When using metal-oxygen distance instead of ionic radii (18), the Goldschmidt's factor t can be defined as:

$t = \dfrac{d(A-O)}{\sqrt{2}d(B-O)}$, where d (A-O) and d (B-O) are the distances between O and the cation in the A or B site. For example the Co^{3+}-O distance (Co six-fold co-ordinated) is 2.003 Å and stable $LaCo_xFe_{1-x}O_3$ perovskite can be obtained.

The aim of the present study, after testing LaNiFe perovskite, (16,17) is to point out the interest of the LaCoFe defined structures in dry reforming of ethane.

Experimental Part

Preparation

The mixed $LaCo_xFe_{1-x}O_3$ perovskites with x values varying from 0 to 1 (x = 0, 0.3, 0.5, 0.7, 1) were prepared from La acetate, Co acetate and Fe powder via a sol-gel related method (16). The starting materials were separately dissolved in hot propionic acid under stirring and then mixed. After 30 minutes stirring, part of the propionic acid of the boiling resulting solution was evaporated until a resin was obtained. This resin was heated 4 hours at 750°C with a ramping of 3°C.mn^{-1}. BET surface areas range between 3 (x=0) and 6 $m^2.g^{-1}$ (x = 1).

Characterisation

The nature of the obtained phases and the lattice parameters were determined by powder X-Ray diffraction (XRD) recorded on a Siemens D-5000 diffractometer using the CuKα radiation.

Transmission electron microscopy (TEM) was performed on a TOPCON-EM002B apparatus equipped with an energy dispersive X-Ray device (EDS).

Temperature programmed reduction (TPR) was performed on 50 mg samples placed in a U-shaped quartz tube (6.6 mm ID), the temperature was increased from 25 to 900°C with a 15°C mn^{-1} slope. The reducing mixture was 3 vol. % hydrogen in helium (50 mL mn^{-1}). A TC detector quantified hydrogen consumption after trapping the formed water on molecular sieve.

In order to check if a metallic phase occurs during the reduction and if any, to characterise it, dynamic magnetic measurements were performed between 25 and 950°C (10°C/mn) under the same reducing flow used in TPR and a magnetic field of 10000 G.

Reaction Conditions

The operating conditions were the following : fixed bed quartz reactor (6.6 mm ID), inlet temperature : 400-850°C; 0.15 $L.h^{-1}$ ethane and 0.3 $L.h^{-1}$ CO_2 diluted in a 2.55 $L.h^{-1}$ argon flow (measured at 1 atm pressure and room temperature). Catalyst amount : 100 mg. The outlet gas was analysed by two on line chromatographs : one quantifying CH_4, C_2H_6, CO and CO_2; the second for CO and H_2.

The catalytic systems became active under the gas mixture and under heating without previous reduction. Reduction occurred in situ during the test. The temperature program was made of three cycles to have a better approach of the ageing of the catalysts. A continuous heating up to 800°C with a 3°C.mn-1 slope then one hour step at 800°C and a 50°C stepwise temperature decrease down to 400°C formed each cycle. Chromatographic analyses were performed every 60°C. The first cycle was carried out to activate the catalyst. The characteristic catalytic activity was obtained in the temperature increase of the second cycle.

The third cycle validated the results and confirmed the stability of the catalyst. The overall duration of one test was 30 hours except for ageing tests experiments (200h).

Results

Characterisation of $LaCo_xFe_{1-x}O_3$ Before Catalytic Test

XRD pattern of $LaCo_xFe_{1-x}O_3$ (x = 0.3; 0.5; 0.7 and 1) are given figure 1.

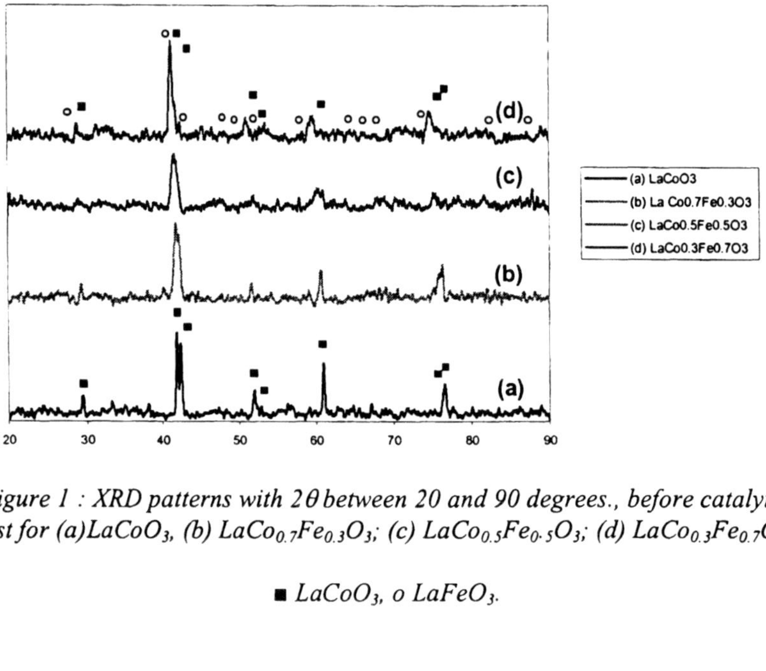

Figure 1 : XRD patterns with 2θ between 20 and 90 degrees., before catalytic test for (a)$LaCoO_3$, (b) $LaCo_{0.7}Fe_{0.3}O_3$; (c) $LaCo_{0.5}Fe_{0.5}O_3$; (d) $LaCo_{0.3}Fe_{0.7}O_3$.

■ $LaCoO_3$, o $LaFeO_3$.

The XRD patterns show a split of the main diffraction lines for $x \geq 0.5$ characteristic of the rhombohedral (R) system of $LaCoO_3$. Compounds without split reflection crystallise in the orthorhombic (O) system such as $LaFeO_3$. Cell parameters (a,b,c) of all catalysts in both crystallographic systems rhombohedral or orthorhombic and the normalised cell volume were calculated. The evolution of the crystalline unit cell with x is presented in figure 2, which plots the reduced cubic parameter as a function of x. The reduced cubic parameter of non-cubic cells corresponds to the cubic parameter $r_{reduced}$ of the cubic cell of same volume : $a_{reduced} = \sqrt[3]{abc}$

In addition to the diffraction lines shape (split or not), we can notice a shift toward small angles of their position from $LaCoO_3$ until $LaFeO_3$. This shift reflects an increasing cell parameter of the perovskite due to the larger ionic radius of Fe^{3+} than the Co^{3+} one. This phenomenon is clearly evidenced in figure 2 where the reduced cell parameter is 3.827 Å for $LaCoO_3$ and reaches 3.931 Å

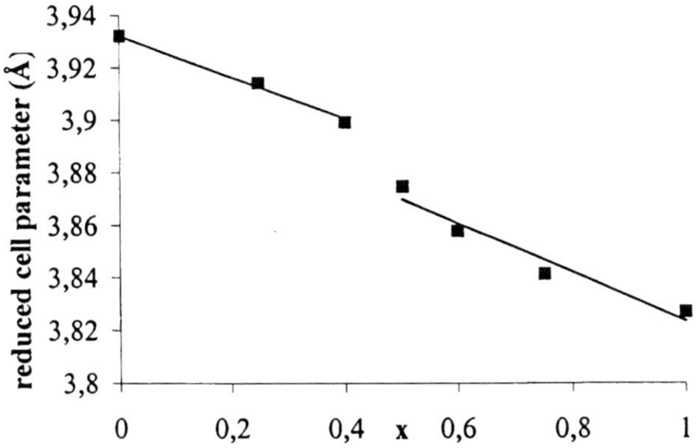

Figure 2 : Evolution of the $LaCo_xFe_{1-x}O_3$ perovskites reduced cell parameter versus x.

for $LaFeO_3$. Furthermore, this increase is not linear all along the series but two linear variation zones occur. As a consequence, $LaCo_xFe_{1-x}O_3$ series is composed of two solid solutions depending on x. For $0.5 \leq x \leq 1$ the obtained solid solution is rhombohedral, isomorphic to $LaCoO_3$. For lower x values, an orthorhombic solid solution is obtained.

The different catalysts have been analysed by Electron Dispersive X-Ray spectroscopy (EDS) coupled to TEM measurements on different areas of the samples with a large (200 nm) or a fine (14 nm) focussed beam. The use of a fine focussed beam shows the good homogeneity of each sample (19).The large focussed beam gives only an assessment of the global composition due to the presence of lanthanum. However, the atomic absorption gives the expected composition of La, Fe and Co for each preparation. Results for $LaCo_{0.3}Fe_{0.7}O_3$ and $LaCo_{0.7}Fe_{0.3}O_3$ are presented in figures 3a and 3b.

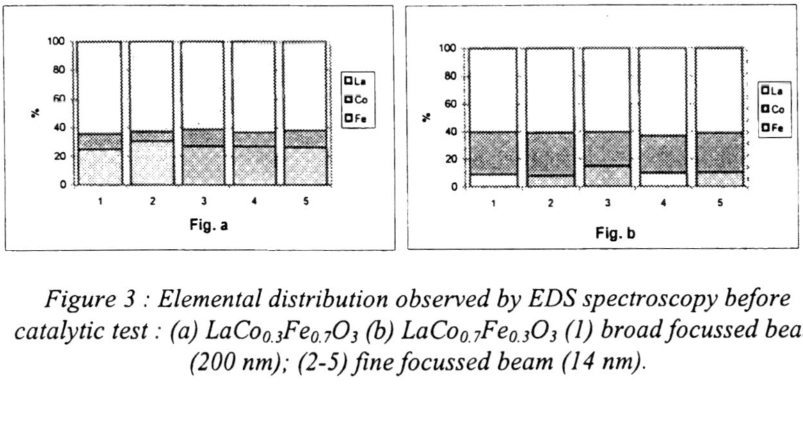

Figure 3 : Elemental distribution observed by EDS spectroscopy before catalytic test : (a) $LaCo_{0.3}Fe_{0.7}O_3$ (b) $LaCo_{0.7}Fe_{0.3}O_3$ (1) broad focussed beam (200 nm); (2-5) fine focussed beam (14 nm).

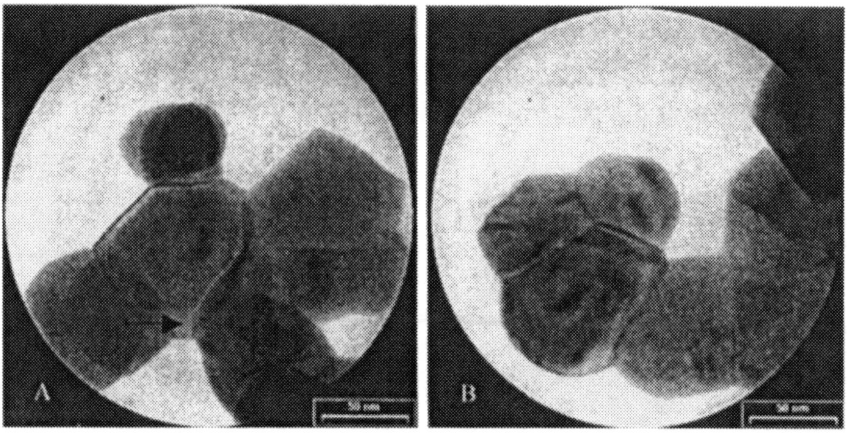

Figure 4 : Transmission electron micrographs : (a) $LaCo_{0.3}Fe_{0.7}O_3$ (b) $LaCo_{0.7}Fe_{0.3}O_3$ before catalytic test.

XRD and TEM-EDS clearly evidenced the formation of La-Co-Fe solid solutions and the good homogeneity of the obtained systems. Grains of around 50 nm each as shown in figure 4 form each preparation.

Study of the Reducibility of the La-Co-Fe Perovskites

The active species in methane reforming being the metal present at the catalytic surface (8, 20), it is of particular interest to study the reduction of the trimetallic structure and the effect of iron addition on the temperature of cobalt reducibility.

Figure 5 gives the curve of thermoprogrammed reduction (TPR) for different catalysts.

For all the perovskites (except $LaFeO_3$) the TPR shapes reveal two reduction areas. The first between 350 and 500°C, the second with a maximum between 620 ($LaCoO_3$) and 810°C ($LaCo_{0.3}Fe_{0.7}O_3$). The temperature of the second maximum increases with iron content. For $x \leq 0.3$, the reduction temperature is similar to the reaction temperature. During the second reduction step, cobalt and part of the iron are reduced to metals and could form a Co-Fe alloy. So, as for LaNiFe, it is possible to control the reduction temperature of the perovskite by varying the Co/Fe ratio and to correlate it with the dry reforming reaction temperature.

In order to have more information concerning the reduction of cobalt and iron, magnetic measurements were carried out from 300 to 1250 K under the same reductive flow as used in TPR.

An example is given in figure 6 for x = 0.3, 0.4, 0.7 and 1.

The magnetisation curve of $LaCoO_3$ shows two increases at 435 and 525°C corresponding to the formation of Co° (ferromagnetic, magnetisation value at room temperature : 160 uem/g (*21*)). The two zones of Co° formation are the same as the two reduction areas observed in TPR. The magnetisation value remains stable between 440 and 520°C, accounting for the stability of the

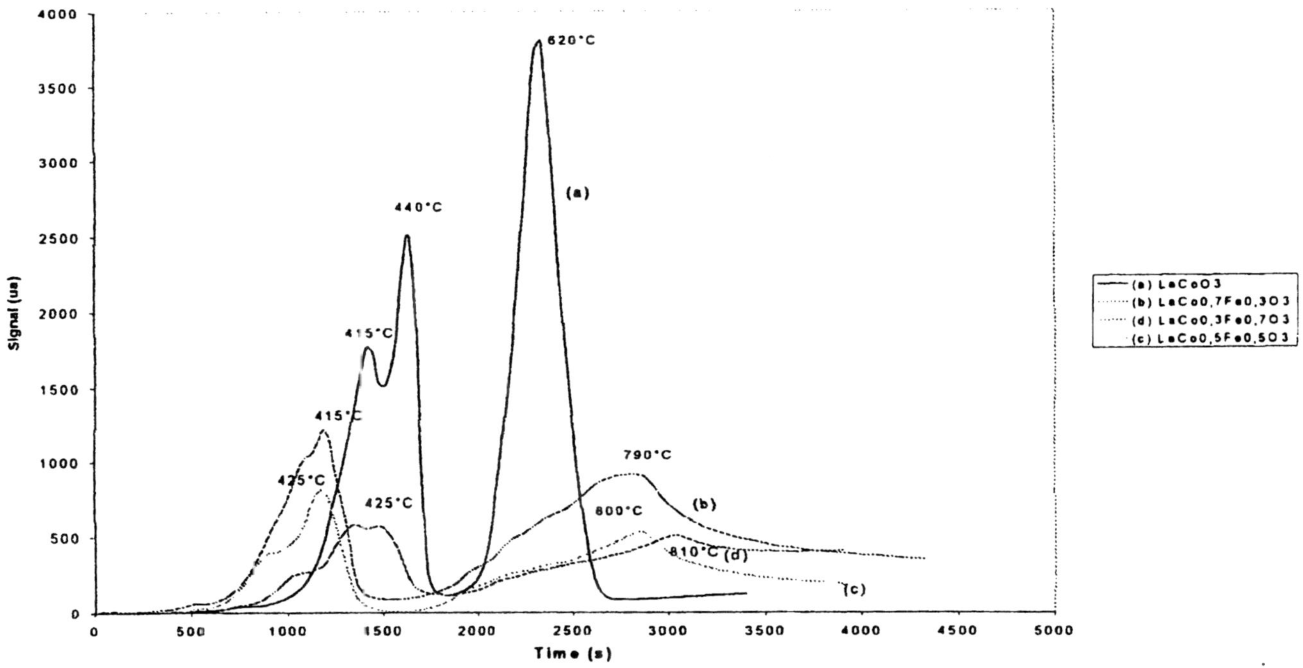

Figure 5 : TPR before catalytic test for (a) $LaCoO_3$, (b) $LaCo_{0.7}Fe_{0.3}O_3$, (c) $LaCo_{0.5}Fe_{0.5}O_3$ and (d) $LaCo_{0.3}Fe_{0.7}O_3$.

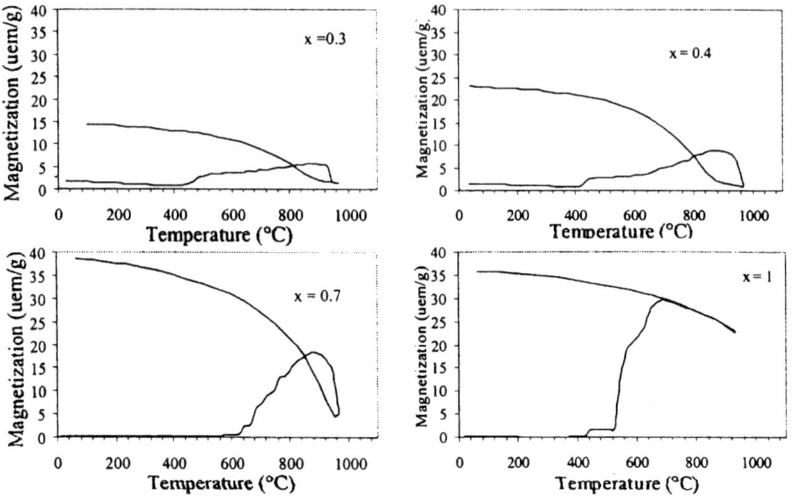

Figure 6: Magnetisation of $LaCo_xFe_{1-x}O_3$ under diluted H_2 from 20 to 950°C.

partially reduced system. All other rhombohedral perovskites (x = 0.7, 0.6 and 0.5) present only one increase of the magnetisation value at higher temperature (almost 600°C for $LaCo_{0.7}Fe_{0.3}O_3$) corresponding to the second TPR area. No cobalt metal is formed during the first TPR reduction area, which mean that the reduction of Co^{III} to Co^{II} and Co° occurs only during the second TPR reduction peak. For the orthorhombic $LaCo_{0.4}Fe_{0.6}O_3$ and $LaCo_{0.25}Fe_{0.75}O_3$ perovskites, cobalt metal is obtained during the first reduction since their magnetisation curve shows an increase at 410 and 425°C respectively. Around 950°C, magnetisation of all trimetallic perovskites decreases dramatically. This is the behaviour of ferromagnetic samples studied at higher temperatures than their Curie temperature (1115°C for Co, 770°C for Fe). So at 950°C, some iron oxide is reduced in order to form a Co-Fe alloy with a Curie temperature between 880 and 950°C depending on the Co/Fe ratio.

Catalytic Tests

Figures 7 and 8 show respectively the ethane conversion and the CO yield versus x for temperatures between 600 and 800°C. Ethane conversion increases with increasing x values and temperature. There is an important change in conversion and CO yield when x is equal or higher than 0.5. This value corresponds to a change of the structure from orthorhombic to rhombohedral. $LaCoO_3$ shows excellent results too but deactivated very fast. Catalysts with cobalt and iron have a very good stability in addition to performance. Ageing tests of more 200 hours indicate no deactivation at all for x = 0.7 or 0.3. The amount of carbon deposit is very low for trimetallic orthorhombic structure (0.40

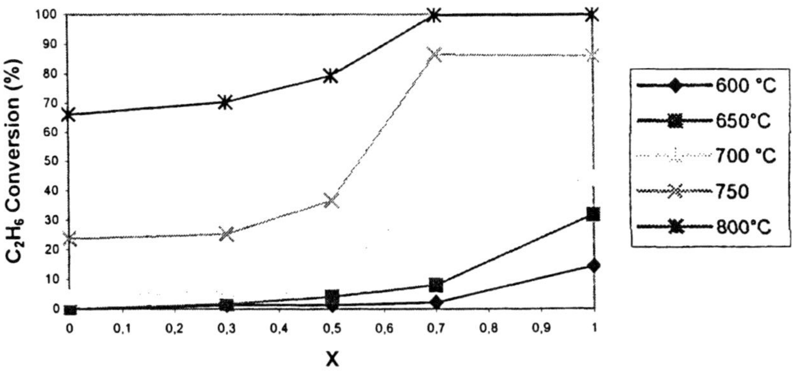

Figure 7 : C_2H_6 conversion at various temperature versus x with a CO_2/C_2H_6 ratio = 2 for $LaCo_xFe_{1-x}O_3$ perovskite.

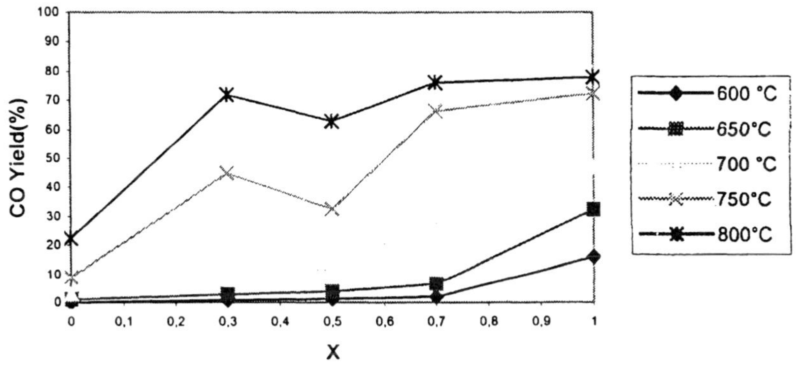

Figure 8 : CO yield (%) at various temperatures versus x with a CO_2/C_2H_6 ratio= 2 for $LaCo_xFe_{1-x}O_3$ perovskites.

and 0.41 wt%/g of catalyst for $LaCo_{0.3}Fe_{0.7}O_3$ and $LaCo_{0.5}Fe_{0.5}O_3$ respectively) but increases for rhombohedral structure (1.49). For $LaCoO_3$ percentage of carbon after 30 hours is high (5.07 wt%/g of catalyst) and deactivation is fast and irreversible.

The interest of the trimetallic perovskite La-Co-Fe compared to the bimetallic La-Co is pointed out by the characterisation study of the catalysts after reaction.

Characterisation After Reaction

Table I gives the different phases characterised by XRD for various x values.

Table I : Crystalline phases detected by XRD after test with ethane for $LaCo_xFe_{1-x}O_3$ perovskites

X	*Crystalline phases detected by XRD after catalytic test*					
0	$LaFeO_3$					
0.3		$LaCo_wFe_{1-w}O_3$	La_2O_3			
0.5		$LaCo_zFe_{1-z}O_3$	La_2O_3			
0.7			La_2O_3	$LaFeO_3$	Co	
1		$La_2O_2CO_3$	La_2O_3		Co	CoO

For the higher cobalt contents (x > 0.5) the trimetallic perovskite is destroyed and the most stable $LaFeO_3$ perovskite is formed together with La_2O_3 and cobalt. $LaFeO_3$ has also been characterised by FTIR (555 cm^{-1}). For low cobalt content, the LaCoFe structure is preserved, although the cobalt content is lower than in the initial structure.

The LaCoFe elemental distributions (EDS) are given in figure 9a and 9c and the La/Fe ratios in figures 9b and 9d for x = 0.3 and 0.7. Figure 9a and 9c show a higher heterogeneity for x = 0.7 than for x = 0.3. For x = 0.3, almost the same La/Co/Fe or La/Fe are observed in different parts of the sample, no free cobalt has been detected confirming that the trimetallic structure was preserved. However analyses show clearly that part of cobalt has left the structure and w and z (table 1) are lower than the initial value of x (0.09 for w; 0.25 for z). For x= 0.7, the LaCoFe structure is destroyed, free cobalt particles evidenced and the ratio La/Fe changed. Free lanthanum (as oxide or hydroxide) has been characterised, however $LaFeO_3$ is always present.

In the latter case, in agreement with XRD results, it can be concluded that $LaFeO_3$ perovskite behaves as a support for cobalt metal particles of 10 to 50 nm (TEM).

Discussion

In the present study, we have seen that whatever the x value up to 0.7, the perovskite structure is preserved : as trimetallic LaCoFe for $x \leq 0.5$ and as a bimetallic LaFe system for $x \sim 0.7$.

The catalytic results have shown an increase of the reactivity with the increase of both x value and reaction temperature. Moreover a sharp increase of activity is due to a change of the perovskite structure from orthorhombic to rhombohedral. Such a change was also been evidenced in Fischer-Tropsch reaction (*22*) but is relatively new. The good ageing of the catalysts (200 h) suggests that coke formation is limited as confirmed by elemental analyses.. Different explanations could be given : formation of small Co metal particles in strong interaction with the perovskite support (*17*), formation of a Co-Fe alloy (*23*), effect of the basic supports presenting a high affinity for CO_2 (*24*), ensemble effects (*6, 25*).

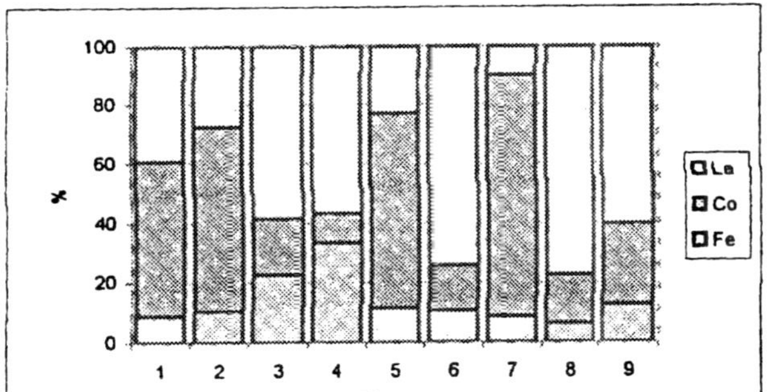

Fig. a

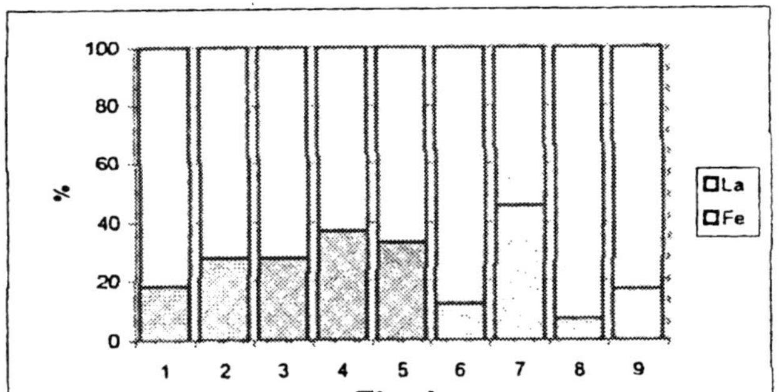

Fig. b

Figure 9 : Elemental distribution observed by EDS after catalytic test for $LaCo_{0.7}Fe_{0.3}O_3$ ((a) ratio La/Co/Fe (b) ratio (La/Fe)) and $LaCo_{0.3}Fe_{0.3}O_3$ ((c) ratio La/Co/Fe) (d) ratio (La/Fe)) (1) Broad focussed beam 200 nm (2-9) fine focussed beam 14 nm.

Continued on next page.

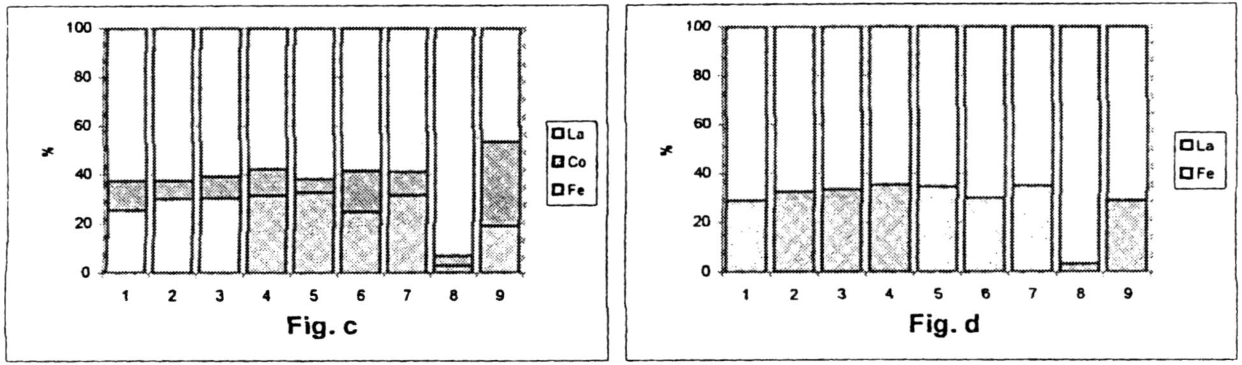

Figure 9. *Continued.*

A priori, the LaCoFe perovskites can combine, at least partially, some effects quoted above: basicity of La_2O_3, small size of cobalt particles due to a more difficult reduction of cobalt when included in the perovskite, strong metal support interaction. Alloy formation was not seen after reactivity as for LaNiFe (23) but only after TPR and magnetisation and cannot be invoked as responsible for the low carbon content. For $0.3 \leq x \leq 0.7$ we suggest that the cobalt inside the partially reduced perovskite network is in strong interaction with the free cobalt of the surface and then limits the migration and the sintering of the cobalt surface metal particles and finally an excessive coke formation. The strong interactions are confirmed by the used catalyst regeneration ability of the cobalt containing perovskites only by heating at high temperature (900°C). However the amount of incorporated cobalt after regeneration is less than in the initial perovskite (x = 0.21 instead of 0.3; and 0.35 instead of 0.5). For Co deposited on bimetallic LaFe perovskite, a mixture of the two perovskites ($LaCoO_3$ and $LaFeO_3$) is obtained after regeneration.

For cobalt rich systems, such interactions are not possible due to the destruction of the trimetallic structure, however a good ageing of $LaCo_{0.7}Fe_{0.3}O_3$ has been seen until 200h. The activity of $LaFeO_3$ in total oxidation of hydrocarbons (26) could be important to prevent coke. For $LaCoO_3$, deactivation is fast and irreversible.

Conclusions

$LaCo_xFe_{1-x}O_3$ catalysts are very efficient in dry reforming of ethane. The evolution of the catalysts depends on x value. For orthorhombic catalysts ($x < 0.5$), under gas mixture, small cobalt metal particles in strong interaction with the remaining trimetallic LaCoFe perovskite are formed. For x=0.7 cobalt is totally reduced and the remaining oxide is a bimetallic LaFe perovskite. For $x \leq 0.5$, catalysts have lower activity than x=0.7 however the strong metal-support interaction prevents the sintering of the metal particles and coke formation. A cobalt metal in strong interaction with a LaCoFe perovskite seems to be the key for a good ageing of the catalyst and the ability to regenerate it.

Acknowledgement

The authors are deeply grateful to French ECOS-NORD programme and to European Community through RUCADI and ALFA program for providing fellowship to G. Rodriguez.

References

(1) Chaumette, P. *Rev. I.F.P.* **1996**, *51*, 653.
(2) Rostrup-Nielsen, J.R. *Catal. Today* **1993**, *18*, 305.
(3) Ashcroft, A.T., Cheetham, A.T., Green, M.L.H., Vernon, P.D.F. *Nature* **1991**, *352*, 225.

(4) Hickman, D.A., Schmidt, L.D., *Science* **1993**, *259*, 343.
(5) Holmen, A., Olsvik, O., Rockstad, O.A. *Fuel Proc. Technol.* **1995**, *42*, 249.
(6). Rostrup-Nielsen, J.R. *Catal. Sci. Techn.* **1983**, *5*, 1 Eds. Springer Berlin
(7) Frusteri, F., Arena, F., Calogero, G., Torre, T., Parmaliana, A. *Catalysis Communications* **2001**, *2*, 49.
(8) Jones, R.H., Ashcroft, A.T., Wallen, D., Cheetham, A.K., Takehira, K. *Catal. Lett.* **1991**, *8*, 169.
(9) Kuminori, K., Umeda, S., Nakamura, J., Uchijima, T. *Bull. Soc. Chim. Jpn.* **1992**, *65*, 2562.
(10) Stojanovic, M., Mims, C.A., Moudahlal, H., Jacobson, A.J. *Proceedings of the 11th Int. Cong. Catalysis Baltimore*, **1996**, 348.
(11) Han, S., De Caul, L.C., Palermo, R.E., Walsh, D.E. *U.S. Patent 5149516*, **1992**, assigned to Mobil Oil Corporation.
(12) Moriga, T., Usaka, O., Imamura, T., Nakabayashi, I., Matsubara, I., Kinouchi, T., Kikkawa, S., Kanamaru, F. *Bull. Soc. Chim. Jpn.* **1994**, *67*, 687.
(13) Hayakawa, T., Andersen, A.C., Shimizu, M., Suzuki, K., Takehira, K. *Catal. Lett.* **1993**, *22*, 307.
(14) Klavna, D., Kirchnerova, J., Gauthier, P., Delval, J., Chaouki, J. *Can. J. Chem. Eng.* **1997**, *75*, 509.
(15) Parvary, M., Jazayeri, S.H., Taeb, A., Petit, C., Kiennemann, A. *Catalysis Communications* **2001**, *2*, 357.
(16) Provendier, H., Petit, C., Estournes, C., Libs, S., Kiennemann, A. *Appl. Catal. A* **1999**, *180*, 163.
(17) Provendier, H., Petit, C., Kiennemann, A. *C.R. Acad. Sci. Paris* **2001**, *4*, 57.
(18) Poix, P. *Bull. Soc. Ceram.* **1966**, *72*, 1.
(19) Houalla, M., Delannay, F., Matsuurai, I., Delmon, B. *J.C.S. Faraday I* **1980**, *76*, 2128.
(20) Slagtern, A., Olsbye, U. *Appl. Catal. A* **1994**, *110*, 99.
(21) Lin, X.M., Sorensen, C.M., Klabunde, K.J., Hadjipanayis, G.C. *Langmuir* **1998**, *14*, 7140.
(22) Bedel, L., Roger, A.C., Kiennemann, A., Estournes, C. *Prep. Pap. Am. Chem. Soc., Div. Petroleum Chem.*, **2000**, *45*, 236.
(23) Provendier, H., Petit, C., Estournes, C., Kiennemann, A. *Stud. Surf. Sci. Catal.* **1998**, *119*, 741.
(24) Yamazuki, O., Nozaki, T., Omata, K., Fujimoto, K. *Catal. Lett.* **1992**, 1953.
(25) Slagtern, A., Oldsbye, U., Blom, R., Dahl, I.M. *Stud. Surf. Sci. Catal.* **1997**, *107*, 497.
(26) Zhang, Z., Verykios, X.E., Mac Donald, S.M., Affrosman, S. *J. Phys. Chem.* **1996**, *100*, 744.

Chapter 5

Combined Carbon Dioxide and Steam Reforming with Methane in Low-Temperature Plasmas

K. Supat[1], S. Chavadej[1], L. L. Lobban[2], and R. G. Mallinson[2,*]

[1]The Petroleum and Petrochemical College, Chulalongkorn University, Bangkok 10330, Thailand
[2]Institute for Gas Utilization Technologies and School of Chemical Engineering and Materials Science, University of Oklahoma, Norman, OK 73019

The combination of carbon dioxide reforming and steam reforming with methane is investigated in a corona discharge reactor with moderate power consumption. The plasma is used to overcome the energy barrier of these two endothermic reactions instead of high temperature catalytic processes. Simultaneous carbon dioxide and steam reforming produces higher methane conversions and CO/C_2 ratios than only steam reforming or carbon dioxide reforming. The H_2/CO ratio of 1.9:1 is close to that desired for Fischer Tropsch or methanol synthesis was achieved at a CO_2/CH_4 ratio of 1:1 with 50% water-vapor in the feed stream. In this condition, the energy consumption is about 12 eV/molecule of carbon converted. Using electricity to drive the reaction at low temperature is counter-balanced by lower costs and simpler operation.

With increasing population and rapid technology development, it is possible that we have begun to encounter alteration of the climate. Greenhouse gases that include carbon dioxide, methane, ozone, halocarbons and nitrous oxide, to a significant extent come from the combustion of fossil fuels (coal, oil, and natural gas) for power generation and transportation. A potentially useful reaction for reducing the impact of carbon dioxide (reducing net emissions), a major component in many natural gas resources, especially in Asia, is carbon dioxide reforming with methane (Reaction 1) for production of synthesis gas (carbon monoxide and hydrogen) at a lower H_2/CO ratio compared to steam reforming of methane (Reaction 2), which is a major conventional process for synthesis gas production. A typical industrial steam reformer operates at 1,120-1,170 K and 15-30 atm over a Ni/Al_2O_3 catalyst (1). However, as with steam reforming, the highly endothermic reaction requires significant high temperature energy input that is usually provided by combustion that releases carbon dioxide. Also, the 1:1 H_2/CO ratio produced from carbon dioxide reforming requires, for GTL applications, the use of the water gas shift reaction, in which much of the potential net consumption of carbon dioxide is lost. However, a major problem with carbon dioxide reforming to this point in time is the deactivation of the catalyst due to carbon deposition (Reactions 3 and 4) at the desired reaction conditions.

$$CH_4 + CO_2 \rightarrow 2CO + 2H_2 \qquad \Delta H = 247 \text{ kJ mol}^{-1} \qquad (1)$$

$$CH_4 + H_2O \rightarrow CO + 3H_2 \qquad \Delta H = 206 \text{ kJ mol}^{-1} \qquad (2)$$

$$CH_4 \rightarrow C + 2H_2 \qquad \Delta H = 75 \text{ kJ mol}^{-1} \qquad (3)$$

$$2CO \rightarrow C + CO_2 \qquad \Delta H = -172 \text{ kJ mol}^{-1} \qquad (4)$$

So far, There is no effective commercial catalyst which can operate without carbon formation (2). The alternative is to combine carbon dioxide reforming with steam reforming. A thermodynamics study has shown that the presence of water in carbon dioxide reforming allows increased methane conversion (3) and probably decreased carbon formation at a specific CO_2/CH_4 ratio. Moreover, for several feed ratios of CH_4/CO_2, the desired H_2/CO ratio at equilibrium is controllable by the amount of water added in feed steam, for example, the desired ratio of H_2/CO of 1:1 can be obtained from a Terrell natural gas by using up to 3.28% of water in the feed (3). Both carbon dioxide reforming and steam reforming have similar low net carbon dioxide reduction due to combustion needed to provide the required heat of reaction. Thus any net reduction of carbon dioxide impact from the process must be compared with partial oxidation or combined/autothermal reforming which are exothermic or self heating, but require the production of oxygen that is itself energy intensive. Therefore, rather than considering the net carbon dioxide reduction as the driver, other factors must be considered. For example, high carbon dioxide natural gas resources may be stranded by the high cost of carbon dioxide removal, which may be reduced or eliminated by use of carbon dioxide reforming. Additionally, the possible lower capital investment and operating costs due to the elimination of the requirement for oxygen makes consideration of this technology worthwhile. Due to the fact that carbon dioxide reforming and steam reforming are endothermic reactions, the high reaction temperature requires an intensive energy input and a high operating cost. One may uses non-equilibrium plasma to reduce energy cost instead of thermal process.

The non-thermal, also called cold, plasma, and its non-equilibrium properties, high electron temperature and low bulk gas temperature, provide the capability for initiating chemical reactions at low temperature and potentially with low energy input. The non-equilibrium plasma is used to overcome the energy barrier to start the reaction by promoting radical production (4).

Corona discharge and dielectric barrier discharge (DBD) techniques are two of the commonly used methods for producing non-equilibrium plasmas at atmospheric pressure (5). The utilization of the non-equilibrium plasma techniques for carbon dioxide reforming with methane has been studied in the dielectric barrier discharge (2, 6-8).

The spectroscopic characteristic of a plasma created in a methane and carbon dioxide mixture by a high-voltage, steep front-voltage, very-short-pulse triggered dielectric barrier discharge in a tubular cell was investigated. The very steep front voltage ($>10^{12}$ V/s) allowed the reactor to operate at higher reduced field values (E/n) than in conventional DBD systems. To confirm the properties of the non-equilibrium plasma, it was shown that the temperature inside the streamers was around 3,000 K and the mean gas temperature in the active volume was close to room temperature (6).

The effects of CO_2/CH_4 ratio and adding helium to a 1:1 CO_2/CH_4 ratio feed stream were studied. The results indicated that CO_2/CH_4 ratio strongly affects the results. Increasing the partial pressure of carbon dioxide from a 1:2 to 1:1 CO_2/CH_4 ratio caused increased carbon dioxide conversion to carbon monoxide but reduced the ethane and hydrogen selectivities. A further increase in the CO_2/CH_4 ratio from 1:1 to 2:1 caused an increase in ethane and hydrogen selectivities. Despite lowering methane partial pressure, helium acts as a third body in enhancing methane conversion and carbon monoxide selectivities (7).

A numerical simulation was performed to investigate the conversion of methane and carbon dioxide to synthesis gas in a DBD reactor. Using curve-fitting, the H_2/CO ratio can be presented as a simple function of a mixing ratio of CH_4/CO_2 which can be approximated over a wide range. Specific electric energy, gas pressure, and temperature hardly influence syngas composition (2).

The combination of solid catalysts with a dielectric-barrier discharge also has been studied for carbon dioxide reforming of methane. It was shown that the nickel and the nickel-calcium coated foams showed nearly identical activities for the reforming reaction in the discharge in the temperature range of 40 - 230 °C and gave significantly higher carbon monoxide production than the rhodium coated catalyst and the uncoated foam. The reaction was operated over the stoichiometric ratio at a CO_2/CH_4 ratio of 3:1 to minimize hydrocarbon formation. After the reaction, small amount of carbon or carbonaceous deposits were found on the foams (8).

Because the corona discharge is particularly easy to establish, it has had wide application in a variety of processes such as electrostatic precipitation, electro-photography, static control in semiconductor manufacture, ionization instrumentation, destruction of toxic compounds, generation of ozone, and synthesis of chemicals (9).

In this work, a corona discharge reactor was used to facilitate carbon dioxide reforming by itself and in combination with steam, at low temperatures. This has been shown to be possible in other methane conversion studies without carbon dioxide, such as helping in the conversion of methane over Sr/La_2O_3 (10) and other materials (11). The disadvantage of using electrical energy, that at best may only be about 60 percent efficient, may be offset by process simplicity and cost, lower carbon formation and less excess energy that must be recovered and utilized only with considerable cost.

Experimental

The flow rate of feed gases were controlled by a set of mass flow controllers supplied by Porter Instrument, Inc. Water-vapor was introduced in the reactor by passing the feed gases through a water-bath the temperature of which was controlled to adjust the concentration of water-vapor in the feed gas mixture. The feed gases were well mixed and then flowed downward through the reactor. The reactor temperature was controlled to be constant 100 °C to avoid water condensation during reaction. All the experiments were operated at atmospheric pressure. The feed gases and the exhaust gases from the reactor were introduced into two condensers in series cooled by ice water that were used to remove the condensable products as well as water and then analyzed on-line by a Carle AGC 400 gas chromatograph with thermal conductivity detector. However, no organic products were collected from the condensers when analyzed with a Varian 3300 GC with a Porapak Q Column. The power supply unit consisted of an AC power supply for converting domestic AC power 120 V, 60 Hz, using a function generator to vary the frequency in the range of 300 to 600 Hz with a sinusoidal waveform. The output was then transmitted to a high voltage alternating current (HVAC) transformer. The HVAC transformer could step up the low side voltage to the high side voltage by a nominal factor of 125 at 60 Hz. This factor may not be constant with changes in frequency due to changes in the power factor because of the capacitive nature of the reactor system. The discharge occurred between two stainless steel electrodes in a quartz tube with an inside diameter of 7 mm. The gap distance between these two electrodes was 1 cm. The feed gas flow rate was the same for every experiment at 100 sccm corresponding to a residence time of 0.23 s. The upper wire electrode is centered axially within the reactor tube, while the lower electrode is a circular plate with holes to allow gas to pass through the reactor and is positioned perpendicular to the reactor axis. An Extech power analyzer was used to measure the power, power factor, current, frequency, and voltage at the low side of the power circuit.

For this system, eV/m_C is defined as the energy consumption for conversion of carbon in the feed gas consisting of methane and carbon dioxide of in units of electron-volt/molecule of carbon converted (eV/m_C). S_X is defined as the selectivity based on converted carbon in both carbon dioxide and methane. S_{total} is defined as the total selectivities of all products which contain C (carbon atom).

All experiments had a carbon balance greater than 84%. Water was not quantified.

Results and Discussion

The electron energy, which corresponds to the electron temperature, is restricted to less than 6 eV in the corona discharge (12). These relatively low-energy electrons have insufficient energy to ionize methane, which has an ionization potential greater than 12 eV (13), but high enough to dissociate carbon dioxide molecules with dissociation energies of 5.5 eV, respectively (Reaction 5) (2, 5, 14).

$$e^- + CO_2 \rightarrow CO + O + e^- \qquad (5)$$

H_2O also requires only a relatively low electron energy to be dissociated to produce O, H and OH radicals as shown in Reactions 6 and 7 (15):

$$e^- + H_2O \rightarrow O^- + H_2 \qquad (6)$$

$$e^- + H_2O \rightarrow OH^- + H \qquad (7)$$

These radicals, H, OH and O, further abstract hydrogen from methane to form methyl radicals (CH_3) leading to a significant rate of methane conversion in the corona discharge (Reactions 8-10).

Methyl radical formation :

$$CH_4 + O \rightarrow CH_3 + OH \qquad (8)$$

$$CH_4 + H \rightarrow CH_3 + H_2 \qquad (9)$$

$$CH_4 + OH \rightarrow CH_3 + H_2O \qquad (10)$$

These methyl radicals react further to either combine with another CH_3 to form ethane (Reaction 11) or break down into CH_2, CH and C and subsequently form ethylene, acetylene and carbon oxides (16). Ethane can be dehydrogenated to form ethylene and acetylene and ethylene can be dehydrogenated to form acetylene. These reaction pathways were also observed in the conversion of methane via microwave plasmas (17). The coupling of hydrogen radicals forms hydrogen. Beyond these main reactions, there maybe many contributing reaction pathways in the corona discharge environment. Some of the possible ethane, ethylene, acetylene and hydrogen formation reactions are shown in Reactions 11-39. Water may be formed in the discharge via Reactions 40-42.

Ethane formation :

$$\mathrm{CH_3 + CH_3 \rightarrow C_2H_6} \qquad (11)$$

$$\mathrm{CH_4 + CH_2 \rightarrow C_2H_6} \qquad (12)$$

$$C_2H_6 + H \rightarrow C_2H_5 + H_2 \qquad (13)$$

$$C_2H_5 + C_2H_5 \rightarrow C_2H_6 + C_2H_4 \qquad (14)$$

Ethylene formation :

$$C_2H_6 \rightarrow C_2H_4 + H_2 \qquad (15)$$

$$C_2H_6 \rightarrow CH_3 + CH_3 \qquad (16)$$

$$CH_3 + CH_3 \rightarrow C_2H_4 + H_2 \qquad (17)$$

$$CH_3 + CH_2 \rightarrow C_2H_4 + H \qquad (18)$$

$$CH_2 + CH_2 \rightarrow C_2H_4 \qquad (19)$$

$$CH_3 + CH_3 \rightarrow C_2H_5 + H \qquad (20)$$

$$C_2H_6 + H \rightarrow C_2H_5 + H_2 \qquad (21)$$

$$C_2H_6 + CH_3 \rightarrow CH_4 + C_2H_5 \qquad (22)$$

$$C_2H_5 + H \rightarrow CH_3 + CH_3 \qquad (23)$$

$$C_2H_5 + H \rightarrow C_2H_4 + H_2 \qquad (24)$$

Acetylene formation :

$$C_2H_4 \rightarrow C_2H_2 + H_2 \tag{25}$$

$$CH_2 + CH_2 \rightarrow C_2H_2 + H_2 \tag{26}$$

$$C_2H_4 + H \rightarrow C_2H_3 + H_2 \tag{27}$$

$$C_2H_3 + H \rightarrow C_2H_2 + H_2 \tag{28}$$

$$CH + CH \rightarrow C_2H_2 \tag{29}$$

$$CH_2 + CH \rightarrow C_2H_2 + H \tag{30}$$

CO_X formation :

$$CH_3 + O \rightarrow HCHO + H \tag{31}$$

$$HCHO + O \rightarrow OH + CHO \tag{32}$$

$$CHO + O \rightarrow OH + CO \tag{33}$$

$$CHO + O \rightarrow CO_2 + H \tag{34}$$

$$HCHO + H \rightarrow H_2 + CHO \tag{35}$$

$$CHO + H \rightarrow H_2 + CO \tag{36}$$

$$HCHO + OH \rightarrow CHO + H_2O \tag{37}$$

$$CHO + OH \rightarrow CO + H_2O \tag{38}$$

Hydrogen formation :

$$2H \rightarrow H_2 \tag{39}$$

Water formation:

$$2OH^- \rightarrow O^{2-} + H_2O \tag{40}$$

$$2H + O \rightarrow H_2O \tag{41}$$

$$H_2 + O \rightarrow H_2O \tag{42}$$

Moreover, the radicals of H, OH, and O generated from dissociation of water are essential ingredient for the inhibition and removal of carbon deposits which have a negative effect on gas discharges, shown in Reactions 43 and 44 (18).

$$C_{(S)} + H \rightarrow CH \tag{43}$$

$$CH_{(S)} + H \rightarrow CH_2 \tag{44}$$

In Table I, results of a feed stream of 50% methane or carbon dioxide or carbon monoxide reacted with 50% water are shown. Many products, CO, CO_2, C_2H_6, C_2H_4, C_2H_2, and H_2, were produced from methane with water but C_2 products were not produced from carbon dioxide with water or from carbon monoxide with water. Methane and water in the corona discharge not only steam reforms methane but other reactions also occur. The H_2/CO ratio is quite high;

more than the stoichiometric ratio for steam reforming. In the industrial catalytic process, the syngas ratio (H_2/CO) obtained with steam reforming is higher than 3 caused by a surplus of steam (19, 20). High hydrogen production here results from the net dehydrogenation with the formation of C_2 hydrocarbon products and hydrogen also can be produced directly from the dissociation of water in the corona discharge as shown in Reaction 6. Steam reforming of methane in the low temperature plasma environment is an alternative way to produce synthesis gas instead of the conventional process.

Table I. Reactions of methane, carbon dioxide, carbon monoxide with water

Carbon (%)	50% CH_4	50% CO_2	50% CO
Water-vapor (%)	50	50	50
Low side Voltage (V)	57	85	63
Frequency (Hz)	600	400	600
Power (W)	13	10	7
Power Factor	0.48	0.27	0.49
Current (A)	0.48	0.44	0.23
C conversion (%)	$X_{CH4} = 27$	$X_{CO2} = 1$	$X_{CO} = 6$
S_{CO} (%)	35	112	-
S_{CO2} (%)	1	0	105
S_{C2H6} (%)	5	0	0
S_{C2H4} (%)	8	0	0
S_{C2H2} (%)	39	0	0
S_{total} (%)	88	112	105
Energy consumption (eV/m_C)	14	353	35
H_2/CO_x	$H_2/CO = 4.7$	$H_2/CO = 0.0$	$H_2/CO_2 = 0.9$
CO/C_2	1.4	-	-
CO/O_2	-	2.3	-

Carbon dioxide with water is essentially unreactive under the conditions of this experiment, where of course both species are strongly favored by thermodynamics. Even though methane with water and carbon dioxide with water have the same current, or an equal the number of electrons, to initiate reactions, carbon dioxide conversion is much lower than methane conversion. The products were only oxygen and carbon monoxide with a CO/O_2 ratio of 2.3. Carbon monoxide with water is converted to some extent with the expected production of hydrogen and carbon dioxide. The H_2/CO_2 was about 0.9, that is close to the stoichiometric ratio of the water gas shift reaction. Energy consumption of carbon dioxide with water and carbon monoxide with water is very high under these conditions.

In Table II, it may be seen that in an excess of carbon dioxide, without water, the methane conversion is low and little carbon dioxide is converted under these conditions. With increasing frequency, keeping power constant at 11 W, the current is nearly constant while the power factor decreased as the low side voltage increased. The power factor decreased with increasing frequency due to the lag between current and voltage waveforms due to the capacitive nature of the reactor system. Frequency affects not only the electrical parameters but also affects conversion and product distribution as well as energy consumption. At these conditions, the maximum conversion of methane and carbon dioxide are found at 400 Hz and consequently the lowest power consumption is found at the same frequency. H_2/CO ratio is constant at 1.3, however, the product distribution changes with frequency. At 400 Hz, CO/C_2 ratio is highest because of decreasing of ethane and ethylene selectivities.

Table II. Carbon dioxide reforming of methane with/without water at a 3:1 CO_2/CH_4 Ratio

CO_2/CH_4	3	3	3	3
Water-vapor (%)	0	0	0	50
Low side Voltage (V)	57	68	79	82
Frequency (Hz)	300	400	500	400
Power (W)	11	11	11	14
Power Factor	0.56	0.48	0.42	0.39
Current (A)	0.35	0.34	0.33	0.44
CH_4 conversion (%)	12	19	14	53
CO_2 conversion (%)	2	4	2	20
S_{CO} (%)	63	53	48	89
S_{C2H6} (%)	14	6	12	0
S_{C2H4} (%)	14	9	13	1
S_{C2H2} (%)	19	18	15	5
S_{total} (%)	110	86	88	95
Energy consumption (eV/m_C)	41	21	33	15
H_2/CO	1.3	1.3	1.3	0.6
CO/C_2	2.7	3.2	2.4	27.5

In the last column, with addition of water, much higher conversions of both carbon dioxide and methane are achieved. Adding water provides more reactive H, OH, and O radicals in the reaction zone. The reactor could be operated at higher power to obtain higher conversion since these radicals inhibit and remove carbon formation. The net energy consumption (eV/molecule of carbon converted) is lower despite a 50 percent dilution of carbon dioxide and methane by the water. In addition, H, OH, and O radicals initiate methyl radical formation (Reactions 8-10) resulting in increased methane conversion and decreased energy consumption. H_2 and H produced from dissociation of water (Reactions 6-7). Little C_2 formation occurs evidently due to dehydrogenation and oxidation of methane such that coupling is inhibited. Most of the C from methane and carbon dioxide form carbon monoxide rather than C_2 products as suggested by the high CO/C_2 ratio. Simultaneous carbon dioxide and steam reforming produces higher methane conversions and CO/C_2 ratio than only steam reforming (Table I) or carbon dioxide reforming (Table II). Oxygen species from decomposition of water has a negative effect on C_2 hydrocarbon formation because the newly formed hydrocarbons are still in excited states and then further to be oxidized with oxygen. In addition, decreasing CH_X concentration with water in the feed stream, makes the coupling of methane less favorable. The H_2/CO ratio drop from 1.3 to 0.6 when the system was operated at 50 % of water in feed stream, even though there is more H atom in the feed gas. The result probably reflecs using high power, H from methane and water reform water.

At a CO_2/CH_4 ratio of 5:1, methane conversion is substantially increased with a more modest increase in carbon dioxide conversion with increasing power, but there are no significant changes in methane and carbon dioxide conversions with more water content in the feed, as shown in Table III. With less water, at 30%, energy consumption per molecule of carbon converted is substantially reduced compared to 50% water. One of the factors is the dilution and therefore lower throughput of methane at the same power level. As previously mentioned, the radicals from the dissociation of water abstract hydrogen from methane to form methyl radicals resulting in increasing methane conversion. However, a greater number of oxygen radicals could speed up the oxidation reaction to produce carbon dioxide, carbon monoxide, and water. For this reason, some input energy is used to convert recovered carbon dioxide so energy consumption per molecule of carbon converted increases with increasing water from 30% to 50%. Also, there is a substantial suppression of C_2 production with increased water content. Consequently, CO/C_2 ratio is very high (~68.8%) at 50% water vapor in the feed gas. This result is similar to the results in Table II. The H_2/CO ratio is nearly constant at 0.4 over these three conditions.

Table III. Effect of power and water on carbon dioxide reforming of methane with water at a 5:1 CO_2/CH_4 Ratio

CO_2/CH_4	5	5	5
Water-vapor (%)	30	30	50
Low side Voltage (V)	62	75	79
Frequency (Hz)	300	300	400
Power (W)	11	13	14
Power Factor	0.43	0.35	0.43
Current (A)	0.41	0.50	0.41
CH_4 conversion (%)	51	57	58
CO_2 conversion (%)	13	17	18
S_{CO} (%)	85	86	100
S_{C2H6} (%)	0	0	0
S_{C2H4} (%)	2	2	0
S_{C2H2} (%)	13	10	3
S_{total} (%)	100	98	103
Energy consumption (eV/m_C)	12	12	17
H_2/CO	0.5	0.4	0.4
CO/C_2	11.1	15.0	68.8

The results at the stoichiometric CO_2/CH_4 ratio of 1:1 and 50% water vapor are shown in Table IV. Increased power at the 1:1 ratio shows increased conversion except when the frequency is increased. The increase in frequency essentially "detunes" the system, reducing the power factor and the power actually applied to the plasma. Power is actually measured on the low voltage side and therefore includes all high voltage system losses. A system designed to properly match impedances and operate at a high power could significantly lower the real power consumption from that reported here.

Table V shows the effect of CO_2/CH_4 ratio on carbon dioxide reforming with water. The methane conversion increases with increasing CO_2/CH_4 ratio while carbon dioxide conversion is constant because of the decreasing methane partial pressure. Energy consumption per carbon converted is lower due to decreased carbon dioxide dilution since methane conversion is the higher of the two. The lower CO_2/CH_4 ratio results in greater C_2 production from the coupling reaction of methane since a decrease in CO_2/CH_4 ratio increases the probability of a methyl radical reacting with another methyl radical. C_2's may be undesirable components of synthesis gas, but the olefinic nature of those produced here are reactive on Fischer Tropsch catalyst systems and therefore may be suitable for feeds to a Fischer Tropsch Synthesis Reactor. The H_2/CO ratio is close to that desired for Fischer Tropsch or methanol synthesis and may be controllable by altering the water content.

Table IV. Effect of power on carbon dioxide reforming with water at a 1:1 CO_2/CH_4 Ratio

CO_2/CH_4	1	1	1	1
Water-vapor (%)	50	50	50	50
Low side Voltage (V)	57	66	88	97
Frequency (Hz)	300	300	400	600
Power (W)	11	13	14	14
Power Factor	0.54	0.45	0.34	0.40
Current (A)	0.36	0.43	0.47	0.36
CH_4 conversion (%)	41	44	47	30
CO_2 conversion (%)	20	21	21	12
S_{CO} (%)	56	60	56	72
S_{C2H6} (%)	2	1	1	3
S_{C2H4} (%)	5	5	5	5
S_{C2H2} (%)	23	24	26	21
S_{total} (%)	86	90	88	101
Energy consumption (eV/m_C)	11	12	12	20
H_2/CO	1.9	1.9	1.9	1.7
CO/C_2	3.9	4.1	3.5	4.9

Table V. Effect of CO_2/CH_4 ratio on carbon dioxide reforming with water at 50% water-vapor

CO_2/CH_4	1	3	5
Water-vapor (%)	50	50	50
Low side Voltage (V)	88	82	79
Frequency (Hz)	400	400	400
Power (W)	14	14	14
Power Factor	0.34	0.39	0.43
Current (A)	0.47	0.44	0.41
CH_4 conversion (%)	47	53	58
CO_2 conversion (%)	21	20	18
S_{CO} (%)	56	89	100
S_{C2H6} (%)	1	0	0
S_{C2H4} (%)	5	1	0
S_{C2H2} (%)	26	5	3
S_{total} (%)	88	95	103
Energy consumption (eV/m_C)	12	15	17
H_2/CO	1.9	0.6	0.4
CO/C_2	3.5	27.5	68.8

In high temperature catalytic steam reforming of methane, the endothermic heat of reaction must be provided at quite high temperatures with very intense heat fluxes. Although the direct fired furnaces used for this are quite efficient (>90%), they are quite expensive including the cost of heat recovery systems. A plasma discharge reactor system uses electricity to drive the reaction and can be assumed to have an efficiency no greater than that of the power generation system (~60% for modern combined cycle combustion turbine plants). However, for the low temperature plasma, little energy is expended in gas heating and heat recovery systems are not needed. Thus, for the plasma system, somewhat higher energy consumption maybe be offset by simple process operation and lower cost for investment. This could be true especially for relatively small scale plants, where fuel gas is inexpensive, or where excess (off peak) power generation is available.

Conclusions

Carbon dioxide combined with steam reforming at moderate conversion levels has been demonstrated at near ambient conditions with moderate power consumption. Under most conditions, carbon formation was not an operational problem. If long term operation with high conversions can be demonstrated, a process can be envisaged wherein the cost and efficiency of using electricity to drive the reaction is counter-balanced by lower costs and simpler operation (without oxygen). For smaller resources and more remote locations, these tradeoffs may be highly advantageous. Whether there is a net reduction in Carbon dioxide impact compared to more conventional reforming/partial oxidation synthesis gas generation processes can only be determined by a detailed life cycle analysis and not simply by the apparent reaction stoichiometry.

Acknowledgements

ChevronTexaco, Inc. and The Thailand Research Fund are gratefully acknowledged for supporting this research. Helpful discussions with Terence A. Caldwell are also acknowledged.

References

1. Walker, A. V.; King, D. A. *J. Phys. Chem. B* **2000**, 104, 6462-6467.
2. Zhou, L. M.; Xue, B.; Kogelschatz, U.; Eliasson, B. *Energy & Fuels* **1998**, 12, 1191-1199.
3. Cancino, J. F. Master Thesis, University of Oklahoma, Norman, OK, 2001.
4. Mutaf-Yardimci, O.; Saveliev, A. V.; Fridman, A. A.; Kennedy, L. A. *Int. J. Hydrogen Energy* **1998**, 23, 1109-1111.
5. Eliasson, B.; Kogelschatz, U.; Xue, B.; Zhou, L.-M. *Ind. Eng. Chem. Res.* **1998**, 37, 3350-3357.
6. Motret, O.; Pellerin, S.; Nikravech, M.; Massereau, V.; Pouvesle, J. M. *Plasma Chem. Plasma Proc.* **1997**, 17, 393-407.
7. Larkin, D. W.; Leethochawalit, M.; Chavadej, S.; Caldwell, T. A.; Lobban, L. L.; Mallinson, R. G. In *Greenhouse Gas Control Technologies*; Riemer, P.; Eliasson, B.; Wokaun, A.; Eds.; Elsevier Science Ltd.: New York, 1999; pp 397-402.

8. Kraus, M.; Eliasson, B.; Kogelschatz, U.; Wokaun, A. *Phys. Chem. Chem. Phys.*, **2001**, 3, 294-300.
9. Chang, J.-S.; Lawless, P. A.; Yamamoto, T. *IEEE Trans. Plasma Sci.* **1991**, 19, 1152-1165.
10. Marafee, A.; Liu, C.; Xu, G.; Mallinson, R.; Lobban, L. *Ind. Eng. Chem. Res.* **1997**, 36, 632-637.
11. Liu, C.; Marafee, A.; Mallinson, R.; Lobban, L. *Appl. Catal. A.* **1997**, 164, 21-33.
12. Eliasson, B.; Kogelschatz, U. *IEEE Trans. Plasma Sci.* **1991**, 19, 1063-1077.
13. Sorensen, S. L.; Karawajczyk, A.; Stromholm, C.; Kirm, M. *Chem. Phys. Lett.* **1995**, 232, 554-560.
14. Eliasson, B.; Liu, C.-J.; Kogelschatz, U. *Ind. Eng. Chem. Res.* **2000**, 39, 1221-1227.
15. Liu, C.; Marafee, A.; Hill, B.; Xu, G.; Mallinson, R.; Lobban, L. *Ind. Eng. Chem. Res.* **1996**, 35, 3295-3301.
16. Huang, A.; Xia, G.; Wang, J.; Suib, S. L.; Hayashi, Y.; Matsumoto, H. *J. Catal.* **2000**, 189, 349-359.
17. Suib, S. L.; Zerger, R. P. *J. Catal.* **1993**, 139, 383-391.
18. Liu, C.-J.; Mallinson, R.; Lobban, L. *J. Catal.* **1998**, 179, 326-334.
19. Tjatjopoulos, G. J.; Vasalos, I. A. *Ind. Eng. Chem. Res.* **1998**, 37, 1410-1421.
20. Kraus, M. Ph.D. Thesis, TU Karlsruhe, Germany, 2001.

Chapter 6

Influence of Electrode Configuration on Direct Methane Conversion with CO_2 as a Co-Reactant Using Dielectric–Barrier Discharges

Yue-ping Zhang[1-3], Yang Li[1,2], Chang-jun Liu[1,2], and Baldur Eliasson[1,4]

[1]Tianjin University-ABB Plasma Greenhouse, Gas Chemistry Laboratory,
[2]State Key Laboratory of C1 Chemistry and Technology, and
[3]Department of Chemistry, Tianjin University, Tianjin 300072, China
[4]Corporate Research, ABB Switzerland Ltd., Baden, CH5405, Switzerland

The effect of the electrode configuration of dielectric-barrier discharge (DBD) reactor on the conversions of CH_4 and CO_2 has been investigated. Two configurations have been applied. The first one is with the high voltage metal electrode covered by the dielectric-barrier material (quartz), while the another one is with the grounded electrode covered by the quartz. Differences both in the Lissajous figures and in the product distribution have been observed, while both reactors can lead to a significantly high conversions. The investigations have suggested a at least two-step plasma methane conversion using DBDs: the first is the initiation of discharge reactions to generate the necessary active species; then the further discharge reactions of such formed species produce the higher hydrocarbons and oxygenates. The future development has also been discussed.

Introduction

The conversion of methane directly into valuable products, including higher hydrocarbons and oxygenates, has been exploited for several decades. Because of its high stability of C-H bond of methane molecule, there is still no established technology for such direct methane conversion. On the other hand, it is well-known that non-equilibrium plasma can be applied for the activation of the stable small molecules, like methane. Such non-equilibrium plasma has been shown to be a promising technology in the chemical utilization of methane thanks to its high non-equilibrium, which means the electrons are with high energy level (up to 10 eV), while the bulk of gases are still at low temperature (as low as room temperature)[1,2]. Light hydrocarbons [3-21], syngas [3,4,21-26], liquid fuels [27-32] and oxygenates [33-43]can be the products from the direct plasma methane conversion under the conditions of corona , microwave, radio frequency discharge and dielectric-barrier discharge (DBD), respectively. Eliasson *et al* [2] has compared the different non-equilibrium plasma chemical processing previously. Most of products produced using corona are some little molecules, such as ethylene, acetylene, syngas (hydrogen and CO)[3,5-8, 16-19]. Under the conditions of DBDs, complex molecules could be produced for its large volume of discharges, which makes multi-step reaction possible. Furthermore, some operation parameters in DBDs can be easily changed with different dielectric materials and discharge gaps. The distribution of products could be very different with the different parameters in DBDs. However, there are some typical advantages with using DBDs, including: (1) DBDs have been successfully industrialized for the synthesis of ozone, (2) the DBDs can be easily operated and scaled-up and (3) the DBDs remain very stable during discharge reactions. On the other hand, the chemical utilization of carbon dioxide is another interesting subject. As the co-feed in methane conversion, CO_2 is a soft oxidant that could provide a carbon atom during the reactions. The synthesis of acetic acid, ethanol, other acids and alcohols with more than one carbon atom directly from methane has been investigated in the presence of carbon dioxide [43]. The effect of electrode materials on the product distribution of the utilization of methane and carbon dioxide has been investigated previously[21]. In this paper, the effect of electrode configurations on the conversions of CH_4 and CO_2 using DBDs will be investigated.

Experimental and Setup

The first electrode configuration used in this work is similar to that used previously[43]. The second electrode configuration of the plasma reactor is

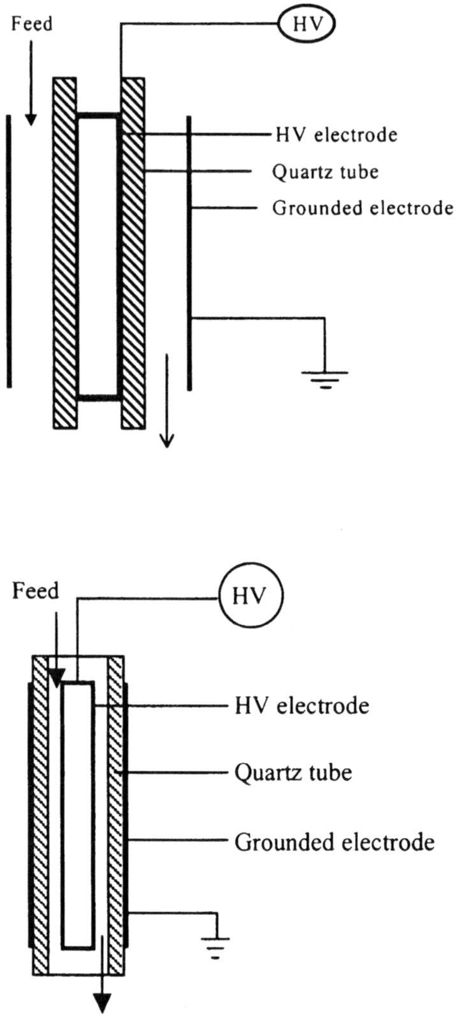

Figure 1. Schematic representatives of DBD reactors applied

shown in Figure 1. With the second electrode configuration, the feed gases passed though the high voltage side other than the grounded side in the previous investigations [29-32,43]. The quartz tube with the inner diameter of 14 mm served as the dielectric barrier. A transparent conductive metal film on the outside of the quartz tube was connected the ground. The high voltage electrode, located at the center of the quartz tube, was with an outer diameter of 12 mm and was made of aluminum foil, which connected with a high voltage (HV) generator. The discharge zone was a gap of 1 mm between the aluminum electrode and the quartz tube with the length of 200 mm. The high voltage (HV) generator provided about an up to 10 kV sinusoidal signal at a frequency of 25 kHz. The voltage and current were measured with a high voltage probe (Tektronix P6015 A) and a pulse current transformer (Pearson Electronics 411) via a digital oscilloscope (Tektronix 2440). The discharge power was conducted via a digital multimeter (Keithley 2000). The discharge power was fixed at 100 W in this investigation.

The reactants, CH_4 and CO_2, were introduced into reactor via mass flow controllers. The total flow rate was fixed at 60 ml/min in the experiments. The products were analyzed with a GC (HP 5890) equipped with a thermal conductivity detector (TCD) and a flame ionization detector (FID) via a HP-PLOT Q (30 m × 0.53 mm × 40 μm) column. A mass selective detector (HP 5971) was used to qualify the products. Liquid products were collected in a cold trap cooled with a mixture of ice and water.

Results and discussion

1. Characteristics of DBDs with different electrode configurations

The voltage versus charge Lissajous figure is a helpful tools to analyze the dielectric-barrier discharges and the discharge power. Figures 2 and 3 show the voltage/charge Lissajous figure recorded with the oscilloscope. The four lines in the parallelogram stand for the periodic change of voltage versus charge at the same frequency to the input voltage. Process of 1 to 2 means that the capacity between the two electrodes is charged by the rising voltage. And the reverse phasic process happens from 3 to 4. The gas discharges take place during the two processes from 2 to 3 and from 4 to 1. The points of 1 and 3 represent the minimum and the maximum of input voltage in one period, respectively. The area in the parallelogram is proportional to the discharge power. The analysis of the Lissajous figure is helpful for clarifying the discharge process and raising the energy efficiency. Comparing Figure 2 to Figure 3, more charge and lower voltage were found with the reactor in this investigation than the case of gases through the grounded electrode side. The obvious difference between the results with the two different reactors was that less liquid hydrocarbons were produced. A higher input voltage was required by the first electrode configuration, which means the generation of electrons with higher energy, leading to more productions of higher hydrocarbons, as observed experimentally.

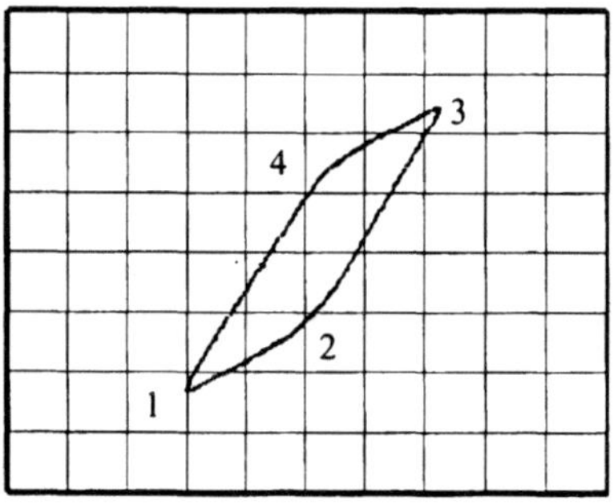

Horizontal: voltage [200 mV/div]; vertical: charge [20 mV/div]; CH_4 concentration of 66.8 v% in the feed; discharge power of 100 W;

Figure 2. Lissajous figure recorded using oscilloscope with the first electrode configuration: gases through the high voltage side

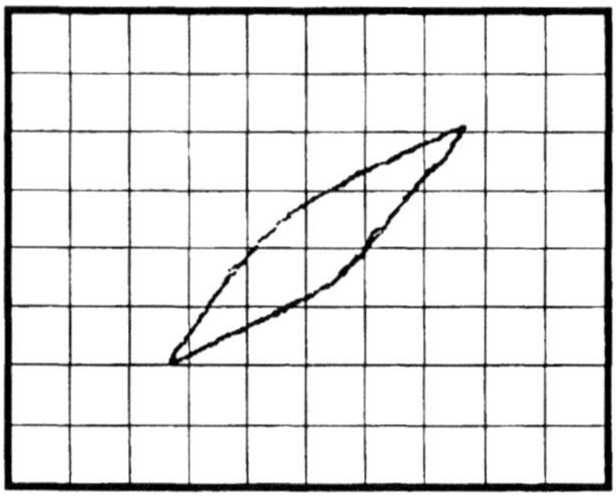

Horizontal: voltage [200 mV/div]; vertical: charge [20 mV/div]; CH_4 concentration of 66.8 v% in the feed; discharge power of 100 W

Figure 3. Lissajous figure recorded using oscilloscope with the second electrode configuration: gases through the grounded electrode side

2. Effect of the electrode configuration on methane conversion

Table 1 shows the effect of the electrode configuration on the conversions. The different electrode configuration presents different trends in the conversions of methane and carbon dioxide. As shown in Table 1, the conversion of carbon dioxide with the second electrode configuration nearly increased proportionally with increasing of methane concentration in the feed. Methane conversion increased with increasing of carbon dioxide conversion when methane concentration was less than 60 v% in the feed. However, methane conversion reached a saturated value when methane was more than 60 v% in the feed. The trend of the change of methane conversion is different from that observed with the first electrode configuration. One reason for these different results was the effect of electrode materials on the conversion of carbon dioxide. Carbon dioxide conversion with the aluminum electrode (as the high voltage one with the second electrode configuration) was much higher than that with the steel electrode (as grounded one with the first electrode configuration)[21]. Therefore, the conversion of carbon dioxide would play a more important role in reactions with the second electrode configuration. The another reason for these different results would be related to the microdischarge characteristics. Figure 2 has shown the first electrode configuration leads to a higher input voltage. Since the operating frequencies for both electrode configurations are the same, therefore the high voltage operation tends to spread the microdischarges through the whole length of the discharge, while the lower applied voltage with the second electrode configuration tends to reignite the old microdischarge channels every half period. As discussed below, this memory effect will be effective to the microdischarge reactions after the plasma-polymerized film is formed on the dielectric surface.

Table 1. Effect of methane feed concentration on conversions

CH_4 feed concentration (%)	Reactor #1		Reactor #2	
	CH_4 conversion (%)	CO_2 conversion (%)	CH_4 conversion (%)	CO_2 conversion (%)
82.8	49.3	44.5	64.2	52.7
75.1	52.7	41.9	64.3	46.2
67.4	54.5	37.4	64.2	43.4
50.4	54.5	32.9	58.3	29.2
34.0	58.0	28.8	50.0	19.1

Figures 4 and 5 present the effect of the electrode configuration on product distribution for plasma methane conversion using DBDs. In these figures, we use "reactor #1"to represent the reactor with the first electrode configuration as

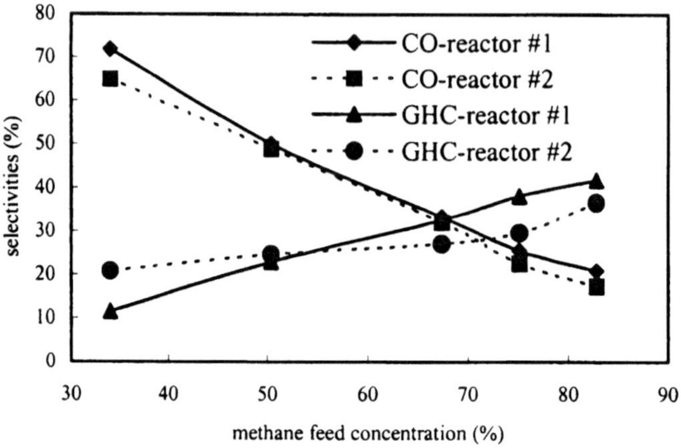

Figure 4. The effect on selectivities of CO and gaseous hydrocarbons

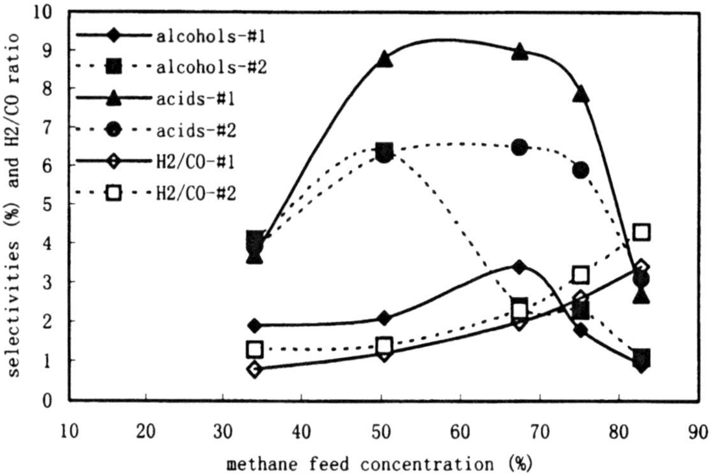

Figure 5. The effect on selectivities of alcohols and acids and the ratio of hydrogen/carbon monoxide

shown in Figure 1 and "reactor #2" to represent the reactor with the second electrode configuration. From these figures, we can see that, at the lower methane feed concentration, the first electrode configuration presents a higher selectivity of CO. With the increasing methane feed concentration, the selectivity of acids obtained with the first electrode configuration is higher than that obtained with the second electrode one until the methane feed concentration becomes more than 80%. Regarding the selectivity of alcohols, the second electrode configuration shows a better activity when the methane feed concentration is below 60%. Once again, we think the surface of the electrode materials will have a significant effect. In addition to the different metal electrode materials applied here with the two reactors, we have also observed some polymers formed on the electrodes that would present some effects. Figure 6 presents a picture that shows a plasma polymerization in the downstream part of the dielectric surface. From the Figure 6, we can divide the whole dielectric surface area into two parts: the first part represents the initiation step of plasma reaction (some carbon black formed as shown in Figure 6). The another part represents the propagation of the radical chain growth to produce oxygenates and higher hydrocarbons (more than C_5 hydrocarbons and even the plasma polymerized film). According to an investigation with the shorter discharge area, less oxygenates are produced. We could think that, in the first part, the initiation of plasma reactions will generate a lot of radicals with the energetic electrons that are so strong that methane molecule can be totally converted into carbon black. In this part, light hydrocarbons, like ethane, propane and acetylene, are the major products. With the discharge reactions, water is produced by the hydrogenation of carbon dioxide as the following:

$$CH_4 + e \rightarrow CH_3 + H + e' \qquad (1)$$

$$CO_2 + 2H \rightarrow CO + H_2O \qquad (2)$$

e' means the electron with less energy.

With the presence of water, the discharge performance starts to change. If the discharge region is long enough, liquid hydrocarbon and oxygenate products are produced. From the visualization of discharge reactions with the second electrode configuration reactor, we observe some liquid drops are hold steady within the discharge region, probably due to the electrostatic forces in the discharge phases. The liquid phase in the discharge region would induce a significant change in the reactions since a significant amount of chemically active species can be generated from water [44]. To get more liquid oxygenate products, it is very necessary to shorten the initiation step of plasma reactions via adjusting the reactive conditions and make the water be generated as early as possible. Further investigation is being conducted to analyse *in situ* the components of the liquid drops for the better understanding the reaction mechanisms, esp., for the formation of oxygenates.

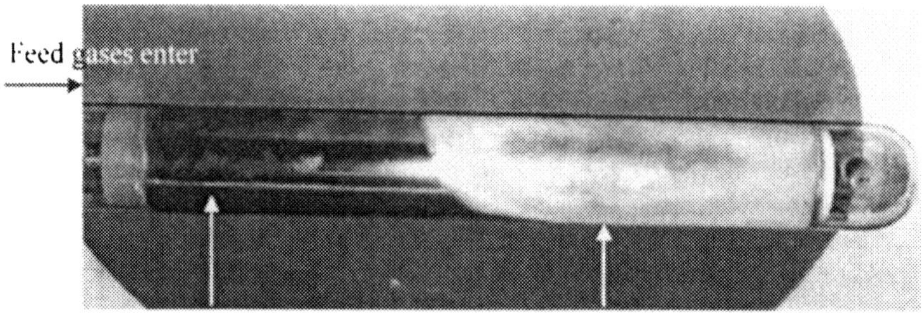

The first step: Initiation of discharge reactions in the upstream

The second step: Propagating of discharge reactions to generate oxygenates and higher hydrocarbons

Figure 6. The dielectric tube after reactions

3. Characteristics of microdischarges with the plasma-polymerized film

The most interesting characteristics of DBDs for most gases is that the breakdown at about atmospheric pressure is initiated in a large number of independent current filaments or microdischarges. The microdischarge channel diameter is roughly 0.1 mm. However, a plasma-polymerized film is normally obtained with the DBD plasma methane conversion with or without carbon dioxide as co-feed. Both electrode configurations will generate such plasma-polymerized film. Normally, such polymer film is likely formed on the high-voltage side. Therefore, for the first electrode configuration, the polymer film will present on the surface of the dielectric (quartz) tube. For the second electrode configuration, the polymer film will mostly present on the high voltage electrode surface. If the plasma polymerized film is formed, the channel must show some changes because all the microdischarges will have to go through such plasma-polymerized film. Especially, if water or the liquid phase is formed within the discharge region, the microdischarge channel will become more complex. Figure 7 presents a SEM image of the plasma-polymerised film on the dielectric surface obtained from the DBD plasma methane conversion in the presence of carbon dioxide. The small holes shown in Figure 7 are generated from microdischarges. Figure 8 shows a microdischarge channel image from such plasma-polymerized film or a scale-up picture of Figure 7. We would think the surface of polymer film is in some kind of "wet" status. This induces some messes in the micro-discharges, as shown in Figure 8. We would think such kind of plasma-polymerized film would induce a significant decrease in the electron energy within a microdischarge channel. On the other hand, the formed plasma-polymerized film will be also bombarded by the energetic electrons to generate more radicals for further discharge reactions.

As a comparison, we also present here a "dry"plasma-polymerized film (as shown in Figure 9 and Figure 10) obtaimed from methane and acetylene using DBDs under the similar reaction conditions. The microdischarges are uniform

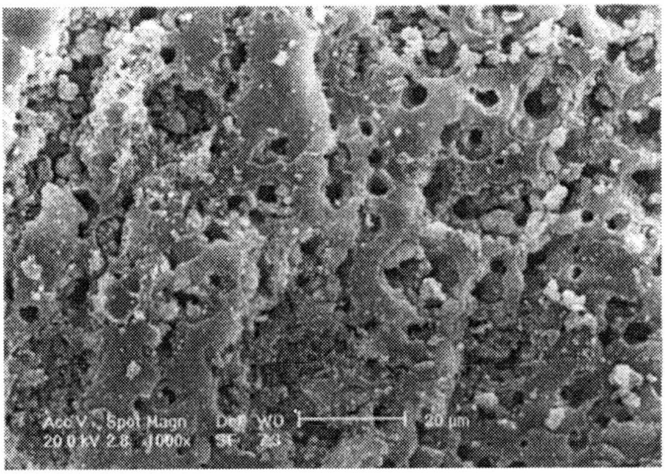

Figure 7. The SEM image of plasma-polymerized film from methane and carbon dioxide

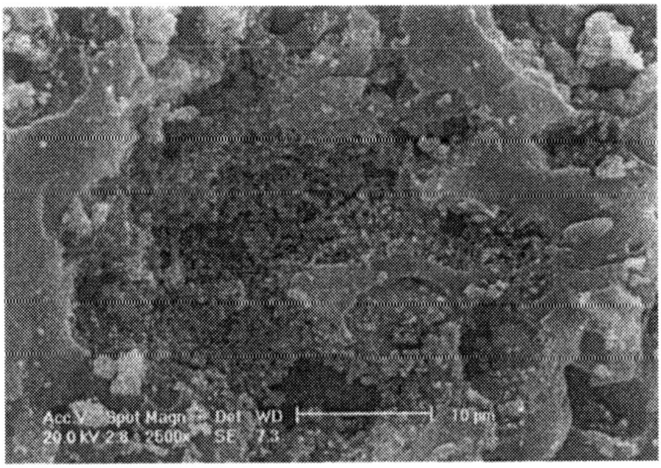

Figure 8. The SEM image of scale-up of the plasma-polymerized film

distributed through the dielectric tube. Especially, the microdischarge channel (see the small holes in the figure) is much more clear. As shown in Figure 9, the diameter of the microdischarge channel is around 10 μm that is around 10 times less than that without polymer film. The mechanism for how the plasma-polymerized film affect on the discharge reactions remains unclear. Normally, for the co-feed of methane and carbon dioxide, the polymers take around 30 to 40% in the total carbon balance.

Figure 9. The SEM image of plasma-polymerized film from CH_4 and C_2H_2

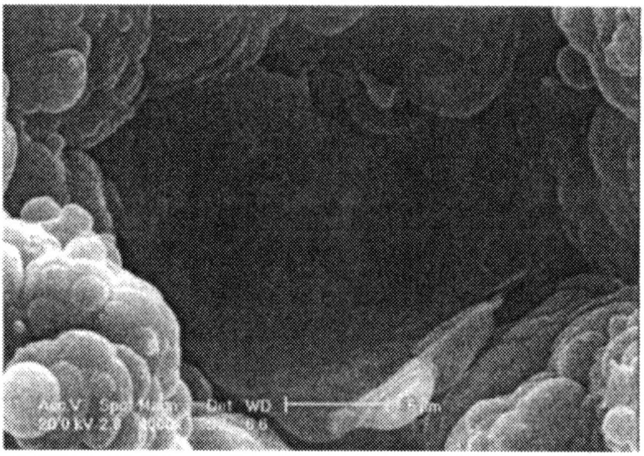

Figure 10. The SEM image ofscale-up of the plasma-polymerized film from CH_4 and C_2H_2

4. Future developments

The conversion of methane into higher hydrocarbons or oxygenates together with the co-feed of carbon dioxide has led to a much higher conversions with both electrode configurations, compared to the conventional catalytic methane conversion. The yield of objective products is also sufficiently high for a potential application of DBD methane conversion technology. Such developed technology is easily scaled-up. The only limitation for this technology is that the energy consumption is still high or the energy yield is still low. For the regions with cheep hydraulic electricity, like Southwest China, it is feasible already in the economics for the present DBD conversion technology. However, for most of areas in the world, we still need a further improvement in the DBD methane conversion. More fundamental investigations are required since the plasma chemistry is still new or young. Compared to those intensive methane conversion via conventionally, the input to the fundamental investigations on plasma methane conversion is very small. The potential of plasma technologies is still not realized worldwide. According to the present understanding, the following improvements in DBD methane conversion would enhance the economics. The first is to use both sides of the dielectric for plasma methane conversion. At present, all the investigations on DBD plasma methane conversion just employed one-side for the discharge reactions. For example, for the first electrode configuration applied in this work, the energy yield could be doubled with using the inner high voltage electrode surface for methane conversion too. The second is that some innovative DBD reactor designs are needed. With the new design, we will be available to better use the applied energy and reduce the generation of species with low excitation status since these species just consume energy with no productivity. The new design of the DBD reactor should be also favored to remove the valuable objective products as soon as they are produced. The third is that the use of some additives or co-reactants for the DBD plasma methane conversion will be helpful to reduce the initiation region of DBD reactions. We have tested various zeolite catalysts, coal and starch. We have found the use of coal or starch as co-reactant can significantly increase the yield of objective liquid products (liquid fuels or oxygenates). On the other hand, we would mention one more point here: the plasma preparation of catalysts will lead to a great improvement in the comprehensive performance of catalysts for methane conversion catalytically. The highly dispersion of active species, stability, activity and selectivities can be totally enhanced by the use of plasma catalyst preparation that we will report elsewhere.

Conclusion

The effectiveness of two reactors with high-voltage electrode covered by a dielectric material (quartz) or by no dielectric material has been investigated in this work. Both electrode configurations employed in this work can lead to a significantly high conversion of methane and carbon dioxide, while the second one leads to a around 10% higher conversion. The first electrode configuration leads to a higher selectivity of acids, while the second one is more favored for the formation of alcohols. It has been found that the DBD plasma methane conversion to higher hydrocarbons and oxygenates is a multi-step reaction. The first step is a necessary initiation of discharge reactions, with which is not favored for the production of higher hydrocarbons (like liquid fuels) and oxygenates. If the objective products are not light hydrocarbons, the first step must be as short as possible. For the future development, the combination of both electrode configurations within a single DBD reactor will be very necessary to improve the efficiency to use both sides of the dielectric for the synthesis.

Acknowledgement

The supports from ABB Switzerland Ltd. and the Key Fundamental Research Project of Ministry of Science and Technology of China (G1999022402) are very appreciated.

References

1. Liu C.-J.; Xu, G.-H.; Wang, T. Non-thermal Plasma Approaches in CO_2 Utilization. *Fuel Processing Technology*, **1999**, *58*, 119.
2. Eliasson B.; Kogelschatz, U. Nonequilibrium volume plasma chemical processing. *IEEE Tran. Plasma Sci.* **1991**, *19*, 1063-1077.
3. Yao, S.L.; Ouyang, F.; Nakayama, A.; Suzuki,E.; Okumoto, M.; Mizuno, A. Oxidative Coupling and Reforming of Methane with Carbon Dioxide Using a High-Frequency Pulsed Plasma. *Energy & Fuels* **2000**, *14*, 910-914.
4. Savinov, S.Y.; Lee, H.; Song, H. K.; Na, B.-K. Decomposition of Methane and Carbon Dioxide in a Radio-Frequency Discharge. *Ind. Eng. Chem. Res.* **1999**, *38*, 2540-2547.
5. Kado S.; Sekine, Y.; Fujimoto, K. Direct synthesis of acetylene from methane by direct current pulse discharge. *Chem. Commun.* **1999**, 2485-2486.
6. Liu C.-J.; Mallinson, R.; Lobban, L. Nonoxidative Methane Conversion to Acetylene over Zeolite in a Low Temperature Plasma. *J. Catal.*, **1998**, *179*, 326.
7. Yao, S.L.; Suzuki, E.; Meng, N.; Nakayama, A. Influence of Rise Time of Pulse Voltage on the Pulsed Plasma Conversion of Methane. *Energy & Fuels* **2001**, *15*, 1300-1303.
8. Liu C.-J.; Mallinson, R.; Lobban, L. Comparative Investigation on Plasma Catalytic Methane Conversion to Higher Hydrocarbons over Zeolites. *Appl. Catal. A* **1999**,*178*, 17.
9. Ogata A.; Mizuno, K.; Kushiyama, S.; Yamamoto, T. Methane Decomposition in a Barium Titanate Packed-Bed Nonthermal Plasma Reactor. *Plasma Chem. Plasma Process.*, **1998**, *18*, 363-373.

10. Oumghar A.; Legrand, J.C.; Diamy, A.M.; Turillon, N. Methane Conversion by an Air Microwave Plasma. *Plasma Chem. Plasma Process.* **1995**, *15*, 87.
11. Suib S. L.; Zerger, R.P. A Direct, Continuous, Low-Power Catalytic Conversion of Methane to Higher Hydrocarbons via Microwave Plasmas. *J. Catal.* **1993**, *139*, 383.
12. Yao S. L.; Nakayama, A.; Suzuki, E. Methane Conversion Using a High-Frequency Pulsed Plasma: Important Factors. *AIChE J.*,**2001**, *47*, 413.
13. Legrand, J.-C.; Diamy, A.-M.; Hrach, R.; Hrachova, V. Methane conversion in the flowing afterglow of a dinitrogen microwave plasma. *Contrib. Plasma Phys.* **1997**, *37*, 521-537.
14. Diamy, A.-M.; Hrach, R.; Hrachova, V.; Legrand, J.-C. Influencing of C atom concentration for acetylene production in a CH_4/N_2 afterglow. *Vacuum* **2001**, *61*, 403-407.
15. Hsieh, L.-T.; Lee, W.-J.; Chen, C.-Y.; Chang, M.-B.; Chang, H.-C. Converting methane by using an rf plasma reactor. *Plasma Chem. Plasma Process.* **1998**, *18*, 215-239.
16. Liu, C.-J.; Marafee, A.; Xu, G.-H.; Mallinson, R.G.; Lobban, L.L. The oxidative coupling of methane with ac and dc corona discharge. *Ind. Eng. Chem. Res.* **1996**, *35*, 3295.
17. Marafee, A.; Liu, C.-J.; Xu, G.-H.; Mallinson, R.G.; Lobban, L.L. An experimental study on the oxidative coupling of methane in a dc corona discharge reactor over Sr/La_2O_3. *Ind. Eng. Chem. Res.* **1997**,*36*, 632.
18. Liu, C.-J.; Marafee, A.; Mallinson, R.G.; Lobban, L.L. Methane conversion to higher hydrocarbons over charged metal oxide catalysts with OH groups. *Appl. Catal. A* **1997**, *164*, 21.
19. Liu, C.-J., R.G. Mallinson and L.L. Lobban; "Comparative Investigation on plasma catalytic methane conversion to higher hydrocarbons over Zeolites," *Appl. Catal. A* **1999**, *178*, 17.
20. Yao, S.L.; Okumoto, M.; Nakayama, A.; Suzuki, E. Plasma Reforming and Coupling of Methane with Carbon Dioxide. *Energy & Fuels* **2001**, *15*, 1295-1299.
21. Li Y., Xu, G.-H.; Liu, C.-J.; Eliasson, B.; Xue, B. Co-generation of syngas and higher hydrocarbons from CO_2 and CH_4 using dielectric-barrier discharge: Effect of electrode Mmaterials. *Energy & Fuels* **2001**, *15*, 299.
22. Huang A.; Xia, G.; Wang, J.; Suib, S.L. CO_2 Reforming of CH_4 by Atmospheric Pressure ac Discharge Plasmas. *J. Catal.*, **2000**,*189*, 349.
23. Kado S.; Urasaki, K.; Sekine, Y.; Fujimoto, K. Low temperature reforming of methane to synthesis gas with direct current pulse discharge method. *Chem. Commun.* **2001**, 415-416.
24. Yao, S.L.; Takemoto, T.; Ouyang, F.; Nakayama, A.; Suzuki,E.; Mizuno, A.; Okumoto, M. Selective Oxidation of Methane Using a Non-Thermal Pulsed Plasma. *Energy & Fuels* **2000**, *14*, 459-463.
25. Lesueur, H.; Czernichowski, A.; Chapelle, J. Electrically assisted partial oxidation of methane. *Int. J. Hydrogen Energy* **1994**, *19*, 139-144.
26. Mutaf-Yardimcu, O.; Saveliev, A.V.; Fridman, A.A.; Kennedy, L.A. Employing plasma as catalyst in hydrogen production. *Int. J. Hydrogen Energy* **1998**, *23*, 1109-1111.

27. Liu C.-J.; Xue, B.; Eliasson, B.; He, F.; Li, Y.; Xu, G.-H. Methane conversion to higher hydrocarbons in the presence of carbon dioxide using dielectric-barrier discharge plasmas. *Plasma Chem. Plasma Process.*, **2001**, *21*, 301.
28. Okumoto, M.; Mizuno, A. Conversion of methane for higher hydrocarbon fuel synthesis using pulsed discharge plasma method. *Catal. Today* **2001**, *71*, 211-217.
29. Eliasson, B.; Liu C.-J.; Kogelschatz, U. Direct Conversion of Methane and Carbon Dioxide to Higher Hydrocarbons Using Catalytic Dielectric-Barrier Discharges with Zeolites. *Ind. Eng. Chem. Res.* **2000**, *39*, 1221-1227.
30. He, F.;Liu, C.-J.; Eliasson B.; Xue, B. XPS Characterization of Zeolite Catalyst in Plasma Catalytic Methane Conversion, *Surf. Interface Analy.* **2001**, *32*,198-201.
31. Liu,C.-J.; Eliasson, B.; Xue, B.; Li, Y.; Wang, Y. Zeolite-Enhanced Plasma Methane Conversion to Higher Hydrocarbons with Dielectric-barrier Discharges. *React. Kinet. Catal. Lett.* **2001**, *74*, 71-77.
32. Jiang, T.; Li, Y.; Liu, C.-J.; Xu, G.-H.; Eliasson B.; Xue, B. Plasma Methane Conversion Using Dielectric-barrier Discharges with Zeolite A. Catalysis Today **2002**, *72*, 229~235.
33. Matsumoto, H.; Tanabe, S.; Okitsu, K.; Hayashi, Y.; Suib, S.L. Selective Oxidation of Methane to Methanol and Formaldehyde with Nitrous Oxide in a Dielectric-Barrier Discharge-Plasma Reactor. *J. Phys. Chem. A* **2001**, *105*, 5304-5308.
34. Larkin, D.W.; Lobban, L.L.; Mallinson, R.G. Production of Organic Oxygenates in the Partial Oxidation of Methane in a Silent Electric Discharge Reactor. *Ind. Eng. Chem. Res.* **2001**, *40*, 1594-1601.
35. Larkin, D.W.; Zhou, L.; Lobban, L.L.; Mallinson, R.G. Product Selectivity Control and Organic Oxygenate Pathways from Partial Oxidation of Methane in a Silent Electric Discharge Reactor. *Ind. Eng. Chem. Res.* **2001**, *40*, 5496-5506.
36. Larkin, D.W.; Caldwell, T.A.; Lobban, L.L.; Mallinson, R.G. Oxygen Pathways and Carbon Dioxide Utilization in Methane Partial Oxidation in Ambient Temperature Electric Discharges. *Energy Fuels*, **1998**, *12*, 740-744.
37. Bugaev S. P.; Kozyrev, A.V.; Kuvshinov, V.A.; Sochugov, N.S.; Khryapov, P.A. Plasma-Chemical Conversion of Low Alkanes with Stimulated Condensation of Incomplete Oxidation Products. *Plasma Chem. Plasma Process.* **1998**, *18*, 247-261.
38. Hijikata K.; Ogawa, K.; Miyakawa, N. Methanol conversion from methane and water vapor by electric discharge (effect of electric discharge process on methane conversion) . *Heat Transfer-Asian Research*, **1999**, *28*, 404.
39. Kozlov K. V.; Michel, P.; Wagner, H.-E. Synthesis of organic compounds from mixtures of methane with carbon dioxide in dielectric-barrier discharges at atmospheric pressure. *Plasmas and Polymers* **2000**, *5*, 129-150.

40. Okumoto M.; Su, Z.; Katsura, S.; Mizuno, A. Dilution effect with inert gases in direct synthesis of methanol from methane using nonthermal plasma. *IEEE Tran. Ind. Appl.* **1999,** *35*, 1205-1210.
41. Zhou, L.M.; Xue, B.; Kogelschatz, U.; Eliasson, B. Partial oxidation of methane to methanol with oxygen or air in a non-equilibrium discharge plasma. *Plasma Chem. Plasma Process.* **1998,** *18*, 375-393.
42. Okazaki, K.; Hirai, S.; Nozaki, T.; Ogawa, K.; Hijikata, K. Plasma chemical reactions at atmospheric pressure for high efficiency use of hydrocarbon fuels. *Energy* **1997**, *22*, 369-374.
43. Liu,C.-J.; Li,Y.; Wang, Y.; Zou, J.; Eliasson, B.; Xue, B. Production of acetic acid directly from methane and carbon dioxide using dielectric-barrier discharges. *Chem. Lett.* **2001**, 1304~1305.
44. Sunka, P.; Babicky V.; Clupek, M.; Lukes, P.; Simek, M.; Schmidt, J.; Cernak, M. Generation of chemically active species by electrical discharges in water, *Plasma Sources Sci. Technol.*, **1999**, *8*, 258-265.

Chapter 7

Effect of H_2S on the Reaction of CH_4 with CO_2 over Titania-Supported Noble Metal Catalysts

András Erdőhelyi, Tamás Szailer, and Éva Novák

Institute of Solid State and Radiochemistry, University of Szeged, P.O. Box 168, H–6701 Szeged, Hungary

The reaction of methane with carbon dioxide was studied on different titania-supported noble metal catalysts in the presence of hydrogen sulphide. In addition the hydrogenation of CO_2 and the interaction of H_2S with the catalysts were also studied. It was found that in the presence of H_2S the conversions of methane and carbon dioxide decreased while the ratio of the amounts of CO and H_2 formed in the reaction increased in time. Temperature-programmed desorption studies revealed that adsorbed H_2S desorbed as SO_2 at higher temperature either from the catalysts or from the support. In the hydrogenation of carbon dioxide the rate of product formation was higher in the presence of 22 ppm H_2S that in the absence of it. This influence of H_2S resulted in the changes of the CO/H_2 ratio in the $CO_2 + CH_4$ reaction. An explanation for the effect of H_2S on these reactions is also proposed.

Introduction

Great efforts are made in the last decade in the catalytic transformation of methane and carbon dioxide, the cheapest carbon containing materials, into more valuable compounds. One of the possibilities to achieve this goal is to react CH_4 with CO_2 to produce synthesis gas. All of the Group VIII metals on a variety of supports have been studied as dry reforming catalysts *(1)*. The high efficiency of supported noble metals in this reaction was confirmed by our previous studies *(2-4)*. Relatively little attention was paid to the effect of sulphur on the CO_2 + CH_4 reaction, although natural gases always contain more or less amounts of sulphur compounds.

The effect of sulphur on the CO_2 + CH_4 reaction was studied on nickel catalysts in the SPARG process in which partly sulphur-poisoned catalyst is used *(5)*. This passivation results in the decrease of the surface carbon formation. Adsorbed sulphur atoms cause the deactivation of the neighboring nickel atoms. It was shown that carbon formation rate depends more sensitively on sulphur coverage than on the reforming rate.

The present report gives an account of the effect of H_2S on the CO_2 + CH_4 reaction in the presence of different titania-supported noble metal catalysts. This reaction can be described as the sum of different reactions; for example the steam reforming and the reverse water gas shift reactions. For comparison the CO_2 hydrogenation was also studied in the presence of H_2S. In addition to the catalytic behavior of titania-supported noble metals, these reactions were studied on silica-supported metals, too.

Experimental

Materials. The catalysts were prepared by impregnating the TiO_2 (Degussa P25) and SiO_2 (Cab-O-Sil) support with the solution of Pt metal salts to yield a nominal 1 % metal content. The following salts were used: $H_2PtCl_6*6H_2O$, $PdCl_2$, $RhCl_3*3H_2O$ and $RuCl_3*3H_2O$. The impregnated powders were dried at 383 K. Before any measurements the fragments of catalyst pellets were oxidized at 473 K for 30 min and reduced at 673 K in the catalytic reactor for 1 hour.

Methods. The adsorption and interaction of the reacting gas mixtures were studied by FTIR spectroscopy. The spectra were recorded with a Bio-Rad FTS/135 type FTIR spectrometer, equipped with a diffuse reflectance attachment (Harrick type, with CHC-CHA low temperature reaction chamber).

Catalytic reactions were carried out in a fixed bed continuous-flow reactor (100 mm x 27 mm o.d.). The ratio of CH_4/CO_2 in the reacting gas mixture was 1:1 and that of CO_2/H_2 was 1:4. The inlet gas mixture contained 20 % of Ar or

He as diluents and 22 ppm or 116 ppm of H_2S when the effect of H_2S was studied. The amount of catalysts used was usually 0.3 - 0.5 g. The flow rate of the reactants was 80 ml/min. Analyses of the gases were performed with gas chromatograph (Chrompack 9001) using Porapak QS column. The products were detected simultaneously by TC and FI detectors.

The amount and the reactivity of surface carbon formed in the catalytic reactions were determined by temperature-programmed reduction. After the catalytic run the reactor was flushed with Ar, the sample was cooled to room temperature, then the Ar flow was changed to H_2 and the sample was heated up to 1173 K with a 10 K/min heating rate and the hydrocarbons were determined by flame ionization detector.

Temperature-programmed desorption (TPD) experiments were carried out in a quartz reactor (8 mm o.d.) filled with 0.3 g of samples. The heating rate was usually 20 K/min. The products were analysed by an online mass spectrometer (Balzers QMS 200). The adsorption of H_2S before the TPD was performed by exposing the reduced samples to Ar flow containing 110 ppm of H_2S for 30 min at 373 K. Afterwards the reactor was flushed with Ar for 10 min, cooled down to room temperature and the sample was heated up to 1073 K.

XPS measurements were performed in a Kratos XSAM 800 instrument using Al K_α (1486.6 eV) primary radiation. To compensate for possible charging effects the binding energies were normalized with respect to the position of C (1s), this value being assumed constant at 285.1 eV. The samples were pretreated in a separate equipment which was connected directly to the analyzing chamber.

The dispersities of the supported metals were determined by H_2 adsorption and by H_2-O_2 titration at 298 K in a conventional gas volumetric apparatus at room temperature after the same pre-treatment mentioned above.

Results and Discussion

Infrared spectroscopy measurements revealed that CO is formed in the interaction of $CH_4 + CO_2$ and of $CO_2 + H_2$ far below the reaction temperature, between 473 – 523 K *(3)* and at room temperature, respectively. CO spectra obtained in the interactions differed from those observed after the adsorption of CO. Only linearly bonded CO formed while in the latter case dicarbonyl species are also detectable. In the interactions the CO band appeared at lower wavenumber because probably carbonyl hydrid was produced. In the presence of H_2S, after the adsorption of CO, nearly the same features were registered, only the intensity of the absorption bands decreased and the reductive agglomerization of surface Rh atoms occurred at higher temperatures. Zhang et al. also found that the presence of S does not significantly influence the geometrical structure and chemisorption energy of CO adsorbed on Rh (111) surface *(6)*. In the interaction or in the reaction of CH_4 + CO_2 the CO band appeared at significantly higher temperature (between 573 – 598K) and with lower intensities in both cases on the spectra in the presence of H_2S than in the absence of it, although in the case of Rh/TiO_2 the gas phase spectra of CH_4, which was the main product in the H_2 + CO_2 reaction, were detectable with the same intensities.

The reaction between CO_2 and CH_4 proceeded rapidly above 673 K on TiO_2 supported noble metals to give CO and H_2 with different ratios. The activities of the samples were relatively stable, only a slight deactivation of the catalysts were observed. The conversion of CO_2 exceeded that of CH_4 for every catalyst sample. The CO/H_2 ratio was between 2 and 3 in the first minutes of the reaction and only slightly varied during the reaction. The reaction occurred at the highest rate (rates per unit surface area of metals) in the steady state on Ru followed by Pd, Rh and Pt. This is practically the same as the sequence for the CH_4 + CO_2 reaction on alumina-supported noble metals *(2)*.

In the presence of H_2S the initial rate of H_2 or CO formation and the conversions of the reactants decreased in time in all cases (Fig. 1 and Table 1).

The deactivation rate was the highest on supported Rh catalyst, and the lowest on Ru/TiO_2. The initial conversion of methane at 773 K was about 12 % on Rh/TiO_2 but after 3 hours of the reaction it was only 2.1 %. On Ru/TiO_2 this value changed from 8.8 % to 7.7 %.

While the rate of products formation decreased, the CO/H_2 ratio surprisingly increased continuously in all cases (Fig. 2).

The enhancement of the CO/H_2 ratio at the beginning of the reaction was relatively low, but after a time lag it increased considerably. The length of this period depends on the metal; it was about 50 minutes on Rh/TiO_2 and more than 100 minutes on titania-supported Pt catalyst. In the presence of Rh/TiO_2 after

Table 1. Some Characteristic Data of the CH_4 + CO_2 Reaction on Titania-Supported Noble Metal Catalysts at 773 K in the Presence of 22 ppm H_2S

Samples	*D*	*Conversion*				*CO/H_2*	
		CH_4		*CO_2*			
		15	360	15	360	15	360
	%	*Minutes*					
1%Ru/TiO_2	9.4	8.8	5.5	14.9	10.8	2.32	2.65
1%Rh/TiO_2	41	12.0	2.1*	17.9	4.9*	3.0	8.2*
1%Pd/TiO_2	16	12.9	1.2	20.8	2.2	2.1	6.3
1%Pt/TiO_2	38	5.5	1.7	10.4	4.2	2.7	5.0

* Data were obtained in the 180th minute of the reaction

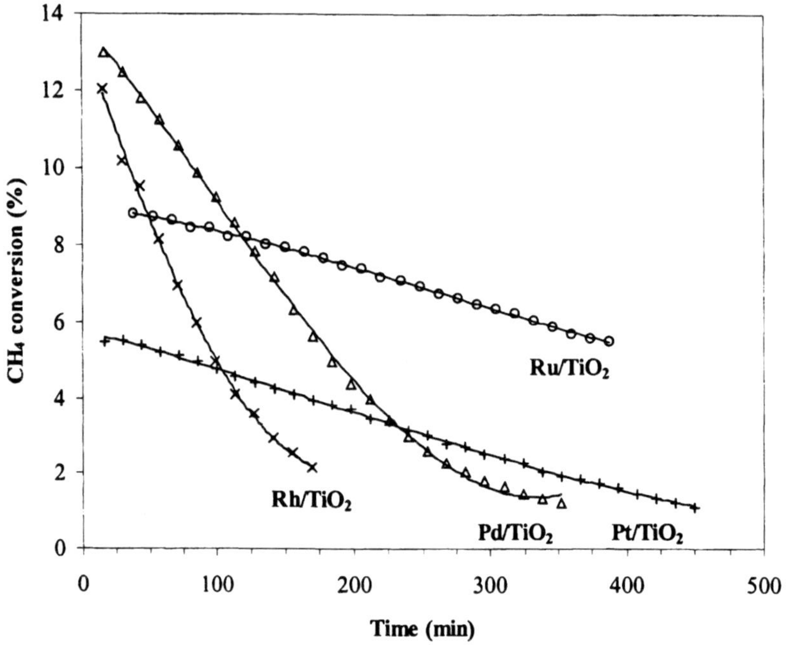

Figure 1. *The conversion of methane in the $CH_4 + CO_2$ reaction on different TiO_2 supported catalysts at 773K in the presence of 22 ppm H_2S*

180 minutes of the reaction the CO/H_2 ratio was more than 8 (Table 1). The most stable catalyst was Ru/TiO_2; on this sample the CO/H_2 ratio was only 2.7 after 400 minutes.

Using more catalyst – 0.5 g instead of 0.3 g – the deactivation rate decreased and the CO/H_2 ratio enhanced after a longer induction period than it was shown on Fig. 2.

When the H_2S concentration was increased in the reacting gas mixture (116 ppm) the conversion of CH_4 or CO_2 decreased more rapidly. After 180 minutes the CH_4 conversion was more than one order of magnitude less than the initial value in all cases. The CO/H_2 ratio increased, too, after 200 minutes of the reaction in the case of Rh/TiO_2 it was more than 10.

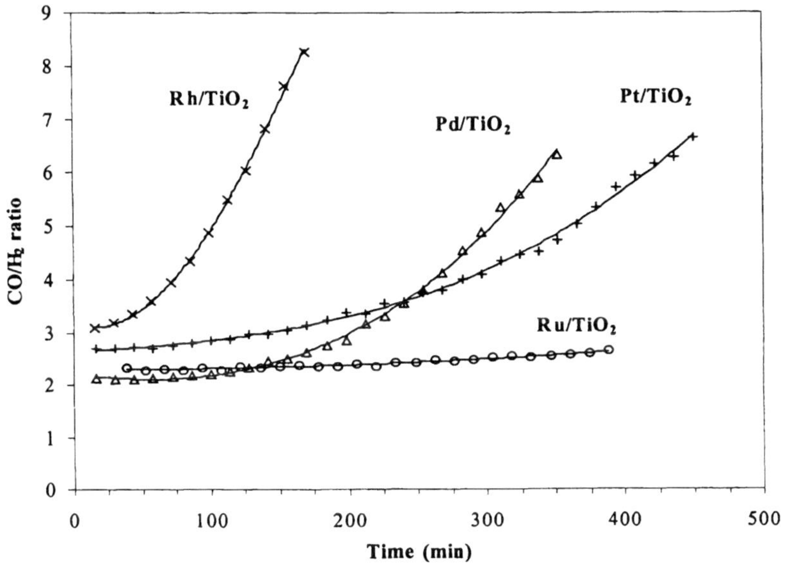

Figure 2. *Changes of the CO/H_2 ratio in the CH_4 + CO_2 reaction on different TiO_2 supported catalysts at 773 K in the presence of 22 ppm of H_2S*

The amount of surface carbon was determined after the catalytic reaction by TPR. In the case of Rh/TiO_2 in the presence of H_2S the amount of surface carbon was about one third (16 μmol/g) of that measured in the absence of it. These results agree well with the observation made in the Sparg process *(5)* when the supported Ni catalyst was pre-sulphided to decrease the surface carbon formation. The reactivity of the carbonaceous deposit did not change significantly.

On silica-supported samples in the presence of H_2S, although the conversion decreased as was observed on titania-supported samples, the CO/H_2 ratio remained nearly the same (Fig. 3). The Pt/SiO_2 catalyst completely loses its activity after 50 minutes of the reaction and so the results measured on this sample were not depicted on Fig. 3.

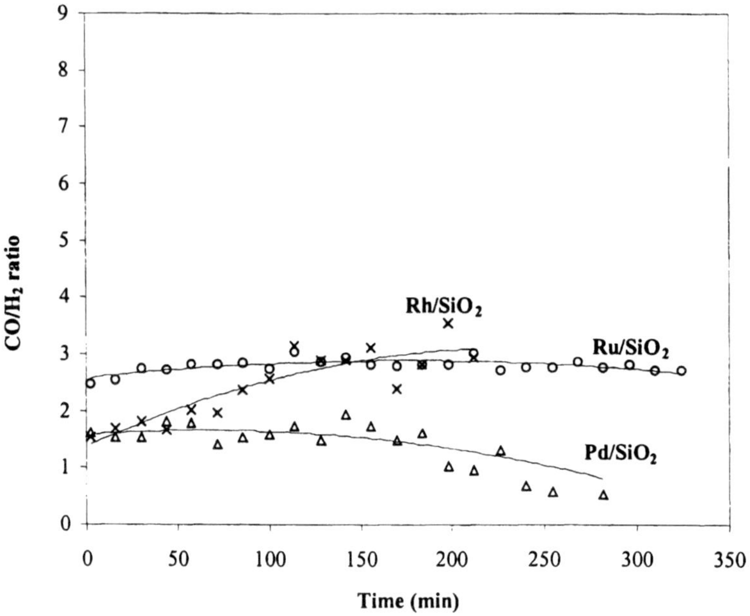

Figure 3. Changes of the CO/H_2 ratio in the $CH_4 + CO_2$ reaction on different SiO_2 supported catalysts at 773 K in the presence of 22 ppm H_2S

It was found that the CO_2 conversion was higher than that of methane in all cases and that the CO/H_2 ratio increased in time in the presence of H_2S. These observations indicated that the reaction

$$CH_4 + CO_2 = 2\,CO + 2\,H_2$$

was followed by several secondary processes, including the hydrogenation of CO and CO_2, the water gas shift reaction and the Boudouard reaction.

CO_2 hydrogenation was studied in the presence of 22 ppm H_2S at 548 K (Fig. 4.). The main product of the reaction was methane on Ru and Rh catalysts; its selectivity was near 100 %. On Pt and Pd samples mainly CO was formed, the methane selectivity was about 20 % at the beginning of the reaction.

These results agree well with our previous findings *(7-9)* obtained on different supported noble metal catalysts. In spite of the 22 ppm H_2S concentration of the reacting gas mixture the rate of products formation or the CO_2 conversion changed only slightly after the first minutes of the reaction in all cases (Fig. 4.). The highest decrease was in the case of Pd catalyst when the initial conversion decreased from 10.0 % to 6.5 % in the 6 hours of the reaction. It has to be mentioned that during the same time in the $CO_2 + CH_4$ reaction the CO_2 conversion decreased from 20.8 to 2.9%. Comparing the rates of products

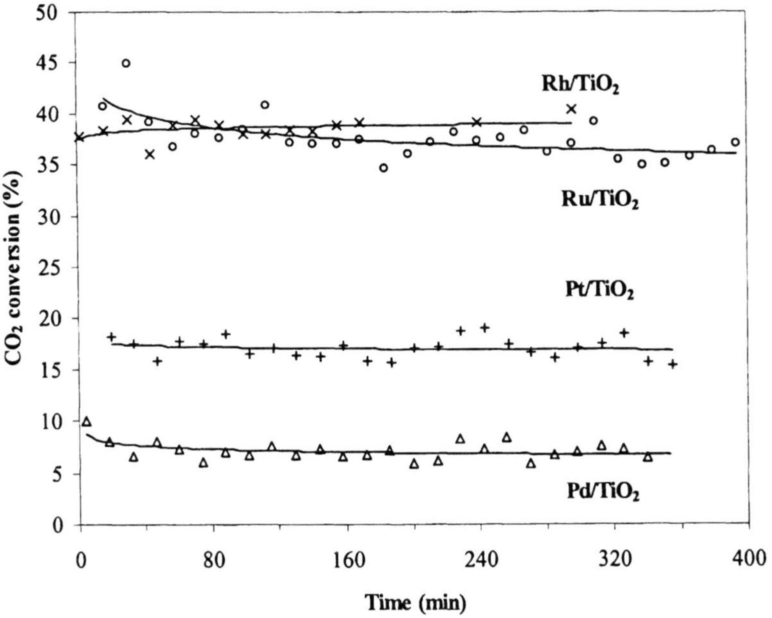

Figure 4. *Changes of the CO_2 conversion in the $CO_2 + H_2$ reaction on different TiO_2 supported catalysts at 548 K in the presence 22 ppm of H_2S*

formation in the $CO_2 + H_2$ reaction in the presence and in the absence of H_2S we can state that the methane evolution was surprisingly higher when H_2S was in the reacting gas mixture, while the product distribution remained nearly the same (Table 2.). When the reaction was studied at 773 K, at the temperature at which the $CO_2 + CH_4$ reaction was followed, the same results were obtained i.e. the rates of CH_4 formation and CO_2 conversion did not decrease even in the presence of H_2S. In this case the contact time was less, so the conversion was in the same range. On Rh/TiO_2 in these experimental conditions CO was also detected, the selectivity of methane formation was only 75.1 %.

Contrary to these results CO_2 conversion decreased continuously in the $CO_2 + H_2$ reaction in the presence of H_2S on the SiO_2 supported samples.

These observations suggest that on titania supported metals H_2S poisons the $CO_2 + CH_4$ reaction, while the $CO_2 + H_2$ reaction was only slightly affected by the same amount of H_2S. This finding may explain the changes of the CO/H_2 ratio. The rate of hydrogen formation in the $CO_2 + CH_4$ reaction decreased while the consumption rate of it in a secondary reaction remained the same in the $CO_2 + H_2$ reaction. In some cases CO is also formed in this process so the CO/H_2 ratio has to increase during the $CO_2 + CH_4$ reaction.

On SiO_2 supported samples in the $CO_2 + CH_4$ reaction the CO/H_2 ratio does not change significantly, and the rate of CO_2 hydrogenation decreased

Table 2. Some Characteristic Data of the CO_2+ H_2 Reaction on Titania-Supported Noble Metal Catalysts at 548 K in the Presence of 22 ppm H_2S

	Rate of product formation									
	In the presence of H_2S					*In the absence of H_2S*				
Samples	*CH_4*		*CO*		*TOF**	*CH_4*		*CO*		*TOF**
	umol/gs				s^{-1}	umol/gs				s^{-1}
	15	360	15	360	360	15	360	15	360	360
	Min					Min				
Ru/TiO_2	10.0	7.5	-	-	0.815	0.7	0.6	-	-	0.07
Rh/TiO_2	5.0	4.8	-	-	0.12	2.8	2.9	-	-	0.07
Pd/TiO_2	0.45	0.25	1.88	1.18	0.017	0.09	0.05	0.3	0.2	0.02
Pt/TiO_2	0.56	0.26	2.0	1.93	0.014	0.16	0.06	0.97	0.73	0.05

* Rate of methane formation related to the amount surface metal atoms

continuously, so the effect observed on TiO_2 supported metals depends on the support. For an interpretation of these results the interaction of H_2S with the titania-supported metals were also examined.

XP spectra of reduced titania-supported noble metals pre-treated with H_2S at room temperature show that the metal was partially oxidized and the extent of it increased with increasing the adsorption temperature of H_2S. For example the binding energy of Rh ($3d_{5/2}$) in the reduced Rh/TiO_2 was 307.2 eV and this value shifted to 307.3 eV when the sample was treated with H_2S at 300 K and to 307.55 eV after adsorption at 773 K. As to the TiO_2 substrate, we could not point out any characteristic or significant changes in the spectra of the Ti(2p) orbital.

The interaction of H_2S with titania-supported metals was studied by temperature-programmed desorption method, too. After H_2S adsorption only SO_2 desorption was detected in two stages. H_2S evolution was not observed. Fig. 5 shows the TPD curve for the Pd/TiO_2 sample.

A small peak was detected above 580 K (T_{Max} = 631 K), but at higher temperature, above 920 K, a great amount of SO_2 desorbed (T_{max} = 988 K). When other titania-supported noble metal was used as adsorbant the same TPD spectra were recorded, the main desorption stage was observed above 870 K. For comparison the TPD spectra of adsorbed H_2S on clean TiO_2 is also presented on Fig. 5. In this case, too, only SO_2 was desorbed, but in one stage in the temperature range 630 – 910 K (T_{Max}= 840 K). Beck and White *(10)* and Chen et al. *(11)* found nearly the same picture on clean TiO_2.

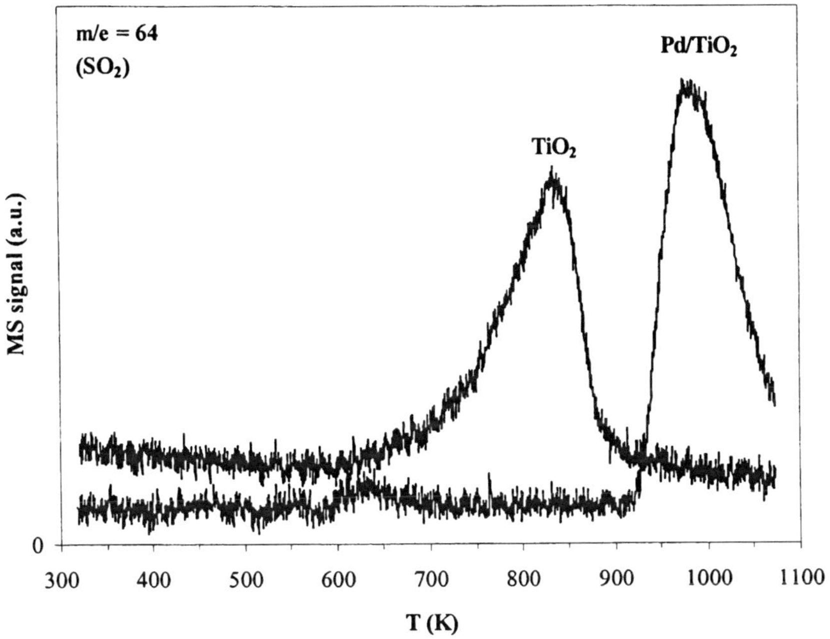

Figure 5. *Temperature-programmed desorption spectra of adsorbed H_2S at 373 K*

These results clearly show that adsorbed H_2S desorbed as SO_2 either from the TiO_2 or from the TiO_2 supported noble metals i.e. TiO_2 can oxidize the surface species formed from H_2S. A possible explanation for the observation that SO_2 evolution was detected on TiO_2 at lower temperature than on supported metal catalysts is that the TiO_2 is reduced in higher extent during the pre-treatment, during the hydrogenation than when metal was supported on its surface. It means that the oxidizing capability of these samples is lower than that of clean TiO_2 support. This resulted that the surface species formed in the H_2S adsorption can be oxidized at higher temperature on titania-supported noble metal catalysts than on TiO_2. The high temperature desorption stage of SO_2 on different titania-supported noble metals hardly depends on the metal. From these results we may conclude that sulphur species are bonded mainly on the support. We cannot exclude that the sulpur is built into the oxygen vacancies of TiO_2 *(11)*.

Recently we have studied the hydrogenation of CO_2 on Rh/TiO_2 previously reduced at different temperatures *(12)*. It was found that the initial rate of methane formation significantly increased with increasing the reduction temperature, but after some seconds it drastically decreased. The promotion effect of the reduction temperature was explained by the formation of oxygen vacancies on the perimeter of the Rh/TiO_2 interface. These sites can be re-oxidized by trace amount CO_2 and H_2O which results in the decrease of the reaction rate. From this observation we may suggest that on titania-supported catalysts the $CO_2 + H_2$ reaction occurred not only on the metal surface, but also on the metal - support interface. The hydrogenation of CO_2 occurred at a higher rate in the presence of H_2S than in the absence of it (Table 2). Based on the TPD measurements we may suppose that sulphur occupied the oxygen vacancies of titania and created new catalytic centers on the metal-support interface. These sites probably have higher catalytic activity in the $H_2 + CO_2$ reaction. The oxygen vacancy can be reoxidized by CO_2 or by water formed in the reaction. When this sites are occupied by sulphur these centers are probably more stable in the presence of CO_2 or water, which results in the permanantly higher activity of the titania supported metals in the CO_2 hydrogenation.

CO_2 hydrogenation is a secondary process of dry reforming. When the rate of this reaction increased, so did the H_2 consumption. It was found that on titania-supported noble metal catalysts in the presence of H_2S the rate of CO_2 + CH_4 reaction decreased but the rate of CO_2 hydrogenation increased. So we can conclude that this is the reason why the CO/H_2 ratio increased in the $CO_2 + CH_4$ reaction in the presence of H_2S.

Acknowledgements

Financial support of this work by OTKA (contract number T 034270) is gratefully acknowledged.

References

1. Bradford, M.C.J., Vannice, M.A., *Catal. Rev.* **1999,** *41*, 1.
2. Solymosi, F., Kutsán, Gy., Erdőhelyi, A., *Catal. Letters* **1991,** *11,* 149.
3. Erdőhelyi, A., Cserényi, J., Solymosi, F., *J. Catal.* **1993,** *141,* 287.
4. Erdőhelyi, A., Cserényi, J., Papp, E., Solymosi, F., *Appl. Catal. A General* **1994,** *108,* 205.
5. Dibbern, H.C., Olesen, P., Rostrup-Nilsen, J.R., Toettrup, P.B., *Hydrocarbon Proc.* **1986,** *65,* 71.
6. Zhang, C.J., Hu, P., Lee, M.-H., *Surface Sci.* **1999,** *432,* 305.
7. Solymosi F., Erdőhelyi, A., *J. Mol. Catal.* **1980,** *8,* 471.
8. Solymosi F., Erdőhelyi, A., Bánsági, T., *J. Catal.* **1981,** *68,* 371.
9. Erdőhelyi, A., Pásztor, M., Solymosi, F., *J Catal.* **1986,** *98,* 166.
10. Beck, D.D., White, J.M., *J. Phys. Chem.* **1986,** *90,* 3123.
11. Chen, Y., Jiang, Y., Li, W., Jin, R., Tang, S., Hu, W., *Catal. Today,* **1999,** *50,* 39.
12. Novák, É., Fodor, K., Szailer, T., Oszkó, A., Erdőhelyi, A., Topics in Catal. **2002,** *20,* 107.

Organic Synthesis with CO_2 as a Reactant

Chapter 8

A Novel Route for Carbon Dioxide Cycloaddition to Propylene Carbonate

W. Wei, T. Wei, and Y. Sun*

State Key Laboratory of Coal Conservation, Institute of Coal Chemistry, Chinese Academy of Sciences, Taiyuan 030001, China

A combination of KI supported catalysts on different supports with a novel continual structural reactor for the production of propylene carbonate from CO_2 cycloaddition was investigated in this paper. With neither solvents nor promoters in this synthesis, both propylene oxide conversion and propylene carbonate selectivity were nearly 100% at moderate conditions over the supported catalysts. In the structural reactor, the lifetime of the catalysts can be highly improved with a much higher yield of propylene carbonate, compared to that obtained from the conventional reactor.

Introduction

The formation of a five-cyclic membered carbonate (such as propylene carbonate or ethylene carbonate) via cycloaddition between CO_2 and epoxides (such as propylene oxide and ethylene oxide) is not only one of the routes for CO_2 chemical fixation (*1*), but also is the first step for the synthesis of dimethyl carbonate (DMC) via transesterification (*2*). Furthermore, propylene carbonate (PC) or ethylene carbonate itself is a nontoxic and versatile intermediate in environmentally benign organic syntheses. In particular, alkylene carbonate is a colorless solvent with a relatively high dielectric constant. It is mainly used as a non-aqueous solvent for electrolytes in high-energy batteries. In addition, it is widely used in polycarbonate syntheses and used as an intermediate in pharmaceutical processes, as an oxyalkylation agent in dyestuff syntheses, and as a solvent in textile production processes. Hence, the green synthesis route of propylene carbonate or ethylene carbonate is expected in this half century.

Nemirovsky first reported the synthesis of alkyl carbonate by phosgenation a century ago (*3*), but the method has gradually been abandoned because of the use toxic and hazardous phosgene. After the 1960s, many patents have been issued for the production of alkyl carbonate (*4-11*). Particularly, much attention has been paid to the cycloaddition between propylene oxide (PO) and CO_2. The main catalysts include organometallic halide-Lewis bases (*1*), organotin halde-tetraalkyphosphonium halides (*1*), alkali metal halide-crown ethers (*12*) and polyethylene glycol-400 (*13*). But the catalytic performance and the severe reaction conditions limit their further utilization. In addition, a variety of solvents or promoters must be used, resulting in high separation costs or low purity of PC. Moreover, the reaction was carried out in an intermission autoclave, leading to a low production capacity. Thus, we have developed a new heterogeneous catalyst with high catalytic activity at moderate conditions (i.e., KI supported catalysts on various supports). In order to improve the lifetime and productivity of the catalysts and avoid the hot spot of the catalysts, a new continual structural reactor was investigated. Combining the new heterogeneous catalysts and the new continual structural reactor, a novel method for cycloaddition between PO and CO_2 as a heterogeneous process was developed in the present work.

Experimental

The purity of CO_2 was higher than 99.5%. Commercially available PO or ethylene oxide (EO) was used without further purification. Catalysts were prepared by impregnating the supports with an aqueous solution of KI, and then dried at 60℃ for 10 h. All supports were commercially obtained.

The catalysts were first evaluated in a conventional intermission autoclave. The cycloaddition between CO_2 and PO is an intense exothermal reaction ($\Delta H \approx 30$kcal/mol), which leads to difficult operations of the heterogeneous catalytic process due to the higher local temperatures and the limitation of the lifetime of catalyst. Thus, a new structural reactor was designed. In this reactor the lifetime of the catalysts was significantly improved.

Results and Discussion

Activity of KI supported catalyst over different supports.

In the heterogeneous process, KI supported on different supports was investigated (see Table 1). First, the reaction conditions (reaction temperature:120 ℃; pressure: 4.0MPa) were moderate, and the pressure was lower than the saturate pressure of carbon dioxide at room temperature. Secondly, both PO conversion and PC selectivity were nearly 100% although neither solvents nor promoters were used. Thus, the product did not need to be separated, which simplified the production of PC. In addition, the yield of PC was much higher than that for the conventional heterogeneous process (< 5g/(g(cat.) • hr)). For activated carbon supported KI catalyst, the yield reached up to 12.0 g/(g(cat.) • hr).

Table Ⅰ. Effect of supports on catalytic performance

Catalysts	*PO Conversion %*	*PC yield %*	*yield kg/(kg(cat.) • hr)*
KI/γ-Al_2O_3	100.0	99.0	7.0
KI/activated carbon	100.0	99.3	12.0
KI/SiO_2	100.0	99.0	7.4

NOTE: T=120℃, P=4.0MPa, CO_2/PO(molar ratio)=3:1.

The structural properties of the catalysts

The BET surface areas of three catalysts are shown in Table 2. Among them, the surface area of the KI/activated carbon was the highest, and that of KI/γ-Al_2O_3 was the lowest. This might be the reason that the supports showed a strong influence. In addition, the acid site of both γ-Al_2O_3 and SiO_2 also might suppress the reaction rate. The detail mechanism is under investigation.

Lifetime of KI/activated carbon

In a 1-L conventional intermission autoclave, the catalytic activity of KI/activated carbon sharply decreased in 100 hours (see Figure 1 and Figure 2). The texture of fresh catalyst and after reaction catalyst greatly changed (see Table 3). Obviously, the fouling of the pores, especially the small pores, was very severe in the reaction in the conventional intermission autoclave, which led to a great decrease in the surface area. In addition, color was observed in the product, indicating that the KI in the catalyst was partly oxidized during reaction. This occurred because the reaction heat could not be removed and thus produced local hot spots.

Table 2. The structural properties of the catalysts

Catalysts	*BET surface* m^2/g	*Average pore diameter* Å	*pore volume.* cm^3/g
KI/γ-Al_2O_3	279.3	43.5	0.32
KI/activated carbon	394.3	22.2	0.22
KI/SiO_2	329.4	79.8	0.64

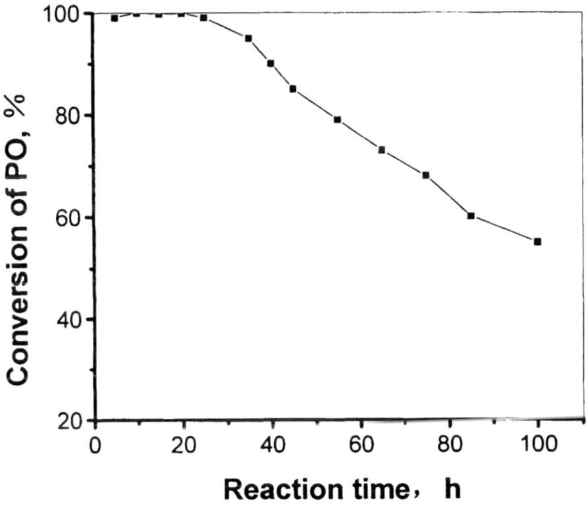

Figure 1. Change of PO conversion with reaction time in 1 L conventional intermission autoclave at 4MPa, 120 ℃ and 3:1 of CO_2 to PO molar ratio.

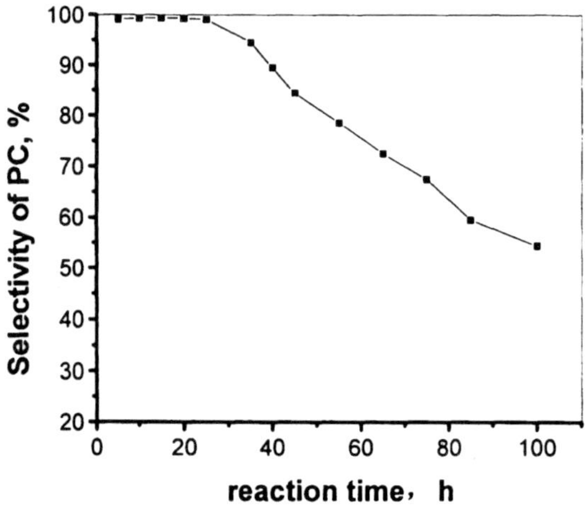

Figure 2. Change of PC selectivity with reaction time in 1 L conventional intermission autoclave at 4MPa, 120 ℃ and 3:1 of CO_2 to PO molar ratio.

Table 3. The structural properties of KI/activated carbon fresh and used catalysts

Catalysts	*BET surface m^2/g*	*Average pore diameter Å*	*Pore volume. cm^3/g*
fresh catalyst	394.3	22.2	0.22
used catalyst	40.7	56.8	0.06

In order to further improve the lifetime and productivity capability of the catalysts and avoid the hot spot in the catalysts, a novel continual structural reactor was designed (*14*). The structural schematic diagram of the reactor is showed in Figure 3. In the reactor the supported catalysts were loaded in the thin baskets that were especially placed. During the reaction, water or nitrogen or air, was introduced into the heat exchanger to remove reaction heat. Furthermore, the reaction system could be operated continually to enhance the productivity. This reactor has the following advantages: (1) PC could be removed from the surface of the catalysts as soon as it was produced; (2) The reaction heat could be efficiently removed. Thus, the lifetime of the catalyst could be improved; (3) the reaction was continual in this reactor, and its productivity was improved. As a result, the activity of catalysts in the new reactor was highly improved. For KI/activated carbon, both PO conversion and PC selectivity hardly decreased after 400 h of continually operation (see Figure 4 and Figure 5). No color in the product was observed. In addition, the structural properties of the catalyst did not change. Thus, in the new reactor the catalyst is stable.

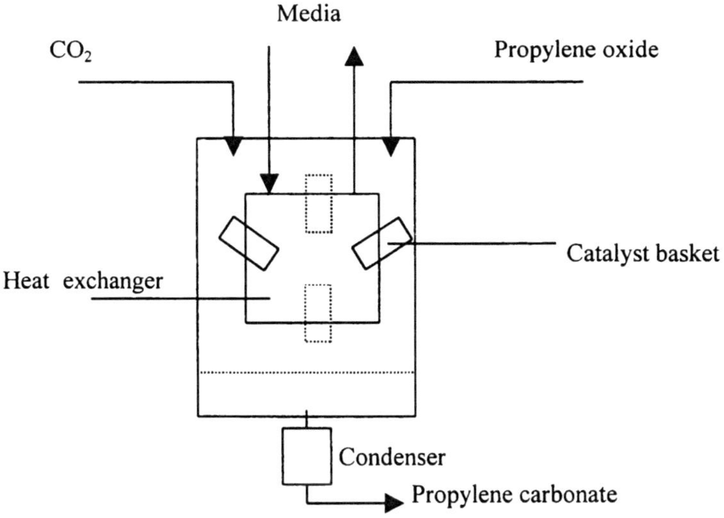

Figure 3. The structural schematic diagram of the new reactor.

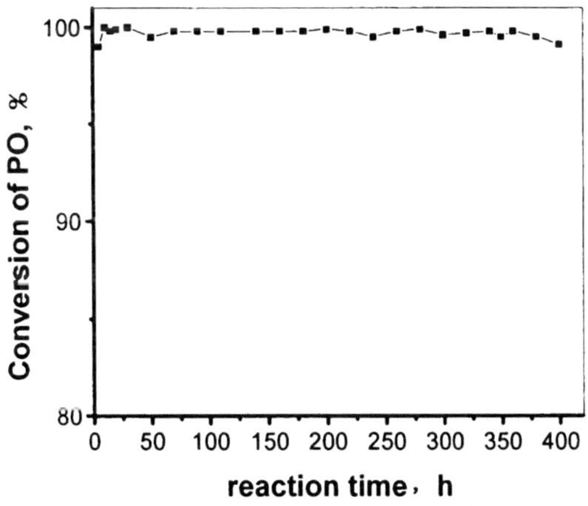

Figure 4. Change of PO conversion with reaction time in 20 L new continual structural reactor at 4MPa, 120 ℃ and 3:1 of CO_2 to PO molar ratio.

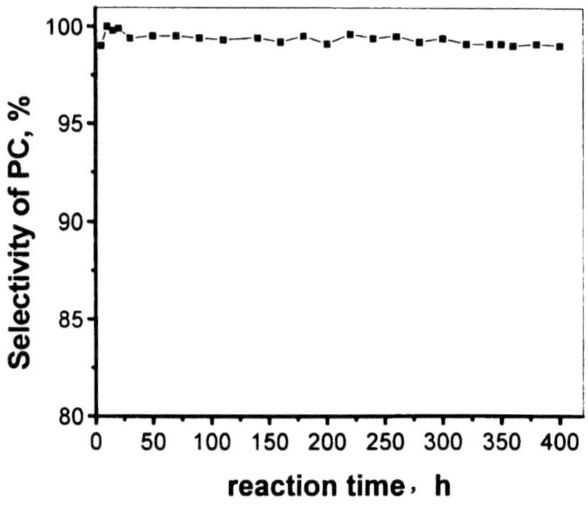

Figure 5. Change of PC selectivity with reaction time in 20 L new continual structural reactor at 4MPa, 120 ℃ and 3:1 of CO_2 to PO molar ratio

Conclusion

A novel structural reactor containing catalyst baskets has been designed to immediately remove the reaction heat. A better performance has been thereby achieved, compared to the conventional reactor. For supported KI catalysts, both PO conversion and PC selectivity approached 100% at moderate conditions, and the space time yield of PC reached up to 12 g/(g(cat.) • hr).

Reference

1. Nomura R., Kimura M., Teshima S , etc., *Bull. Chem. Soc. Jpn.*, 1982,55,3200-3208.
2. Baba A., Nozaki T., Miki T., *Bull. Chem. Soc. Jpn.*, 1987,60,1552-1558.
3. Nemirovsky I., Prakt J. Chem., 1883, 28,439-442.
4. Fumagalli, Carlo; Caprara, Giuseppe; Roffia, Paolo. United States Patent 4,009,183.
5. Gregory A Kao; Jar-lin; Sheng; Ming N. United States Patent 4,224,223.
6. Richard G.; Michaelson; Robert C.; Myers; Richard S. United States Patent 4,824,969.
7. Shikata, Kazuo; Shigemune,Toshiaki. United States Patent 4,217,298.

8. Venturello, Carlo; D'Aloisio, Rino. United States Patent 4,226,778.
9. Robert M. United States Patent 4,931,571.
10. Hoon Sik; Jai Jun; Sang Deuk; Kun You; Hong Gon. United States Patent 6,156,909.
11. Scott J.; Clarence D.; Robert A.; Lorenzo C.; Zhaozhong; Jose G.; Hye Kyung Cho. United States Patent 6,258,962.
12. Rokicji G. and Kuran W., *Bull. Chem. Soc. Jpn.,* 1984,57,1662-1670.
13. Tang Z.Z., Chen Y., Qu Z.J., etc, *J. Peterochem Technol.* (China), 1996, 25, 409-412.
14. Sun, Y.; Chen, X.; Wei, W. China patent: 01108814.1,2001

Chapter 9

Catalytic Behavior of Calcium Oxide for Synthesis of Dimethyl Carbonate from Propylene Carbonate and Methanol Near Room Temperature

Tong Wei, M. Wang, Wei Wei, Yuhan Sun*, and Bin Zhong

State Key Laboratory of Coal Conservation, Institute of Coal Chemistry, Chinese Academy of Sciences, Taiyuan 030001, China

For the first time, DMC was synthesized near room temperature from propylene carbonate (PC) and methanol (MeOH) on CaO with high yields and selectivity. Based on the assumption that hydroxylpropyl methyl carbonate (HPMC) was the intermediate in this reaction, the charge distributions for MeOH, PC and HPMC were calculated and a possible side reaction, the polymerization of PC, was inferred. The effect of temperature, catalyst concentration as well as the ratio of PC to MeOH on DMC yield and DMC selectivity was investigated in detail. The results obtained support the hypothesis of the intermediate and the side reaction. In addition, the effects of the addition method, temperature, catalyst concentration and the ratio of PC to MeOH on reaction rate were also investigated in detail. Based on these results, the possible mechanism of this reaction on CaO is proposed.

Introduction

The development of environmentally benign processes based on the utilization of naturally abundant carbon resources, such as carbon dioxide, has gained considerable attention in recent years (*1, 2*). The synthesis of dimethyl carbonate (DMC) using carbon dioxide is one of the promising reactions from this point of view (*3*). DMC is a unique molecule having versatile chemical reactivity and excellent physical properties (4-*6*). It is an important precursor of polycarbonate resins as well as a useful carbonylation and methylation agent since it can be substituted for toxic phosgene, dimethyl sulfate, or methyl iodide. DMC has about 3 times the oxygen content as methyl t-butyl ether (MTBE) as well as a good blending octane value (R+M/2=105). In addition, it is not only low in toxicity but also quickly biodegrades. Therefore, it is an important alternative to MTBE as a potential gasoline octane enhancer.

Besides being prepared from methanol and hazardous phosgene, there are three main environmentally compatible routes for the production of DMC (*7-9*). It can be prepared with high yields by the oxidative carbonylation of methanol using $CuCl_2$ as a catalyst. The main drawbacks of this method are the deactivation of catalyst at high conversion and the production of corrosive hydrogen chloride. Carbonylation of methyl nitrite over Pd/carbon is also used to produce DMC with high yields. For this method, the expense of the catalyst and the deposition of the unconverted starting material, NO, are the main problems. The other promising route is a transesterification method as shown in Scheme 1, in which a cyclic carbonate is formed from epoxide and carbon dioxide and subsequently trans-esterified to form DMC and alkylene glycol. In the first step, cyclic carbonate can be synthesized with a conversion and a selectivity close to 100% under moderate conditions (*10*). Therefore the key to this route is to improve the conversion of cyclic carbonate and selectivity for DMC in the second step.

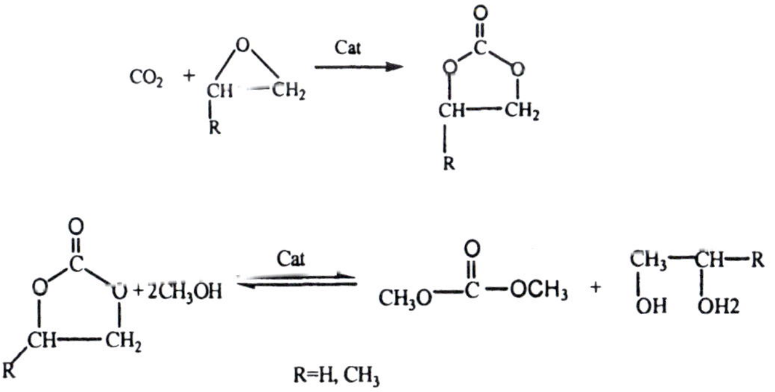

Scheme 1. Synthesis of DMC by transesterification method

There are many reports on the synthesis of DMC from ethylene carbonate and methanol (*11,12*). In these studies, alkali metals, free organic phosphines supported on partially cross-linked polystyrene, and heterogeneous catalysts such as zeolite and hydrotalcite are used for the reaction. To our knowledge, however, there are only a few reports on the transesterification of propylene carbonate and methanol with solid catalysts and most reactions in these reports were carried out above 100℃ (*2*). There has been no paper reporting this reaction being carried out near room temperature with high conversion and selectivity. In the present work, for the first time, we find that CaO calcined from $CaCO_3$ has excellent catalytic activity for the synthesis of DMC from propylene carbonate and methanol near room temperature. The catalytic behavior of CaO, as a solid base catalyst, as well as the reaction mechanism will be discussed in this paper.

Experimental

CaO was prepared from $CaCO_3$ calcined at 1173K for 1hr in a N_2 atmosphere. The BET surface area of CaO was 12.166 m^2/g. The reaction was carried out in a 250mL flask equipped with reflux condenser, water bath and strong magnetic stirring. Except where otherwise noted, the catalyst was added to the flask with vigorous magnetic stirring after the temperature reached the expected one. Reaction products were analyzed by Gas Chromatograph after centrifugal separation of solid catalyst from liquid.

Results and discussion

Possible Reaction in the Process of DMC Synthesis from PC and MeOH

Knifton (*12*) reported that hydroxyethyl methyl carbonate (HEMC), which could be detected only at short reaction time and/or low temperatures, was an intermediate when MeOH reacted with ethylene carbonate, catalyzed by a basic catalyst, to generate DMC and ethylene glycol. Since CaO is a typical solid base, hydroxypropyl methyl carbonate (HPMC) may be the intermediate in the synthesis of DMC from PC and MeOH on CaO. On the basis of this assumption, the charge distributions for MeOH, HPMC and PC were calculated by use of Guassion 98. Table 1 lists the charge of the oxygen and hydrogen in the hydroxyl group and/or the carbon in the carbonyl group of these three molecules.

Table 1. Charge Distribution of MeOH, HPMC and PC

	Charge of atom		
Molecule	O_a	H_b	C_c
MeOH	-0.614	0.393	----
HPMC	-0.620	0.391	0.803
PC	----	----	0.771

NOTE: Atoms in the molecules are illustrated as follow:

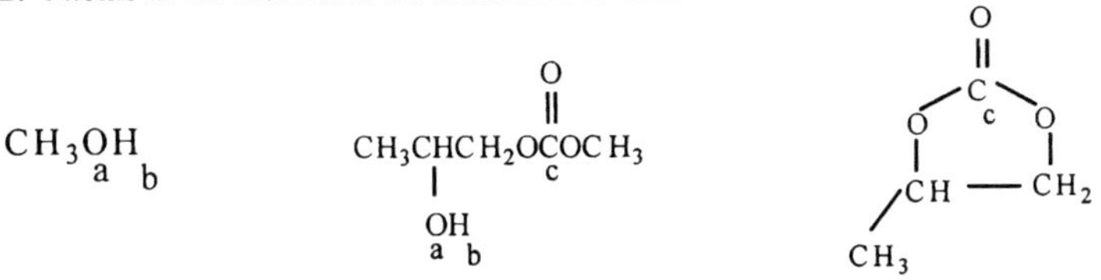

Usually a transesterification reaction catalyzed by a base is taken place when the nucleophilic agent attacks the electrophilic carbonyl carbon of the ester. Thus, when PC transesterifies with MeOH, the oxygen of methanol attacks the carbonyl carbon of PC to initially give HPMC (as shown in Scheme 2). HPMC has both a nucleophlic hydroxyl oxygen and an electrophilic carbonyl carbon. It

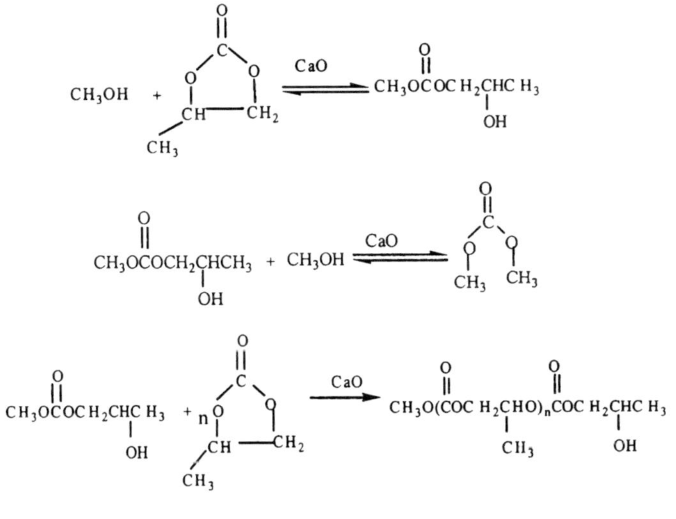

Scheme 2. Possible reactions in the process of DMC synthesis from PC and MeOH on CaO

cannot only react with MeOH to give DMC but also react with PC to give the poly-propylene carbonate (PPC). The two reactions are competitive. From results in Table 1, it can be seen that the charge of the oxygen in MeOH and HPMC is –0.614 and –0.620, respectively, while the carbonyl carbon in HPMC and PC is 0.803 and 0.771, respectively. It is widely accepted that for nucleophilic substitution reactions, a high electron density for the nucleophilic agent and a high positive charge for the attacked atom are beneficial to decrease the energy of the transition state and subsequently increase the reaction rate (*13*). As a result, the free energy for the reaction of HPMC with MeOH is lower than that for MeOH with PC as well as that for HPMC with PC. The reaction rate of HPMC with MeOH is faster than that of MeOH with PC as well as that of HPMC with PC. In summary, for the synthesis of DMC from PC and MeOH, the intermediate is HPMC, which may be detected only at low temperatures and/or short reaction times. This is because the reaction rate of HPMC with MeOH is much faster than that of PC with MeOH. On the other hand, HPMC can continually react with PC to give the side product, PPC, with a rate much slower than that of the transesterification reaction.

Effect of Temperature on DMC Yields and DMC Selectivity

Figures 1 and 2 depict the effect of temperature on the DMC yields and DMC selectivity (reaction condition: MeOH/PC=4/1; catalyst (wt)% =0.90%). It can be seen that CaO shows excellent catalytic ability for the synthesis of DMC from PC and MeOH since even at 10 ℃ the reaction reaches equilibrium in 60 minutes. With the increase of temperature, the reaction rate increases but the equilibrium DMC yields decrease gradually. This is reasonable if the exothermic property of this reaction is considered (ΔH=-7.092 kJ/mol calculated from enthalpies of reactants and products (*14-17*)). The increase of temperature would result in the decrease of the equilibrium constant and consequently decrease the equilibrium DMC yield. As for DMC selectivity, it increases at the beginning of the reaction and then decreases gradually at 10 ℃ and 20 ℃, but at 30 ℃, 40℃ and 50 ℃ the selectivity decreases monotonically. In addition, after reaching equilibrium, DMC selectivity decreases with the rise of temperature. As mentioned above, HPMC is an active intermediate in this reaction and the polymerization of PC is the main side reaction. Therefore, the by-product is mainly the intermediate, HPMC, before the reaction reaches equilibrium, and is mainly PPC after the reaction reaches equilibrium. This also explains the white colloidal solid produced from MeOH and PC after 2-hr reaction at 120 ℃. The increase of temperature not only promotes the transesterification of PC and MeOH but also the polymerization of PC, and consequently, it results in the decrease of DMC selectivity after the equilibrium is reached.

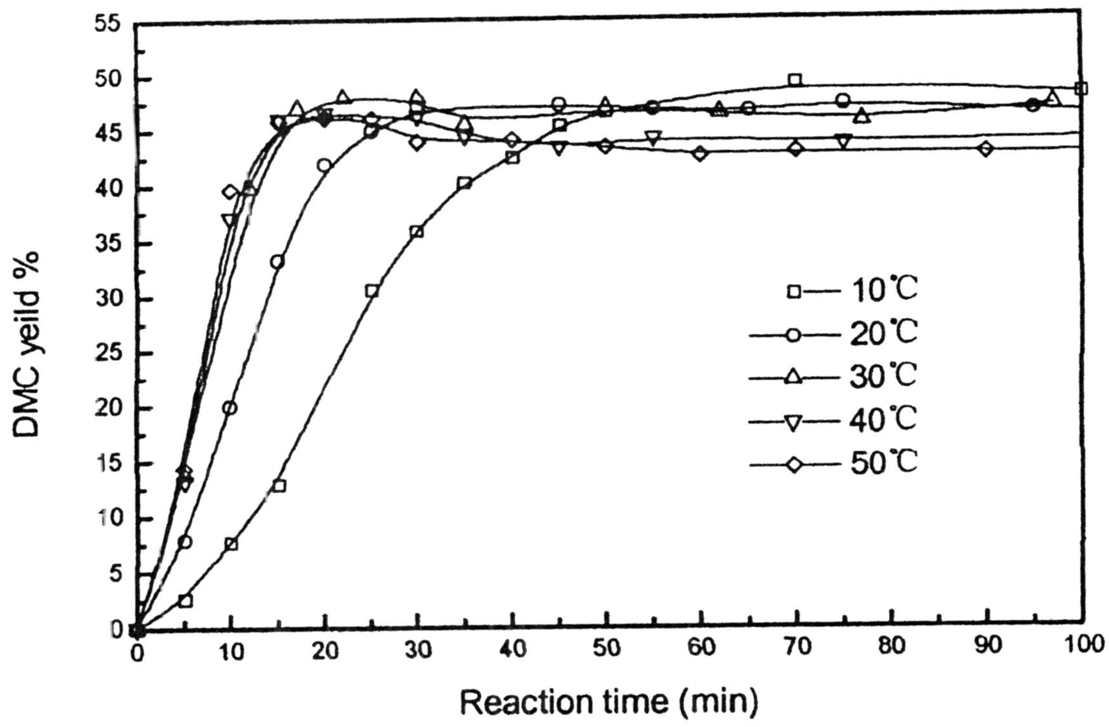

Figure1. The effect of temperature on DMC yield.

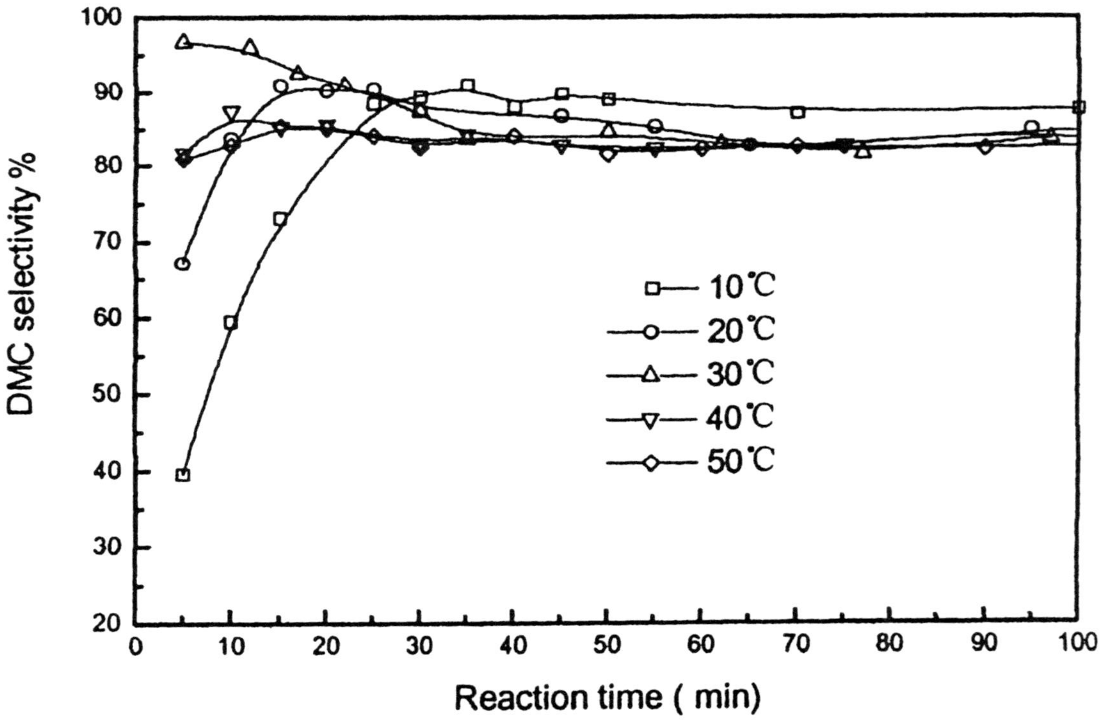

Figure.2 The effect of temperature on DMC selectivity.

Effect of Catalyst Concentration on DMC Yield and DMC Selectivity

The effect of the catalyst concentration on the DMC yield and DMC selectivity is illustrated in Figure 3 and Figure 4 (reaction condition: MeOH/PC=4/1; T=20 ℃). It can be seen that the reaction rate increases initially and then remains constant, but the DMC equilibrium yield hardly changes with catalyst concentration. This means that the reaction has reached thermal equilibrium. On the other hand, when the catalyst concentration is 0.61% and 0.91%, DMC selectivity increases initially, and then decreases as the reaction proceeds with a tendency similar to that seen at 10℃ and 20℃. This may be due to the fact that a decrease in catalyst concentration leads to a lower activated methanol concentration and consequently decreases the reaction rate of HMPC with MeOH. Therefore the concentration of unreacted HPMC possibly increases with a decrease in catalyst concentration at short times. Furthermore, after the reaction reaches equilibrium, DMC selectivity decreases with an increase in catalyst concentration. This is probably due to the promotion of both the transesterification reaction and the PC polymerization by the increase in catalyst concentration. Moreover, the transesterification reaction is reversible but PC polymerization is irreversible.

Effect of the Ratio of MeOH to PC on DMC Yield and DMC Selectivity

The effect of the ratio of MeOH to PC on DMC yield and DMC selectivity is shown in Figure 5 and Figure 6 (reaction condition: T= 20℃; catal. (wt)% (based on PC): 2.76%). It can be seen that the equilibrium DMC yield increases with the increase of the ratio of MeOH to PC. This reaction is reversible. The increase in MeOH concentration definitely promotes PC conversion and improves the equilibrium DMC yield. Figure 6 indicates that the intermediate, HPMC, exists at the beginning when the MeOH to PC ratio is 2:1 and 4:1, since the DMC selectivity initially firstly and then decreases as the reaction proceeds. On the other hand, DMC selectivity increases monotonically with the increase of the ratio of MeOH to PC after the equilibrium is reached. The polymerization of PC competes with the transesterification reaction. The increase of the ratio of MeOH to PC means a decrease of the PC concentration, and thus the PC polymerization is inhibited and DMC selectivity increases after the equilibrium is reached.

Effect of Addition Method, Temperature, Catalyst Concentration and the Ratio of MeOH to PC on Reaction Rate

In order to further understand the mechanism of this reaction on CaO, the effect of addition method, temperature, catalyst concentration and the ratio of

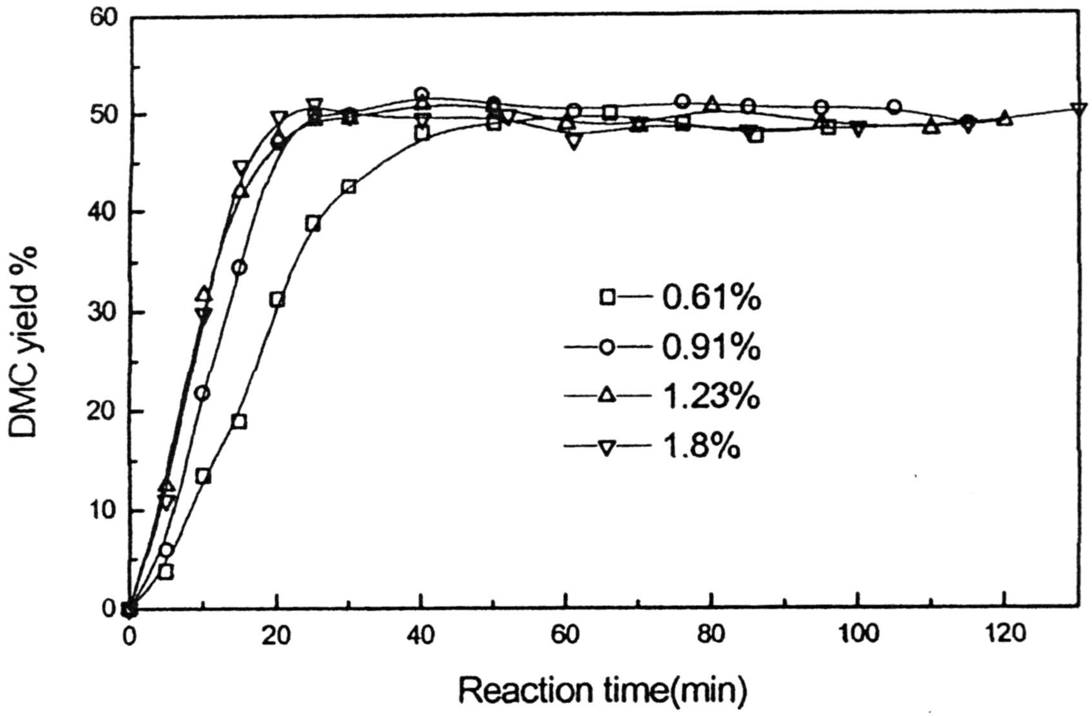

Figure 3. The effect of catalyst content on DMC yield.

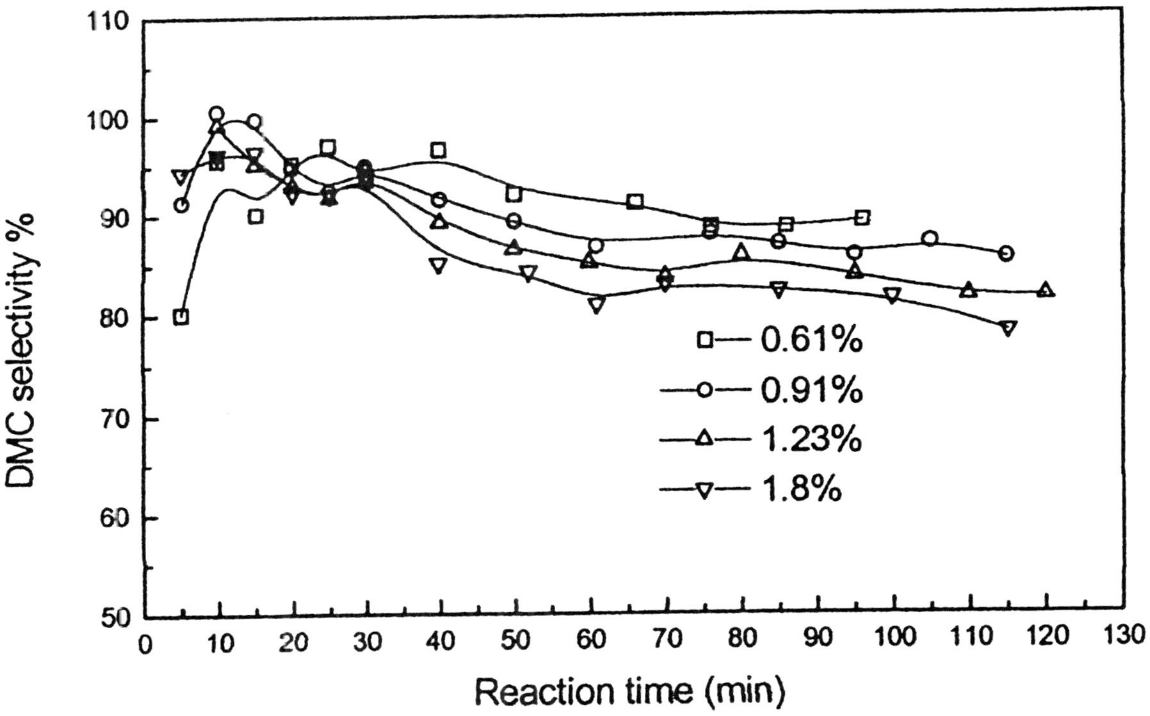

Figure 4. The effect of catalyst concentration on DMC selectivity.

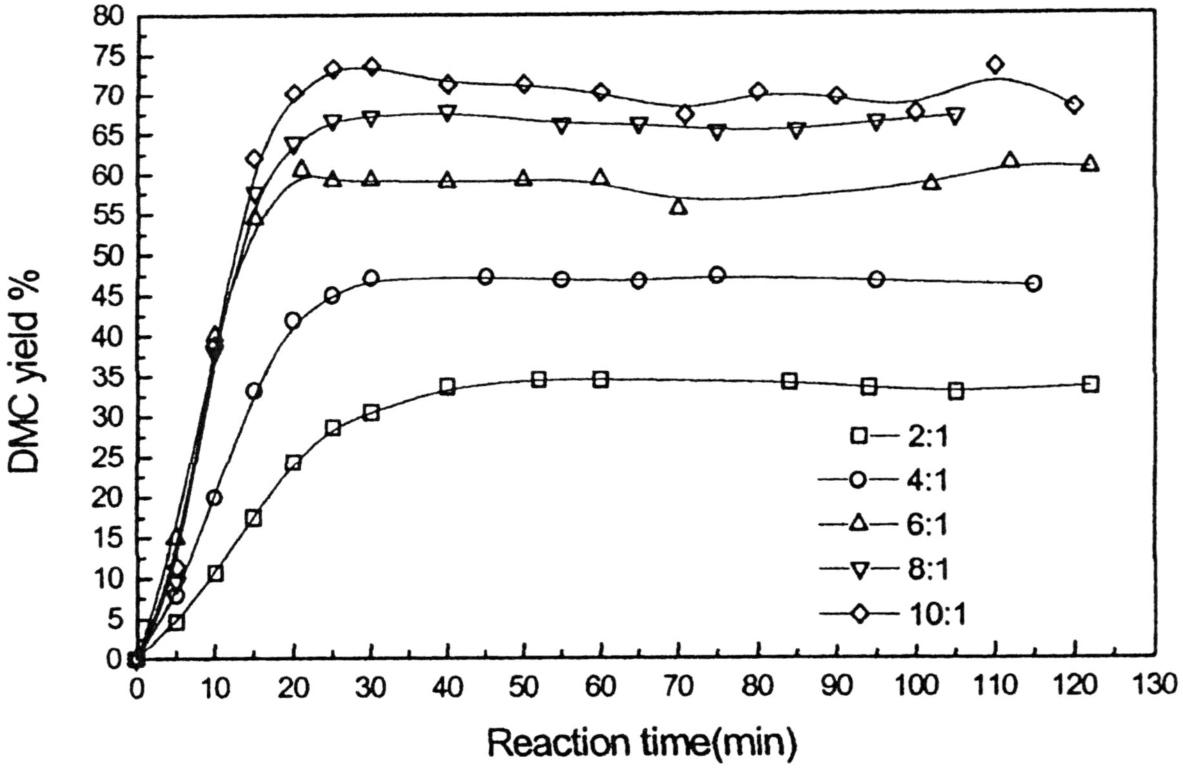

Figure 5. The effect of the ratio of MeOH and PC ratio on DMC yield.

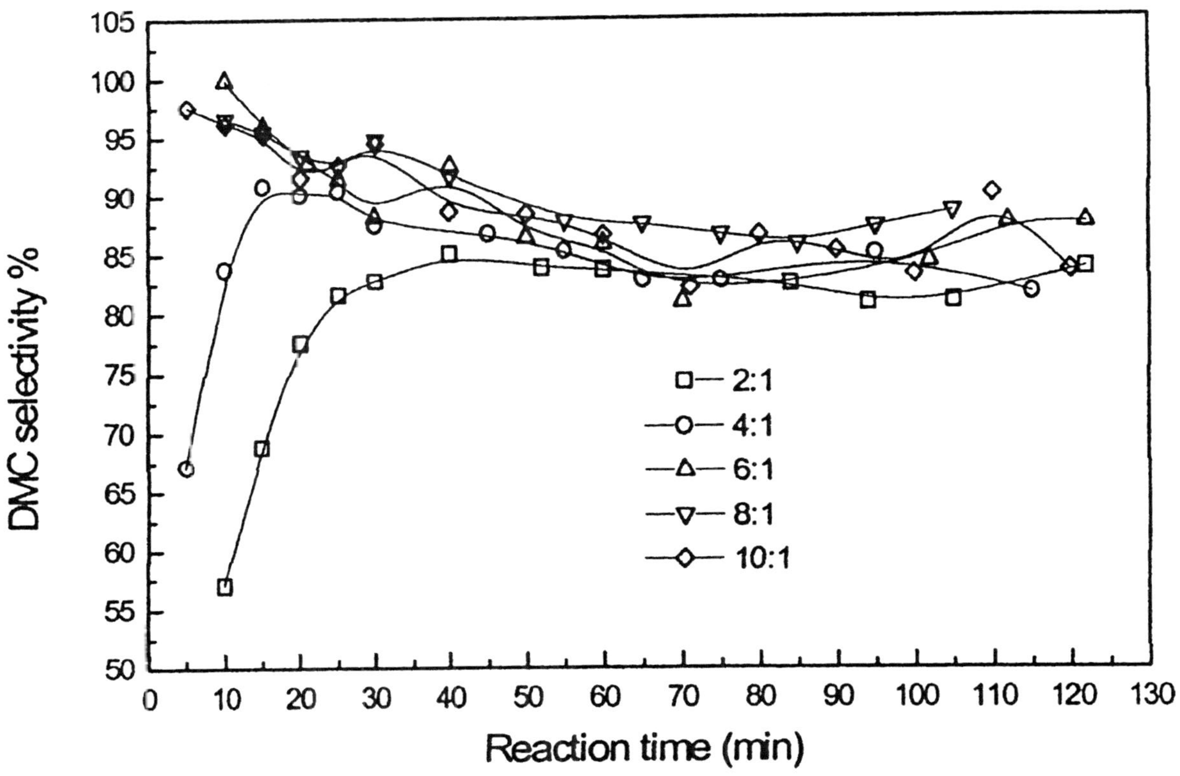

Figure 6. The effect of the ratio of MeOH and PC on DMC selectiviy.

MeOH to PC on the reaction rate were investigated in detail. Figure 7 shows that there is an induction period in the rate-time curve when MeOH and PC are added first as well as when PC and the catalyst are added first. This implies that the induction comes from the activation of MeOH on the solid base catalyst and that the main function of the catalyst is to activate methanol and not PC. On the other hand, the reaction rate is faster when MeOH and PC are added first than that when PC and the catalyst are added first. This may be caused by the activation of the carbonyl carbon in PC through H-bonding between PC and MeOH, since Comelli and Park et al. (18, *19*) reported that when PC and MeOH were mixed, the reaction showed endothermic behavior because of the H-bonding between PC and MeOH.

Figure 8 depicts the effect of temperature on the reaction rate. It can be seen that the induction period is shortened and the maximum rate increases with the increase of temperature. At the beginning, the concentration of activated MeOH is low, so the activation of MeOH has an important effect on the DMC production rate. After the activation step reaches equilibrium, the surface reaction of activated MeOH with PC becomes the important factor that influences the rate of DMC production. The rise of temperature promotes the rate of both steps and consequently increases the maximum rate and shortens the induction time.

As for the effect of catalyst concentration, shown in Figure 9, the maximum rate increases and the induction period is shortened when the catalyst concentration increases from 0.61% to 1.23%, but the rate-time curve is almost unchanged when the catalyst concentration increase from 1.23% to 1.84%. Obviously, this is due to the fact that the increase of catalyst concentration increases the rate of both MeOH activation and the reaction of activated MeOH with PC caused by the increase of the equilibrium activated MeOH concentration. However, the catalyst concentration is not an important factor that affects either the MeOH activation rate or the equilibrium activated MeOH concentration after it reaches a certain amount. This further demonstrates that the main function of the catalyst is to activate MeOH.

Figure 10 shows the effect of the ratio of MeOH to PC on the reaction rate. When the ratio increases from 2:1 to 6:1, the induction period is shortened and the maximum rate increases. But when the ratio increases further from 6:1 to 10:1, the induction period is prolonged a little and the maximum rate decreases. While the shortening of induction time indicates an increase of the MeOH activation rate, the increase of maximum rate indicates an increase of the reaction rate of activated MeOH and PC. With the increase of the ratio of MeOH to PC, the MeOH concentration increases, but both the catalyst concentration and the PC concentration decrease. Therefore, there will be an extreme in the rate of both methanol activation and the reaction of activated MeOH and PC with the increase of the ratio of MeOH to PC. As a result, the induction period is shortened initially and then prolonged a little, while the maximum rate increases first and then decreases with the increase of the ratio.

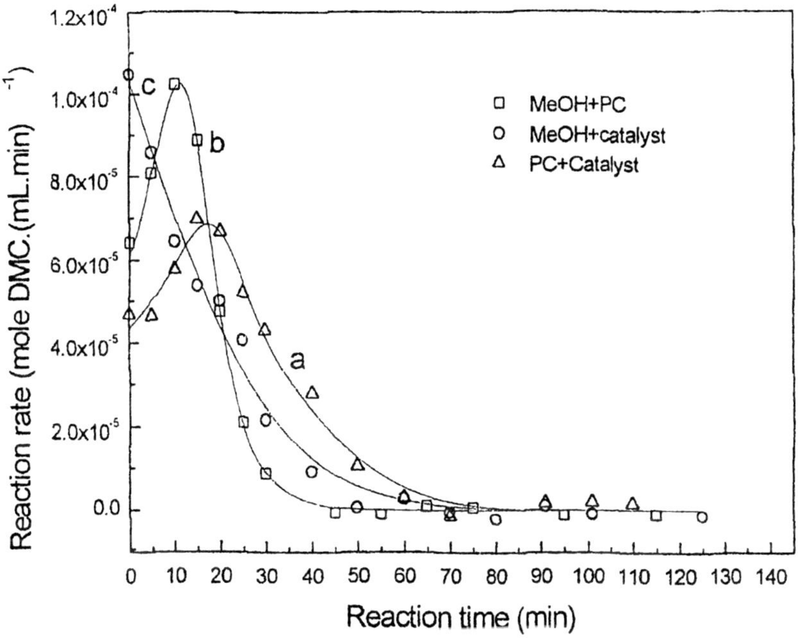

Figure 7. The effect of addition method on reaction rate.

Note: Reaction condition: PC/MeOH=1/4; T=20℃; catal. (wt)%= 0.90%. a: MeOH was added 30 min later after the addition of PC and catalyst; b: catalyst was added 30 min later after the addition of MeOH and PC; c: PC was added 30 min later after the addition of MeOH and catalyst

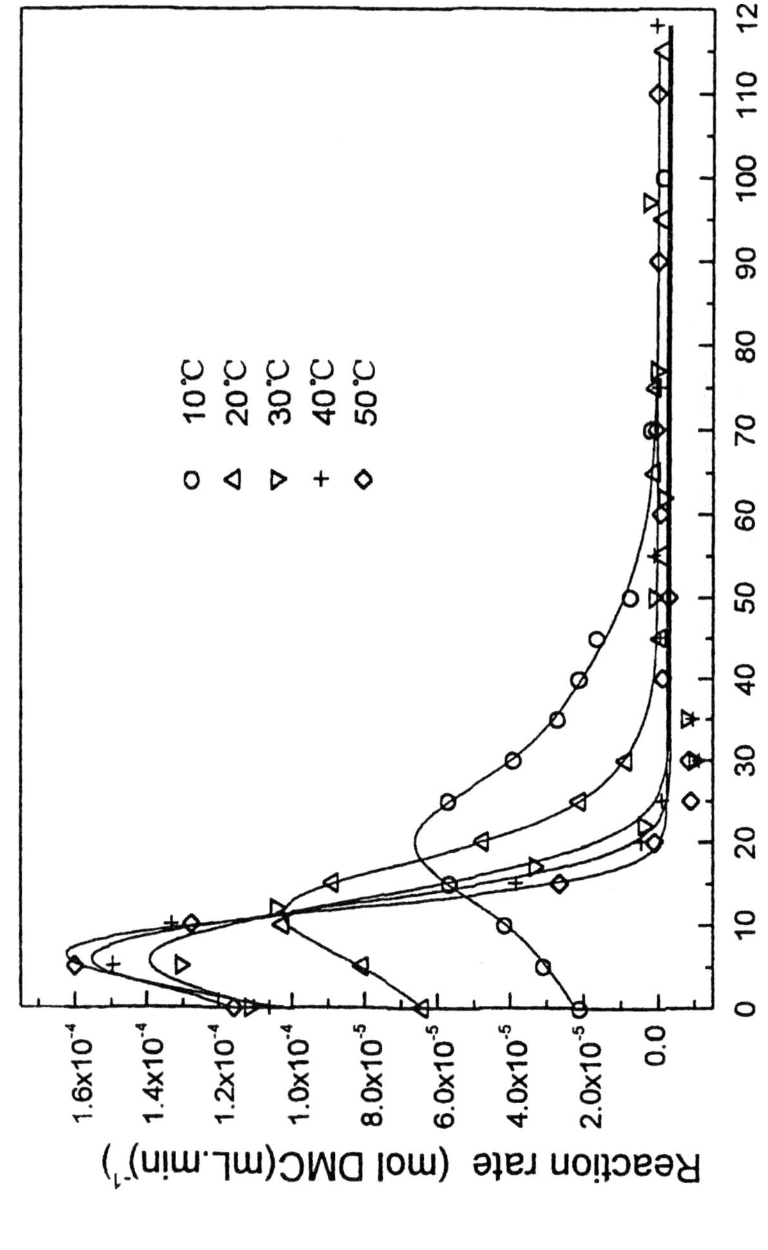

Figure 8. The effect of temperature on reaction rate

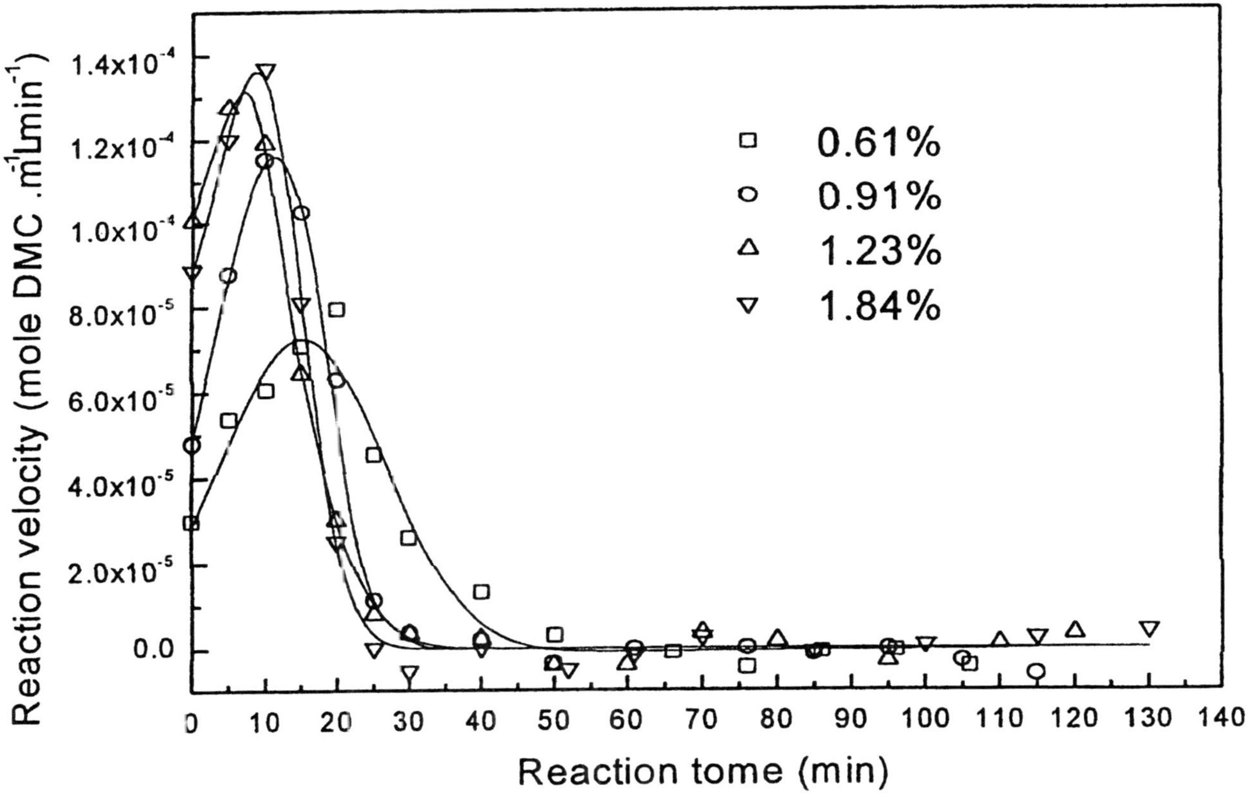

Figure 9. The effect of catalyst concentration on reaction rate

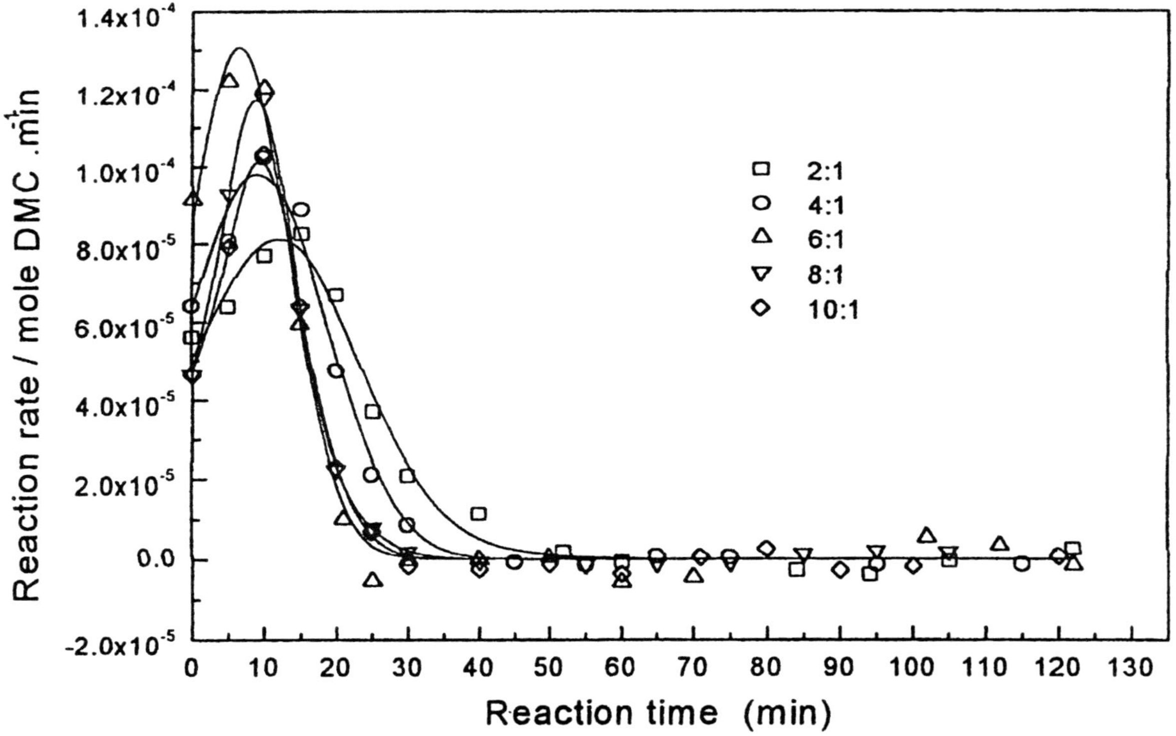

Figure10. The effect of the ratio of MeOH and PC on reaction rate.

Possible mechanism

As mentioned above, the transesterification reaction catalyzed by base is a nucleophilic substitution reaction. For the present reaction, the nucleophilic agent is MeOH and the main function of CaO is to activate MeOH. Iizuka (20) reported that alcohol could be adsorbed on the surface of CaO with $H^{\delta+}$ being dissociated by surface O^{2-} and the anion being stabilized by surface Ca^{2+}. Therefore, based on the results illustrated above, the mechanism of synthesis for DMC from PC and MeOH on CaO is proposed as shown in Scheme 3. First, MeOH was activated due to the abstraction of $H^{\delta+}$ by O^{2-}. Then, $CH_3O^{\delta-}$, which is stabilized by Ca^{2+}, reacts with PC to generate $CH_3OC(O)OCH(CH_3)O^{\delta-}(B^{\delta-})$ stabilized by Ca^{2+}. Afterwards, $CH_3O^{\delta-}$ stabilized in the neighborhood of $B^{\delta-}$ would react with $B^{\delta-}$ to give both DMC and $O^{\delta-}CH_2CH(CH_3)O^{\delta-}$ ($C^{2\delta-}$) stabilized by two consecutive sites. Finally $C^{2\delta-}$abstracts two neighbor $H^{\delta+}$ from O^{2+} to form propylene glycol and desorbs from surface of CaO.

Conclusion

CaO calcined from $CaCO_3$ shows excellent catalytic activity for synthesis of DMC from MeOH and PC. The reaction even reaches equilibrium in 60 minutes at 10℃. The by-product is mainly the intermediate--HPMC before the reaction reaches equilibrium, and is mainly PPC after the equilibrium is reached. The main function of CaO is the activation of MeOH and the subsequent promotion of the reaction of MeOH and PC. The H-bound between MeOH and PC can activate PC and increase the reaction rate. The reaction rate is affected by both the activation of MeOH on CaO and the reaction of activated MeOH and PC. Relatively low temperatures and high ratios of MeOH and PC are preferable for high DMC yield and selectivity. An increase in catalyst concentration has no effect on DMC yield but decreases DMC selectivity.

References

1. Ikedal, Y.; Sakaihoril, T.; Tomishige, K. et al. *Catal. Lett* **2000**, *66*, 59-62.
2. Bhannage, B.; Fujita, S.; Ikushima, Y. et al. *Appl. Catal. A* **2001**, *219*, 259-266.
3. Tundo, P. *Pure Appl. Chem.* **2001**, *73*, 1117-1124.
4. Ono, Y.; Fu, Z. *J. Catal.* **1994**, *145*, 166-170.
5. Kim, W. B.; Lee J. S. *Catal. Lett.* **1999**, *59*, 83-88.
6. Pacheco, M. A.; Marshal C. L. *Energy Fuels* **1991**, *11*, 2-29.
7. Shaikh, A.G. *Chem. Rev.* **1996**, *96*, 951-976.
8. Ono, Y. *Appl. Catal. A: General* **1997**, *155*, 133-166.
9. Sato, Y.; Yamamoto, T.; Souma, Y. *Catal. Lett.* **2000**, *65*, 123-126.

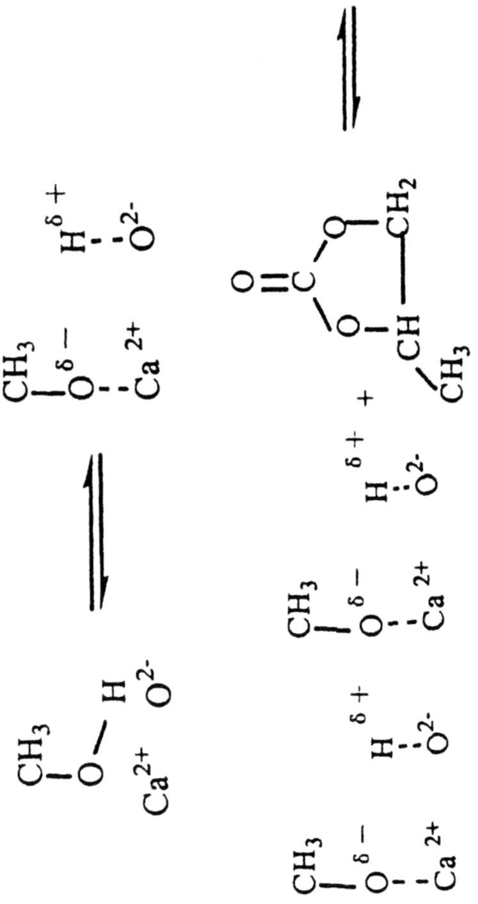

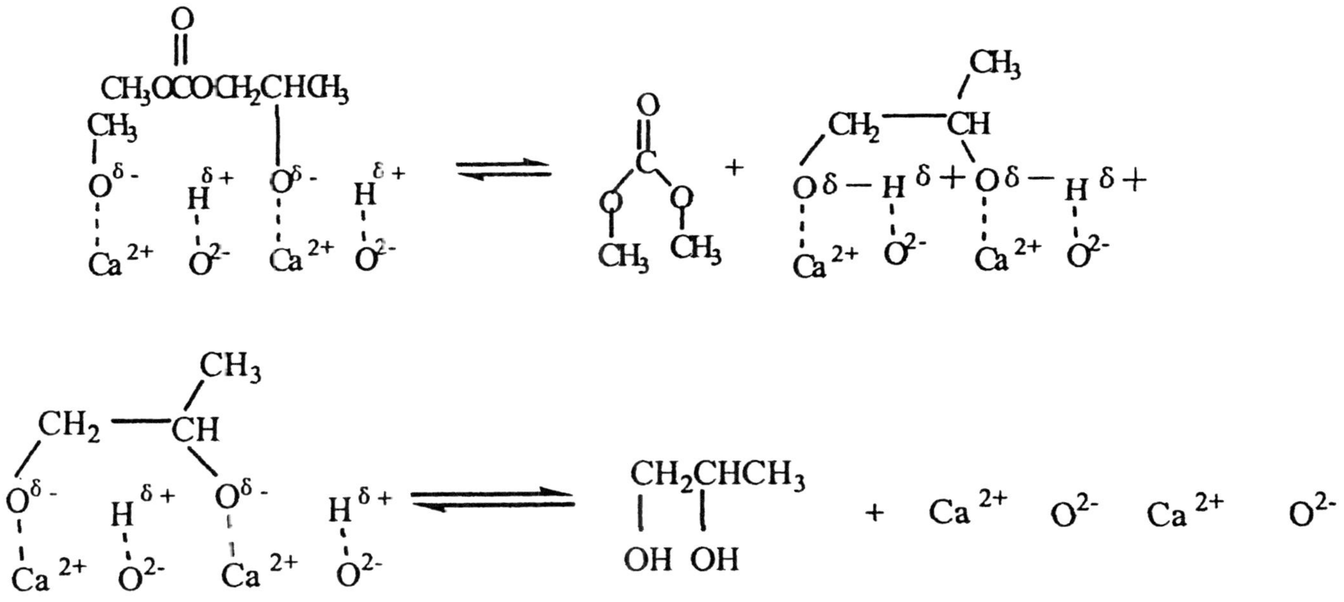

Scheme 3. Mechanism of DMC synthesis from PC and MeOH on CaO

10. Sun, Y.; Chen, X.; Wei, W. China patent: 01108814.1, 2001
11. Watanabe, Y.; Tatsumi, T. *Microporous and Mesoporous Materials* **1998**, *22*, 399-407.
12. Knifton, J. F.; Duranleau, R.G. *J. Mol. Catal.* **1991**, *67*, 389-399.
13. Tang, P. *Fine Organic Synthesis Chemistry and Technology*, 1st ed.; Tianjin University Press, Tianjin, 1993, pp. 22.
14. Yin,Y.; Xi, Z.; Li, D. *Physical Chemistry*, 3rd ed.; Higher Education Press, Beijing, 1994, pp. 569.
15. Dong, Z. *Hangzhou Chemical Engineering* **1979**, 39-96.
16. Reid R.C.; Prausnitz J.M.; Poling B.E. *The Properties of Gases and Liquids*, 4th ed.; McGraw-Hill Company, New York, 1987, pp. 683.
17. Steele, V.; Chirico, R. P.; Knipmeyer, S. E. et al. *J. Chem. Eng. Data.* **1997**, *42*, 1031-1052.
18. Francesconi, R.; Castellari, C. *J. Chem. Eng. Data.* **1999**, *44*, 144 –147.
19. Park, S. B.; Kim, J. S.; Lee, H. *J. Thermodynamics* **1999**, *31*, 1265-1271.
20. Iizuka, T.; Hattori, H.; Ono, Y. et al. *J. Catal.* **1971**, *22*, 130-139

Chapter 10

Catalytic Esterification of Carbon Dioxide and Methanol for the Preparation of Dimethyl Carbonate

Fa-hai Cao, Ding-ye Fang, Dian-hua Liu, and Wei-yong Ying

Department of Chemical Engineering, East China University of Science and Technology, Shanghai 200237, China

Dimethyl carbonate (DMC) is an important methylating agent and potential additive for clean fuel production. Several processes have been developed for the manufacture of DMC. The direct synthesis of DMC from CO_2 and methanol is especially attractive and important. In this work, the continuous synthesis of DMC from methanol and CO_2 in the region near the critical point of CO_2, with methyl iodide and potassium carbonate as the promoters, was investigated. The reactions were performed in a stainless steel autoclave with an inner volume of 500 mL equipped with a magnetic stirrer and an electric heater. The effects of the reaction temperature and pressure were determined first. It was shown that the yield of DMC increases with the increase of the reaction pressure. When the pressure approaches the critical pressure of CO_2, the optimal yield of DMC is obtained. Then the reaction characteristic was discussed from the unique characteristics of supercritical CO_2. Finally, a new process for the production of DMC has been proposed, for which water is used as an extractant for the separation of DMC from the reaction mixture.

Introduction

Carbon dioxide is commonly regarded as a major greenhouse gas. It is mainly produced during the combustion of fossil fuels (coal, oil and natural gas). On the other hand, carbon dioxide can be converted into many useful organic compounds. Therefore, the utilization and recycling of carbon dioxide can be beneficial to the environment. We proposed a process in which dimethyl carbonate (DMC) is produced by the catalytic esterification of carbon dioxide and methanol according to the reaction shown below:

$$2CH_3OH + CO_2 \rightarrow CH_3OC(O)OCH_3 + H_2O$$

DMC has attracted much attention as a non-toxic substitute for dimethyl sulfate and phosgene, which are toxic and corrosive methylating or carbonylating agents. In addition, DMC is considered to be an option for meeting the oxygenate specification for fuels.

The conventional synthesis of DMC is via the reaction of methanol and phosgene. Owing to the high toxicity of the raw materials and severe corrosity, this method has been abandoned gradually. The other two widely used methods for synthesis of DMC are an ester exchange process (*1-2*) and the oxidative carbonylation of methanol (*3-5*). Recently, a more challenging method (*6-9*) is the direct synthesis from carbon dioxide and methanol. Although metallic magnesium powder, Sn(IV) and Tl(IV) alkoxides have been used as the catalysts, unfortunately, the yield of DMC was low even in the presence of chemical dehydrates mainly due to thermodynamic limits.

On the other hand, regions near the critical point are considered to be very important because in these regions the supercritical characteristics have the greatest effect on the reactions. The supercritical conditions will also play a crucial role in the activation and conversion of carbon dioxide. In this work, we studied the process for the continuous synthesis of DMC near the critical regions of carbon dioxide. Carbon dioxide performed not only as a medium of the supercritical fluid but also as one of the reactants.

Experimental

Chemicals. Methanol (99.5% purity), dimethyl carbonate (99.8% purity) and CH_3I and K_2CO_3 (99.8% purity) were all commercially obtained. Compressed carbon dioxide (>99.9% purity) was used.

Apparatus and Analysis. The experimental configuration is shown in Figure 1. The reactor (4) was a stainless steel autoclave (FYX-5A, made in

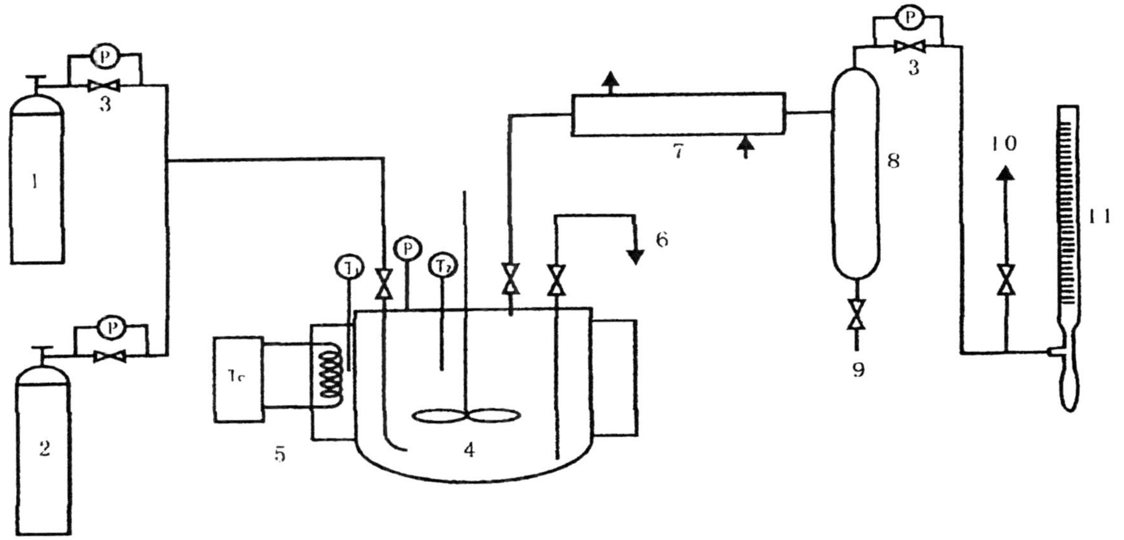

Figure 1. The schematic diagram of the experimental apparatus

1 — Cylinder of feed gas; 2—Cylinder of N_2; 3—Pressure reducing valve; 4—Autoclave; 5—Heater; 6—Tapped hole; 7—Condenser; 8—Gas-liquid separator; 9—Gas chromatograph; 10-Vent; 11—Bulb meter

Dalian, China) with an inner volume of 500 mL, a magnetic stirrer and an electric heater (5). First given amounts of CH_3I and K_2CO_3 were mixed with methanol (150 mL) in the autoclave, and then the autoclave was sealed. The autoclave was then flushed with nitrogen and then purged with CO_2+N_2 until the desired pressure was reached at room temperature. The autoclave was then heated and stirred constantly at the desired temperature and pressure for a given period of time. The products, which were collected to a proper volume from the tapped hole (6), were cooled to ambient temperature. The samples so obtained were analyzed by gas chromatograph (GC-900B, made in Shanghai, China) using a thermal conductivity detector with a TDX-02 column (80-100 mesh). After condensation (7), gas-liquid separation (8) and depressurization, the products were continuously monitored on-line by GC.

Results and Discussions

In all experiments, the molar ratio of feed gases was CO_2:N_2=3:5. The rotation speed of the magnetic stirrer was controlled at 250 rpm. The preliminary tests showed that the optimal dosages of CH_3I and K_2CO_3 were 10.0 mL and 8.0 g, respectively. If one of these two compounds was absent, the other compound had no activity for the catalytic reaction.

x_D is defined as the mole fraction of DMC in the liquid-phase product.

Effects of Reaction Pressure and Temperature

Figures 2 and 3 show the effects of the reaction pressure and the temperature on x_D, respectively. Figure 2 shows that x_D increases initially with the increase of the reaction pressure. When the reaction pressure approaches about 7.3 MPa, x_D reaches the maximum. As pressure is further increased, x_D decreases. The change tendency of Figure 3 is similar to that of Figure 2, with the peak value of x_D obtained at 80-100℃. Obviously, near the critical regions of carbon dioxide, carbon dioxide can be converted effectively into DMC.

Supercritical Phenomena and Reaction Pathway (*10-12*)

There is a conjugated double bond in a molecule of carbon dioxide. Near the critical regions of carbon dioxide, the reaction rate varies with the change of

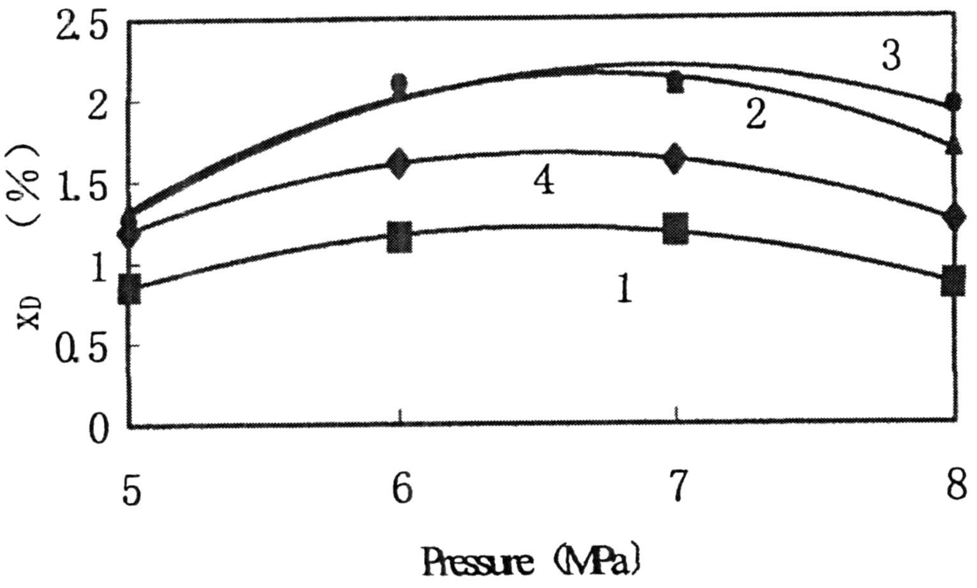

Figure 2. Effects of the reaction pressure on x_D
1-50.0℃; 2-75.0℃; 3-90.0℃; 4-115.0℃

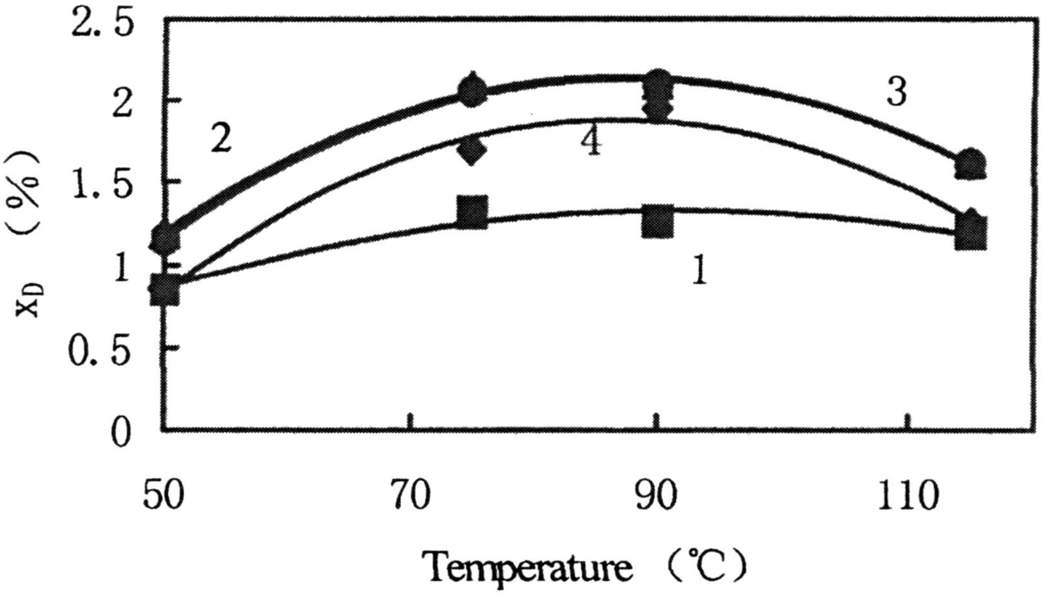

Figure 3. Effects of the reaction temperature on x_D
1-5.0MPa; 2-6.0MPa; 3-7.0MPa; 4-8.0MPa

the electric polarization of the conjugated system of the molecule, so that carbon dioxide can take part in an addition reaction. Furthermore, carbon dioxide has a Lewis acid carbon atom and two weak Lewis base oxygen atoms. It is therefore not surprising that it can undergo various reactions, such as initially binding "abnormally" to form the intermediate. The presence of the basic material, K_2CO_3, is useful for the activation of methanol to form CH_3O^-. The nucleophilic attack of CH_3O^- upon CO_2 leads to the formation of DMC. Methyl iodide is only involved in the catalytic cycle. Based on above analysis, near the critical point of carbon dioxide, x_D reaches the maximum value. The possible reaction pathway is described as follows:

$$CH_3OH \xrightarrow{K_2CO_3} CH_3O^- + H^+$$

$$CH_3O^- \xrightarrow{CO_2} [CH_3OC(O)O]^-$$

$$[CH_3OC(O)O]^- \xrightarrow{CH_3I} CH_3OC(O)OCH_3 + HI$$

$$HI \xrightarrow{CH_3OH} CH_3I$$

A New Tentative Process for the Commercial Production of DMC

In the effluents, in addition to the products, DMC and water, there are large amounts of the feed gases and small amounts of by-products like dimethyl ether. In order to obtain a pure product of DMC, a two-step method can be utilized to separate dimethyl carbonate from the reaction products. The first step is pre-distillation. It uses the azeotropic feature of methanol with dimethyl carbonate. The azeotropic mixture of CH_3OH and DMC can be separated from the by-products and the feed gases in a packed column (2). The second step is a refining step, for which an extractive distillation was adopted. Based on the features that DMC is insolvable in water, and that water does not react with DMC during distillation. We therefore selected water as the extraction reagent to extract and separate DMC from the azeotrope.

Figure 4 shows the schematic diagram of our tentative process for the commercial production of DMC. After pre-distillation, the gasified azeotrope is added into the extractive distillation column (3) and water is sprayed into the top of the column. After cooling (4), the distillate, which contains mostly dimethyl carbonate, flows into the separator (5) and is separated into two parts: one is the organic phase that contained 97% DMC, and the other is the aqueous phase that contained 87% water, 1.2% methanol and 11.8% DMC. The effluent, which contained 95% water, 5% methanol and less than 0.05% DMC, is discharged from the bottom of the extractive distillation column (3) and added into the fractionating tower (6). The aqueous phase is pumped back to the extractive distillation column (3). After fractionating (6), water was returned to the top of the extractive distillation column (3).

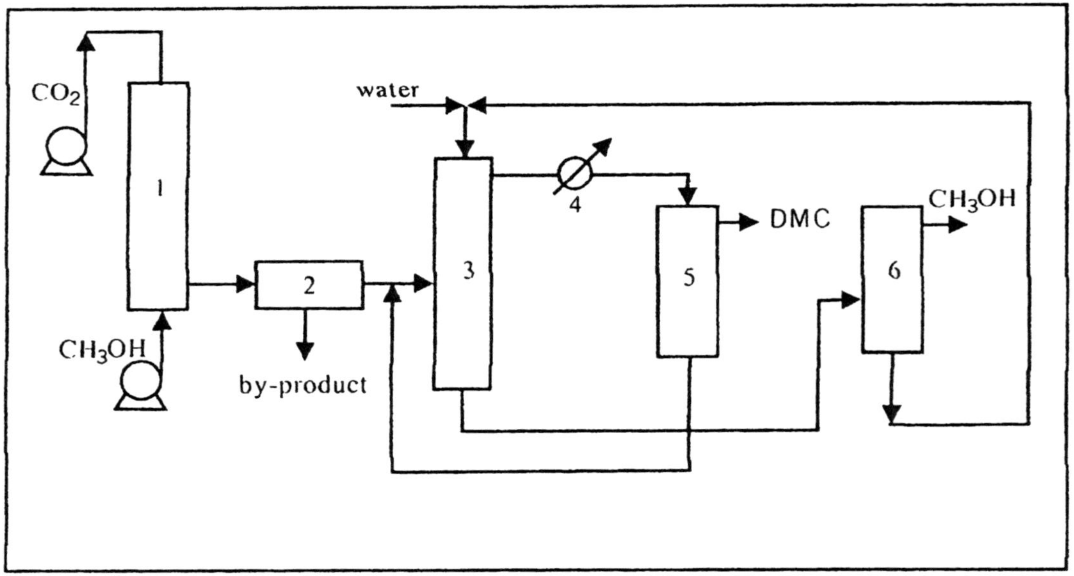

Figure 4. The schematic diagram of the tentative process for the commercial production of DMC

1-reactor; 2-primary tower; 3-extractive distillation; 4-condensor; 5-separator; 6-fractionating town

If it is possible to adopt one more extractive distillation column, we can obtain DMC in higher purity.

The drawback of this method for DMC synthesis from methanol and CO_2 is that the conversion is limited by the reaction equilibrium. But this limitation is avoidable by the removal of water from the reaction system, for example, by introducing trimethyl orthoformate. If it is possible to develop a suitable method for the removal of water, the DMC yield will be improved significantly.

Conclusion

The continuous catalytic synthesis of dimethyl carbonate from carbon dioxide and methanol in the presence of CH_3I and K_2CO_3 with reaction temperature and pressure near the critical point of carbon dioxide was studied in an agitated reactor. The optimal yield of dimethyl carbonate was obtained under a reaction pressure of 7.3 MPa, whereas the reaction temperature is 80-100 ℃. It can be concluded that carbon dioxide is effectively activated in the supercritical state. A possible reaction pathway was presented for the supercritical synthesis of dimethyl carbonate. The results obtained will be useful for the esterification of carbon dioxide and methanol.

Acknowledgments

Supports from the Research Fund for the Doctoral Program of Higher Education of China (97025105) and the Basic Research Project of SINOPEC (598013) are very appreciated.

References

1. Knifton, J. F. U.S. Patent 4,661,609, 1987.
2. Duranleau, R. G.; Nieh, E. C. Y.; Knifton, J. F. U.S. Patent 4,691,041, 1986.
3. Romano, U. *Ind. Eng. Chem. Prod. Res. Dev.* **1980**, 19, 396.
4. Fenton, D. M.; Steinward, P. J. *J. Org. Chem*, **1974**, 39, 701.
5. Curnutt, G. L. W.O. Patent 07601, 1987.
6. Kizlink, J. *J Collect Czech Chem Commun*, **1993**, 58, 1399.
7. Jiang, Q.; Lin, Q. H.; Huang, Z. T. *Chinese Journal of Catalysis* (*Cuihua Xuebao)*, **1996**, 17, 91 (in Chinese).
8. Fang, S.; Fujimoto, K. *Applied catalysis A: General*, **1996**, 142, L1-L3.
9. Gui, X. S.; Cao, F. H.; Liu, D. H.; Fang, D. Y. *Journal of east china university of science and technology*, **1998**, 1, 7 (in Chinese).
10. Savage, P. E.; Gopalan, S.; Mizan, T. I. *AICHE J.* **1995**, 7, 1723.
11. Xu, X. D.; Moulijn J. A. *Energy Fuels* **1996**, 10, 305.
12. Brounstein, P.; Matt, D.; Noble, D. *Chem. Rev.* **1988**, 88, 747.

Chapter 11

Electrochemical Reduction of CO_2 on Cu Electrode in Methanol at Low Temperature

Satoshi Kaneco[1], Hideyuki Katsumata[1], Tohru Suzuki[2], and Kiyohisa Ohta[1]

[1]**Department of Chemistry for Materials, Faculty of Engineering and** [2]**Environmental Preservation Center, Mie University, Tsu, Mie 514–8507, Japan**

The electrochemical CO_2-to-hydrocarbons reduction system with high Faradaic efficiency at copper electrode using methanol media has been developed. Since the cold methanol has been industrially used as the physical absorbent of CO_2 in Rectisol method, a combination technology of the Rectisol CO_2 absorption process and the electrochemical CO_2 conversion method will be able to apply the large scale plant.

Methanol is a better solvent of CO_2 than water, particular at low temperature, because the solubility of CO_2 in methanol is approximately five times that in water at ambient temperature and eight to fifteen times that in water at temperatures below 273 K (1-3). Therefore, methanol has been industrially used as the physical absorbent of CO_2 in Rectisol method at low temperature (3). Currently, over 70 large-scale plants apply the Rectisol process.

For the CO_2 sequestration technologies, the chemical conversion of CO_2 by radiochemical, chemical, thermochemical, photochemical, electrochemical and biochemical procedures has been of significant interest from the practical viewpoints (4). The electrochemical method appears to be a very suitable method for the conversion and reduction of CO_2 (5). Since the present conversion efficiencies of single crystalline, multicrystalline and amorphous silicon photoelectrochemical solar cells are at the respective levels of about 24 %, 18 % and 13 %, respectively, one possible system to industrially convert CO_2 to useful products may be a hybrid method of the solar cell and the CO_2 electrochemical reduction cell (6). Therefore, a combination technology of the Rectisol CO_2 absorption process and the electrochemical CO_2 conversion method will be able to apply the large scale plant. Recently, many investigators have actively studied the electrochemical reduction of CO_2 using various metal electrodes in organic solvents, because organic aprotic solvents dissolve much more CO_2 than water (7-9). It has been described that low reduced products containing carbon monoxide, oxalic acid and formic acid were produced by the electroreduction of CO_2 in dimethyl sulfoxide, N,N-dimethyl formamide, propylene carbonate and acetonitrile. However, even at a cupper electrode, few hydrocarbons have been obtained in these organic solvents. In this article, we describe the electrochemical reduction of CO_2 in methanol-based electrolyte, and especially, the electroreduction on a copper electrode at low temperature.

Experimental

The electrochemical reduction of CO_2 was performed in a custom-made, divided H-type cell. An Aldrich Nafion 117-type ion exchange membrane (0.18 mm thickness) was used as the diaphragm. The cathode potential was measured with respect to an Ag/AgCl sat. KCl electrode that was connected with the catholyte through an agar salt bridge. The methanol (99 %) was purified by double distillation from metallic magnesium Water content in the pure methanol was less than 1 ppm (confirmed by the Karl Fisher test). Various origanic salts and inorganic salts were used as the supporting salt in the methanol-based catholyte. The pH of the catholyte was measured with a glass electrode for nonaqueous solvent, calibrated in water. Mechanical processing of the Cu and Pt electrodes required polishing each surface with successively finer grades of alumina powder down to 0.05 μm, followed by the removal of grease with acetone. The electrodes were activated by anodic polarization at 500 mA for 100 sec in 14.7 mol/dm^3 phosphoric acid. Finally, both electrodes were rinsed with water and ethanol. A discontinuous electroreduction procedure was used. First, CO_2 gas was bubbled into the methanol catholyte for 1 h at a rate of 30 ml/min. Then, the CO_2-saturated solution was reduced electrolytically at cathodic polarization. Stirring of the catholyte was provided by a magnetic bar. The Faradaic efficiencies of formation for the main products were calculated assuming that a total of 50 coulombs of charge passed through the cell. Gaseous products obtained during electroreduction were collected in a gas collector and were analyzed by gas chromatography. Products soluble in the catholyte were analyzed by using high-performance liquid chromatography and gas chromatography.

Results and Discussion

Generally, the product distributions of CO_2 electrochemical reduction in an aqueous solution depend on a number of factors, such as medium, electrode material, electrode shape, a type of supporting salt and its concentration, temperature, pressure, potential, current density and others (5). In the present study, the effects of electrode material and a type of supporting electrolyte on the electrochemical reduction of CO_2 in methanol-based electrolyte were investigated in order to selectively obtain hydrocarbons such as methane and ethylene, which can be used as fuel gases.

Organic salts

Benzalkonium chloride

In order to explore the CO_2 reduction characteristics of metallic electrodes, the electrochemical reduction of CO_2 at various metal electrodes was studied in methanol-based electrolytes with benzalkonium chloride at ambient temperature and pressure. Benzalkonium chloride was used as the ionophore at a concentration of 10 mmol/dm^3. A wide variety of metals, such as Ag, Au, Co, Cu, Fe, Ni, Pt, Sn, Ti, and Zn, were used as the cathode. The treatment of the electrodes was described elsewhere (10). The results are shown in Table I. Carbon monoxide, methane and ethane were the main products. In the water, formic acid was the major product on Sn, Pb, Zn and Cu electrodes (11,12), but in the methanol-based electrolyte with benzalkonium chloride salts no formic acid was produced. CO was selectively produced on Ag, Au and Zn electrodes. Hydrogen production was significant at Co, Fe, Ni, Pt and Ti electrodes. The maximum formation efficiency for hydrocarbons (4.9 %) was observed at Cu electrode. Hori et al. presented that in the electrochemical reduction of CO_2 in water copper was an excellent electrode for formation of hydrocarbons. Therefore, so as to produce hydrocarbons with high efficiency, the electrochemical reduction of CO_2 at the copper electrode in methanol with various supporting salts was examined. The electrolysis was performed at 243 K

Table I. Electrochemical reduction products of CO_2 on various metallic electrode in benzalkonium chloride/methanol at 288 K

Cathode	*Potential (V)*	*Faradaic efficiency*						
		CH_4	*C_2H_4*	*C_2H_6*	*CO*	*HCOOH*	*H_2*	*Total*
Ag	-1.5	0.8	-	-	71	-	41	113
Au	-1.2	1.2	-	-	63	-	10	74
Co	-1.5	2.3	-	-	0.2	-	84	87
Cu	-1.9	1.4	3.5	-	16	-	54	75
Fe	-1.1	1.3	0.2	-	0.2	-	86	88
Ni	-1.9	2.7	0.5	0.3	0.3	-	95	99
Pt	-1.8	0.2	-	-	-	-	94	94
Sn	-1.5	1.8	-	-	28	-	4.4	34
Ti	-1.8	2.0	-	-	0.2	-	114	116
Zn	-1.6	0.6	-	-	64	-	21	86

Catholyte : 10 mmol/dm^3 benzalkonium chloride in methanol.
Anolyte : 100 mmol/dm^3 $KHCO_3$ in water.
SOURCE: Reproduced from reference 10. Copyright 1998.

because temperature in the Rectisol process is customarily in the range of 243 to 263 K.

The CO_2 electrochemical reduction with Cu electrode at 243 K was studied in 80 mmol/dm^3 benzalkonium chloride/methanol. The results are shown in Table II. The reduction reaction began at less than −1.8 V in the electrolyte, the applied potential range was from −1.8 to −2.1 V. The products of the electrochemical reduction of CO_2 were methane, ethylene and CO. The formation of methane increased as the potential decreased when the potential was above −2.0 V. The maximum Faradaic efficiency of methane was obtained at −2.0 V. The Faradaic efficiency for carbon monoxide was 12 % at -1.8 V. The efficiency decreased with decreasing the potential. The maximum Faradaic efficiency of ethylene production was 2.1 %. Azuma et al. (11) investigated the

Table II. Electrochemical reduction products of CO_2 on Cu elelctrode in benzalkonium chloride/methanol at 243 K

Potential (V)	*Current density (mA/cm²)*	*Faradaic efficiency*						
		CH_4	*C_2H_4*	*C_2H_6*	*CO*	*HCOOH*	*H_2*	*Total*
-1.8	5.0	11.0	0.5	-	12.0	-	11.0	34.5
-1.9	6.9	23.8	1.6	-	9.8	-	5.0	40.2
-2.0	9.2	42.5	2.1	-	6.3	-	7.0	57.9
-2.1	11.8	36.0	1.6	-	4.3	-	7.5	49.4

Catholyte : 80 mmol/dm³ benzalkonium chloride in methanol.
Anolyte : 80 mmol/dm³ benzalkonium chloride in methanol.
SOURCE: Reproduced from reference 13. Copyright 1995.

electrochemical reduction of CO_2 at the Cu electrode in 50 mmol/dm³ $KHCO_3$ aqueous soulution, at 293 K, and obtained methane, ethylene and ethane with Faradaic efficiencies of 17.8 %, 12.7 % and 0.039 %, respectively. Moreover, at 273 K, the current efficiency was 24.7 % for metnae, 6.5 % for ethylene and 0.015 % for ethane. Hori et al. (12) reported 38 % Faradaic efficiency for methane in the electrochemical reduction of CO_2 at Cu electrode in 1 mol/dm³ aqueous $KHCO_3$ solution. Kiriacou et al. (14) described the formation efficiencies in the electrochemical reduction of CO_2 on Cu, in 500 mmol/dm³ $KHCO_3$ solution, 298 K, as follows: 16 % for methane; 14 % for ethylene; etc. Therefore, in methanol-based electrolyte, ethane was not produced but the efficiency of methane was much better than those obtained in water.

Tetraethylammonium perchlorate (TEAP)

The electrochemical reduction of CO_2 on a copper electrode in methanol at 243 K was investigated with 5 mmol/dm³ TEAP supporting salts. The results are shown in Table III. The main products from CO_2 were methane, ethylene, ethane, carbon monoxide and formic acid. From the cyclic voltammetry study, the onset (starting) potentials of the cathodic current, i.e., those potential values at which a current density of 0.01 mA/cm² is observed, were −1.1 V and −1.5 V in CO_2-saturated methanol and N_2-pureged methanol, respectively. The reduction current density in CO_2-saturated methanol was larger than those in N_2-pureged methanol. The Faradaic efficiency of methane increased up to 31.4 % as the potential decreased. The pattern for ethylene formation efficiency was convex

Table III. Electrochemical reduction products of CO_2 on Cu electrode in TEAP/methanol at 243 K

Potential (V)	*Current density (mA/cm²)*	*Faradaic efficiency*						
		CH_4	C_2H_4	C_2H_6	CO	$HCOOH$	H_2	*Total*
-2.5	0.46	1.9	2.0	0.10	49.2	13.2	11.9	78.3
-3.0	0.64	10.3	5.1	0.49	40.8	18.9	12.9	88.5
-3.5	1.1	10.0	11.5	0.02	45.7	10.4	6.2	83.8
-4.0	1.3	22.0	9.8	0.01	44.7	11.4	4.1	87.9
-4.5	1.3	31.4	3.2	0.02	18.9	21.8	6.7	82.0

Catholyte : 100 mmol/dm³ TEAP in methanol.

Anolyte : 100 mmol/dm³ KOH in methanol.

SOURCE: Reproduced from reference 15. Copyright 1998.

and the maximum efficiency was 11.5 % at –3.0 V. The current efficiency of formic acid tended to increase with lowering the potential. The formation efficiency for CO was nearly constant at > -4.0V and then the efficiency dropped at < -4.0 V. The current efficiency for ethane was less than 0.5 % at all potentials. The best total Faradaic efficiency for CO_2 reduction was observed at –4.0 V, and the efficiency was 87.9 %. Generally, in the electrochemical reduction of CO_2 in water, hydrogen formation is simultaneous to CO_2 reduction. Therefore, the suppression of hydrogen formation is very important because the applied energy is wasted on hydrogen evolution instead of being used for the reduction of CO_2. The current efficiency for hydrogen formation in the TEAP/methanol-based catholyte was suppressed to below approximately 13 %.

Inorganic salts

Lithium hydroxide

The electrochemical reduction of CO_2 on Cu in methanol at 243 K was investigated with 80 mmol/dm³ LiOH supporting salts. The results are presented in Table IV. Although $LiHCO_3$ may be formed in the methanol while bubbling CO_2 through the solution for several minutes, the amount of hydrogen carbonate is assumed to be negligible, since one cannot observe any precipitate. Therefore, CO_2 can considered to be physically dissolved, i.e., under intact form in the methanol catholyte. The onset potential of the cathodic current in CO_2-saturated

Table IV. Electrochemical reduction products of CO_2 on Cu elelctrode in LiOH/methanol at 243 K

Potential (V)	*Current density (mA/cm²)*	*Faradaic efficiency*						
		CH_4	*C_2H_4*	*C_2H_6*	*CO*	*HCOOH*	*H_2*	*Total*
-2.0	0.84	16.8	7.7	-	35.2	13.7	5.1	78.5
-2.5	2.0	37.1	9.8	-	25.4	13.4	6.5	92.2
-3.0	5.3	58.0	16.5	-	13.5	6.8	1.3	96.1
-3.5	7.6	53.7	13.0	-	15.9	7.3	1.9	91.8
-4.0	10	63.0	14.7	-	14.2	5.2	1.3	98.4
-4.5	12	52.0	13.0	-	13.7	5.7	1.1	85.5
-5.0	14	58.4	18.1	-	10.3	4.7	1.4	92.9

Catholyte : 80 mmol/dm³ LiOH in methanol.

Anolyte : 300 mmol/dm³ KOH in methanol.

SOURCE: Reproduced from reference 16.

methanol was around –0.5 V. In LiOH/methanol-based electrolyte, the onset potential recorded in N_2-purged methanol was shifted to more negative potential values relative to the onset potential recorded in CO_2-saturated methanol. The pH of catholyte after saturation with CO_2 was 6.2 and the pH of N_2-purged methanol was about 14. Both the reduction and onset potentials may be affected by the pH of catholyte. Methane, ethylene, CO, and formic acid were detected as reduction products from CO_2. The current efficiency of methane increased from 20 % to 60 % from -2.0 to -3.0 V. At potentials more nagative -3.0 V, the high efficiency was maintained. The current efficiency of ethylene gradually increased to 18 %, as the potential was lowered. The current efficiency of methane was larger than those of ethylene in the entire investigated potential range. A maximum Faradaic efficiency for hydrocarbons (methane and ethlene, 78 %) was observed at -4.0 V. The formation efficiencies of CO and formic acid gradually increased with the positive direction at < -3.0 V. At the polarizations above -3.0 V, both efficiencies rose abruptly. In LiOH/methanol-based electrolyte, the Faradaic efficiency for hydrogen formation on the Cu electrode at 243 K, was suppressed to less than 2 %, at polarization below –3.0 V.

Table V. Electrochemical reduction products of CO_2 on Cu elelctrode in methanol with various potassium supporting salts at 243 K

Salts (mmol/dm³)	*Potential (V)*	*Faradaic efficiency*						
		CH_4	*C_2H_4*	*C_2H_6*	*CO*	*HCOOH*	*H_2*	*Total*
CH_3COOK (50)	-3.0	8.6	3.2	-	5.4	13.1	53.9	84.2
CH_3COOK (50)	-5.0	18.4	11.4	0.01	19.4	12.7	8.7	70.6
CH_3COOK (500)	-3.0	27.0	7.1	0.02	6.5	13.0	32.5	86.1
CH_3COOK (500)	-4.0	16.1	14.6	0.01	4.2	13.7	18.0	66.0
KBr (50)	-3.0	6.4	9.8	-	35.0	17.4	3.6	72.2
KBr (50)	-5.0	19.1	18.2	0.13	12.0	12.0	4.6	66.0
KI (50)	-3.0	15.8	19.9	0.17	15.2	-[a]	3.5	54.6
KI (50)	-6.0	6.8	12.1	0.13	16.8	-[a]	4.0	40.2
KI (50)	-7.0	19.4	10.5	0.01	21.0	-[a]	1.2	52.1
KI (500)	-3.0	7.0	14.9	-	14.1	-[a]	8.1	44.1
KSCN (50)	-3.0	22.4	12.6	0.23	21.0	-[a]	15.3	71.5
KSCN (50)	-6.0	22.5	9.5	0.01	27.1	-[a]	2.6	61.7
KSCN (500)	-3.0	16.7	16.5	0.34	11.1	-[a]	8.1	52.7
KSCN (500)	-4.0	11.8	16.0	0.01	5.5	-[a]	9.7	43.0

[a] Not measured.

Anolyte : 100 mmol/dm³ KOH in methanol.

SOURCE: Reproduced from reference 17. Copyright 1999.

Potassium supporting salts

The electrochemical reduction of CO_2 with Cu electrode at 243 K was studied in methanol-based catholyte using such potassium supporting salts as CH_3COOK, KBr, KI and KSCN. The results are shown in Table V. The onset potentials for cathodic current were approximately -1.3, -1.3, -1.4, -1.4, -1.5, -1.5 and -1.6 V in 50 and 500 mmol/dm^3 CH_3COOK, 50 mmol/dm^3 KBr, 50 and 500 mmol/dm^3 KI, 50 and 500 mmol/dm^3 KSCN/methanol-based electrolyte, respectively. The reduction products from CO_2 were methane, ethylene, ethane, CO and formic acid. A maximum Faradaic efficiency of methane (27.0 %) was observed in 500 mmol/dm^3 CH_3COOK/methanol electrolyte. The best ethylene formation efficiency (19.9 %) was obtained in 50 mmol/dm^3 KI/methanol-based catholyte. Ethane formation efficiency was less than 0.34 % under the experimental conditions studied. The total current efficiency for hydrocarbons may roughly be independent of the supporting salts. In K^+/methanol-based electrolyte, the Faradaic efficiency for hydrogen formation on the Cu electrode at 243 K was suppressed to less than 15.3 % except for that in the case of acetate. Especially, the efficiency was below 8.1 % when potassium halide salts were used as supporting electrolytes. Only in CH_3COOK/methanol electrolyte, hydrogen formation efficiency was relatively high (53.9 and 32.5 %) at -3.0 V and decreased with the negative potential.

Cesium supporting salts

The electrochemical reduction of CO_2 at Cu electrode was investigated in methanol-based electrolyte using various cesium supporting salts as the ionophore at 243 K. Cesium acetate, chloride, bromide, iodide and thiocyanate were used as the ionophere. The results are shown in Table VI. The potentials were -1.4, -1.3, -1.3, -1.3, -1.3 and -1.1 V in 50 and 500 mmol/dm^3 CH_3COOCs, CsCl, CsBr, CsI and CsSCN/methanol-based electrolyte, respectively. The reduction products from CO_2 were methane, ethylene, ethane, CO and formic acid. The Faradaic efficiency of methane increased in the order of acetate > thiocyanate > chloride > iodide > bromide. A maximum formation efficiency of ethylene (9.9 %) was observed in CsI/methanol-based catholyte at the potential of -3.0 V. Ethane formation efficiency was less than 0.05 % at all experimental conditions studied. Ethylene/methane Faradaic efficiency ratio increased in the sequence of bromide > iodide > chloride > thiocyanate > acetate. The current efficiency for ethylene was larger than those for methane in the methanol-based electrolyte using cesium supporting salts except for acetate. Therefore, the methanol-based electrolyte using cesium supporting salts seems to be effective for the formation of C2 compounds (ethylene). At -3.0 V, CO formation was favorable with the sequence of bromide > iodide > thiocyanate > chloride > acetate. At more negative potential (-6.0 V), the order was bromide > thiocyanate > iodide > chloride > acetate and was nearly similar to those at the

Table VI. Electrochemical reduction products of CO_2 on Cu elelctrode in methanol with various cesium supporting salts at 243 K

Salts (mmol/dm³)	*Potential (V)*	*Faradaic efficiency*						
		CH_4	*C_2H_4*	*C_2H_6*	*CO*	*HCOOH*	*H_2*	*Total*
CH_3COOCs (50)	-3.0	8.3	2.2	-	9.6	23.0	31.2	74.3
CH_3COOCs (50)	-6.0	4.7	7.5	0.01	21.5	14.7	19.3	67.6
CH_3COOCs (500)	-3.0	12.3	3.3	0.01	3.1	15.6	30.1	64.4
CH_3COOCs (500)	-5.0	9.6	7.0	0.01	8.4	16.2	15.8	57.0
CsCl (30)	-3.0	2.8	8.8	0.01	35.6	14.6	8.4	70.2
CsCl (30)	-6.0	2.3	9.0	0.01	23.4	12.6	6.7	54.0
CsBr (25)	-3.0	0.1	5.9	-	54.1	13.0	7.0	80.1
CsBr (25)	-6.0	0.2	6.4	0.007	34.9	14.5	10.1	66.1
CsI (30)	-3.0	1.0	9.9	0.05	40.0	4.8	6.2	61.9
CsI (30)	-6.0	1.3	7.2	0.01	28.1	3.5	7.7	47.8
CsSCN (30)	-3.0	2.4	6.0	-	36.1	16.3	7.2	68.0
CsSCN (30)	-6.0	5.1	9.8	0.009	28.1	9.9	7.9	61.4

Anolyte : 300 mmol/dm³ KOH in methanol.

SOURCE: Reproduced from reference 18. Copyright 2000.

potential of -3.0 V. A minimum current efficiency for formic acid was obtained in CsI/methanol-based electrolyte. The best partial current densities for CO_2 reduction and hydrocarbons (18.5 and 4.3 mA/cm^2, respectively) were observed in CH_3COOCs/methanol-based catholyte at -5.0 V. In cesium supporting electrolytes, the Faradaic efficiency for hydrogen formation on the Cu electrode at 243 K could be suppressed to below 10.1 % except for that in the case of acetate. In CH_3COOCs/methanol-based catholyte, hydrogen formation efficiency was relatively high (31.2 and 30.1 %) at -3.0 V and decreased with the negative potential.

Table VII. Electrochemical reduction products of CO_2 on Cu elelctrode in CsOH/methanol at 243 K

Potential (V)	*Current density (mA/cm²)*	*Faradaic efficiency*						
		CH_4	*C_2H_4*	*C_2H_6*	*CO*	*HCOOH*	*H_2*	*Total*
-3.0	5.4	5.5	25.7	0.02	7.4	13.4	23.3	75.3
-3.5	8.0	4.0	32.3	0.05	9.0	13.1	18.4	76.9
-4.0	11	8.3	23.7	0.02	7.2	11.1	18.1	68.4
-4.5	13	5.5	31.6	0.94	10.1	10.7	17.1	75.9
-5.0	16	6.4	25.9	0.008	7.7	12.7	17.5	70.2

Catholyte : 80 mmol/dm^3 CsOH in methanol.

Anolyte : 300 mmol/dm^3 KOH in methanol.

SOURCE: Reproduced from reference 19. Copyright 1999.

Cesium hydroxide

The electrochemical reduction of CO_2 with Cu electrode at 243 K was investigated in 80 mmol/dm^3 CsOH/methanol-based electrolyte. The results are shown in Table VII. Although $CsHCO_3$ can be formed in the methanol catholyte by bubbling CO_2 gas for several minutes, the amount may be very small because the formation reaction seems to be extremely slow at 243 K. Therefore, CO_2 was considered to be dissolved intactly in the methanol catholyte. The onset potential of the cathodic current was around -1.5 V and the potential was nearly the same as in N_2-purged methanol. Since the pH of catholyte after CO_2-saturation was 7.1 and the pH of N_2-purged methanol was approximately 14, both the reduction and onset potentials may be affected by the pH of catholyte. Methane, ethylene, ethane, CO and formic acid were detected as reduction products from CO_2. The formation efficiency of ethylene was roughly independent of the potential and was in thc range of 23.7 to 32.3 %. The current efficiency for methane had a convex pattern and the maximum efficiency (8.3 %) was observed at -4.0 V. The Faradaic efficiency of ethylene was larger than that of methane in all the potential range. Trace amount of ethane (less than 1 %) was formed. The formic acid formation efficiency gradually decreased with the negative potential at > -4.5 V and then slightly increased at < -4.5 V. The CO formation efficiency curve was nearly flat and the efficiency was approximately 8.3 %. In CsOH/methanol-based electrolyte, the Faradaic efficiency for hydrogen formation on the Cu electrode at 243 K was suppressed to below 23 %.

Reaction mechanism

The mechanism of the electrochemical reduction of CO_2 in methanol-based electrolyte was investigated for a copper electrode. A GC-MS study with deuterated methanol catholyte demonstrated that no reduction product was produced from methanol (20). When the electrolysis was conducted under nitrogen atmosphere, electrolysis yielded exclusively hydrogen. Consequently, the targeted products were produced by the electrochemical reduction of CO_2. These experimental data and literature reports (10-20) suggest the pathway by which methane, ethylene, carbon monoxide, and formic acid are formed are formed as illustrated in Fig. 1. Hydrocarbons are yielded by a series of simultaneous or consecutive electronation/protonation steps. The adsorbed $\cdot CO_2^-$ radical anion formed in the first electronation step undergoes a second electronation/protonation to yield adsorbed CO as the key intermediate. By a succession of four electronation/protonation steps an adsorbed reactive methylene group forms, and this may either stabilize as a methane molecule by a subsequent double electronation/protonation sequence or dimerize to form ethylene and ethane. For the formation of formic acid and CO, we assume the usual pathway, which involves a one-electron reduction followed by a second electronation/protonation to yield formic acid, and by the disproportionation of $\cdot CO_2^-$ radical anions to neutral CO molecules and dinegative carbonate ions. Several researchers (14) have presented that the methane/ethylene Faradaic efficiency ratio increased as the radius of the cation decreased, i.e., $Li^+ < Na^+ < K^+ < Cs^+$, in the electrochemical reduction of CO_2 on Cu in water. Hence, a similar tendency may be obtained in the electrochemical reduction of CO_2 at Cu

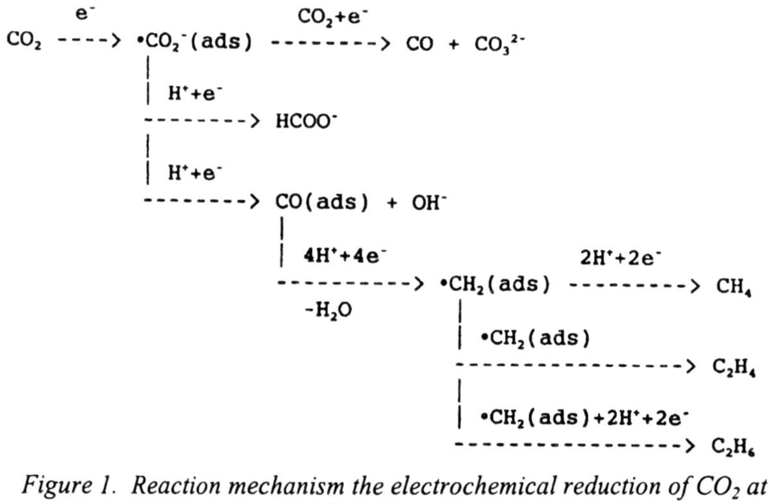

Figure 1. Reaction mechanism the electrochemical reduction of CO_2 at Cu electrode in methanol.

electrode in methanol. Small cations such as Li^+ and Na^+ were not adsorbed at the electrode surface, due to their strong hydration. In addition, small cations carry to the cathode a large number of water molecules, and thus supply protons for the electroreduction. The conversion of intermediate $Cu{=}CH_2$ to methane require the presence of adsorbed hydrogen. Thus, this reaction may be favored at larger surface hydrogen coverage, as is the case of Li^+. The effects of the cationic and anionic species on the electrochemical reduction of CO_2 in methanol-based electrolyte are summarized as follows: 1) The Faradaic efficiency of methane increase in the order of $Li^+ > K^+ > Cs^+$. 2) When Li salts were used, high current efficiency for methane can be obtained. Conversely, high formation efficiency of ethylene can be produced with Cs salts. 3) The hydrogen formation roughly decrease with decreasing the cation size. 4) In the presence of acetate, the formation efficiency of hydrogen is relatively high.

Conclusion

Based on this work, the high efficiency electrochemical CO_2-to-hydrocarbon conversion method appears to be achieved. Future work to advance this technology may include the use of solar energy as the electric energy source. This research can contribute to the large-scale manufacturing of fuel gases from readily available and inexpensive raw materials: CO_2-saturated methanol from industrial absorbers (the Rectisol process). Thus, the synthesis of hydrocarbons by the electrochemical reduction of CO_2 might be of practical interest in fuel production, storage of solar energy and the production of intermediate materials for the petrochemical industry.

Acknowledgments

The present research was partly supported by a Grant-in-Aid for Young Scientists (B) No.14750655 from the Ministry of Education, Culture, Sports, Science and Technology of Japan, as well as the Industrial Technology Research Grant Program in '02 (No. 00B66010c) from the New Energy and Industrial Technology Development Organization (NEDO).

References

1. Lide, D. R. (ed) *Handbook of Chemistry and Physics*, 72nd edn.; CRC press: Boston, MA, 1991; pp 6-4, 3-321.
2. Chihara, H. (ed) *Kagaku Binran-Kiso (Handbook of Chemistry-basic, in Japanese)*, 3rd edn.; Vol II, Maruzen, Tokyo, 1984; pp 158,165.
3. Hochgesand, G. Rectisol and Purisol. Efficient Acid Gas Removal for High Pressure Hydrogen and Syngas Production. *Ind. Eng. Chem.* **1970**, 62, 37.

4. Halmann, M. M. *Greenhouse Gas Carbon Dioxide Mitigation*, Lewis Publ., Washington, D.C.; 1999, pp 1-560.
5. Sullivan, B. P. (ed) *Electrochemical and Electrocatalytic Reduction of Carbon Dioxide*, Elsevier, Amsterdam, 1993; pp 1-299.
6. Gratzal, M. *Nature*, **2001**, 414, 338.
7. Ito, K.; Ikeda, S.; Iida, T.; Nomura, A. *Denki Kagaku*, **1982**, 50, 463-469.
8. Ito, K.; Ikeda, S.; Yamauti, N.; Iida, T.; Takagi, T. *Bull. Chem. Soc. Jpn*, **1985**, 58, 3027.
9. Ikeda, S.; Takagi, T.; Ito, K. *Bull. Chem. Soc. Jpn*, **1985**, 60, 2517.
10. Ohta, K.; Kawamoto, T; Mizuno, T.; Lowy D. L. *J. Appl. Electrochem.*, **1998**, 28, 717.
11. Hori, Y.; Kikuchi, K; Suzuki, S. *Chem. Lett.* **1985**, 1695.
12. Azuma, M.; Hashimoto, K.; Hiramoto, M.; Watanabe, M.; Sakata, T. *J. Electrochem. Soc.* **1990**, 137, 1772.
13. Mizuno, T.; Naitho, A.; Ohta, K. *J. Electroanal. Chem.*, **1995**, 391, 199.
14. Kyriacou, G. Z.; Anagnostopoulos, A. K. *J. Appl. Electrochem.*, **1993**, 23, 483.
15. Kaneco, S.; Iiba, S.; Ohta, K.; Mizuno, T. *J. Energy Envion. Econ.*, **1998**, 7, 153.
16. Kaneco, S.; Iiba, S.; Suzuki, S.; Ohta, K.; Mizuno, T. *J. Phys. Chem. B*, **1999**, 103, 7456.
17. Kaneco, S.; Iiba, S.; Ohta, K.; Mizuno, T. *J. Solid State Electrochem.*, **1999**, 3, 424.
18. Kaneco, S.; Iiba, S.; Ohta, K.; Mizuno, T. *Energy Sources*, **2000**, 22, 127.
19. Kaneco, S.; Iiba, S.; Hiei, N.; Ohta, K.; Mizuno, T.; Suzuki, T. *Electrochim Acta*, **1999**, 44, 4701.
20. Naitoh, A.; Ohta, K.; Mizuno, T.; Yoshida, H.; Sakai, M.; Noda, H. *Electrochim. Acta*, **1993**, 38, 2177.

Chapter 12

CO_2 Hydrogenation over Copper-Based Hybrid Catalysts for the Synthesis of Oxygenates

Son-Ki Ihm[1], Se-Won Baek[1], Young-Kwon Park[2], and Jong-Ki Jeon[3]

[1]National Research Laboratory for Environmental Catalysis, Department of Chemical and Biomolecular Engineering, Korea Advanced Institute of Science and Technology, 373–1 Kusong-Dong, Yusong-Gu, Taejon 305–701, Korea
[2]Faculty of Environmental Engineering, University of Seoul, 90 Jeonnong-Dong, Dongdaemun-Gu, Seoul 130-743, Korea
[3]Department of Chemical Engineering, DongYang University, YoungJou, Kyungbook 750–711, Korea

Hybrid catalysts consisting of Cu-based methanol synthesis catalyst and zeolite were investigated for CO_2 hydrogenation by following the idea that thermodynamic limitation in methanol synthesis from CO_2 hydrogenation could be overcome through *in situ* transformation of methanol to dimethylether(DME) to enhance the yield of oxygenates (methanol + DME). $Cu/ZnO/ZrO_2$ and $Cu/ZnO/Ga_2O_3$ were chosen as methanol synthesis catalysts, and NaZSM-5, HZSM-5, H-Ga-silicate and SAPO-34 as zeolites. The hybrid catalyst of $Cu/ZnO/ZrO_2$ and H-Ga-silicate showed the highest yield of oxygenates. The modification of reducibility or oxygen coverage due to the interactions between methanol synthesis catalyst and zeolite was found to improve the catalytic performance in oxygenates synthesis.

Introduction

Hydrogenation of carbon dioxide is recognized as one of the best methods to fix huge amount of emitted CO_2 in a short time. In order to improve climate conditions as well as to solve the carbon resource problem, it is desirable to develop techniques whereby carbon dioxide can be converted into valuable materials. Methanol synthesis through carbon dioxide hydrogenation has been widely investigated, and direct synthesis of dimethylether or hydrocarbon from carbon dioxide has become one of the recent research interests[1-14].

Since carbon dioxide is one of the most stable compounds, most reactions including CO_2 as a reactant are not favorable thermodynamically. The synthesis of methanol from carbon dioxide ($CO_2 + 3H_2 = CH_3OH + H_2O$, $\Delta G_{250℃} = 46.5$ kJ/mol) is less favored compared to that from carbon monoxides ($CO + 2H_2 = CH_3OH$, $\Delta G_{250℃} = 26.9$ kJ/mol), therefore the equilibrium conversion of carbon dioxide is extremely low even at a commercially desirable operating temperature around 250 °C. One way to overcome the thermodynamic limitation of CO_2 conversion is to combine the process for methanol synthesis with that for methanol conversion by using hybrid catalysts. Copper based catalysts and zeolites have been used for the methanol synthesis and methanol-to-hydrocarbon processes, respectively[12].

For the CO_2 hydrogenation over hybrid catalyst[8], oxygenates synthesis was favorable at lower temperature around 250 °C while hydrocarbon synthesis was favorable at higher temperature around 400 °C. The hydrocarbon distribution could be controlled by selecting adequate zeolite. SAPO-34 was found to be selective to propene in methanol conversion[13]. In the reaction over hybrid catalyst, ethane was the main hydrocarbon with HZSM-5 as zeolite, while C_3 to C_5 hydrocarbons were predominant with SAPO-34. The product distribution and corresponding reaction mechanism strongly depended upon the type of zeolite and its acidity[12]. A new peak appeared in the NH_3-TPD due to the interaction between $CuO/ZnO/ZrO_2$ and SAPO-34, which lead to the simultaneous changes in acidity of zeolite and reducibility of metal oxides. It was found that catalytic performance was well correlated with the peak intensity of the new medium acid site due to hybridization[14].

Dimethylether(DME) is considered not only as a useful building block for making important chemicals but also as an alternative clean fuel for diesel engines. DME can easily be obtained through dehydration of methanol over solid acids, while methanol is commercially manufactured from syngas by using Cu-based catalysts. In this regard, many attempts have been made to combine the methanol synthesis and methanol conversion processes in a single step to obtain oxygenates directly from CO/H_2 or CO_2/H_2[1-7,15]. The *in situ* transformation of methanol to dimethyl ether(DME) would shift the thermodynamic equilibria and thereby improve the total oxygenates yield(methanol + DME). Dubois et al.[5] demonstrated that combination of $Cu/ZnO/Al_2O_3$ catalyst with a solid acid, which enhanced the DME formation, indeed provided a strong driving force for CO_2 conversion. It was also reported that HZSM-5 supported copper-molybdenum oxide catalysts showed an enhanced activity and DME selectivity[15].

In the present work, hybrid catalysts made of Cu-based methanol synthesis catalyst($Cu/ZnO/ZrO_2$ or $Cu/ZnO/Ga_2O_3$) and a zeolite were used for the direct formation of oxygenates(methanol + DME) from CO_2 hydrogenation, and their catalytic behaviors were investigated in terms of reducibility and oxygen coverage.

Experimental

Catalyst preparation

Cu/ZnO based catalysts were prepared by co-precipitation. An aqueous solution of copper acetate, zinc nitrate and metal nitrate(Zr, Ga) and an aqueous solution of sodium hydroxide (1.0M) were mixed at 85°C and pH 7.0 ±0.3. The precipitate was aged for 1h, filtered, washed and dried. The precursors were calcined at 400°C for 12h to prepare $Cu/ZnO/ZrO_2$ and $Cu/ZnO/Ga_2O_3$ catalysts.

ZSM-5 was obtained through a three-day hydrothermal synthesis at 170°C in a Teflon-lined 450ml Parr bomb by following the procedures in a US patent[16]. After washing, the crystallized zeolites were calcined overnight at 550°C. The calcined NaZSM-5 sample was converted into the NH_4^+ form by ion exchange with 1M NH_4Cl solution at room temperature and calcined at 550°C for 12h to obtain HZSM-5. H-Ga-silicate was synthesized according to the procedure similar to that of HZSM-5 except that Al source was substituted with Ga nitrate. SAPO-34 was hydrothermally prepared from reactive gels containing a silicon source (Ludox HS-30), an aluminum source(pseudoboehmite), phosphoric acid, morpholine and water[17]. This reaction mixture was heated in a Teflon-lined Parr bomb at 200°C for 72h. After washing and drying, the resultant product was calcined at 550°C.

Hybrid catalysts were prepared by physically mixing equal weight of a copper based catalyst and a zeolite. They were pelletized and crushed to obtain desired sizes of 60-100 mesh.

Catalyst characterization

The exposed copper surface area was measured by titration with nitrous oxide[18]. The copper surface area(Cu_{react}) of used catalyst after 1hr of reaction was measured first, and the total copper surface area(Cu_{total}) was obtained from the same catalyst after complete reduction of copper. The oxygen coverage was calculated by the following equation[19] : $\Theta_0 = (Cu_{total}-Cu_{react})/(Cu_{total} \times 2)$.

Temperature programmed reduction(TPR) was performed to determine the reduction temperature peak of catalyst. In a typical run, 0.015g of a catalyst was placed in a quartz tubular reactor and heated at 300°C under a helium flow of 200ml/min for 1h. The reactor was then cooled to 100°C and then temperature was increased to 700°C under H_2 flow of 40ml/min for 2h.

Temperature programmed desorption(TPD) measurements by using ammonia as an adsorbate were carried out to determine the acidic properties of zeolites. In a typical run, 0.3g of a calcined sample was placed in a quartz tubular reactor and heated at 500°C under a helium flow of 45 ml/min for 1h. The reactor was then cooled to 100°C and the adsorption was conducted at that temperature by exposing the sample to an ammonia flow of 23 ml/min for 1h. Physically adsorbed ammonia was removed by purging the sample with a helium stream flowing at 45 ml/min for 30min at 100°C.

Activity measurement for CO_2 hydrogenation

Carbon dioxide hydrogenation was performed in a continuous-flow fixed bed microreactor system made of a stainless-steel tube. In a typical experiment, 1.0g of hybrid catalyst was loaded in the reactor. Before reaction, the catalyst was reduced according to the following procedure; heating to 150°C at a rate of 5°C/min and heating to 280°C at a rate of 2°C/min in 5% H_2/Ar(60cc/min) at atmospheric pressure. The hydrogen concentration was then increased stepwise in the sequence to 10,20,40 and 100%(30min per step). Subsequently the catalyst was held in a flow of pure hydrogen(60cc/min) for 1h at the same temperature. After the catalyst bed was reduced with hydrogen, the reactor was pressurized to 28atm with helium, and CO_2 and H_2 were introduced into the reactor. CO_2 hydrogenation products were passed through a heated transfer line to a gas chromatograph(Hewlet-Packard 5890 series II) with TCD and FID detectors. Products were separated in a cross-linked methylsilicon capillary column (i.d. = 0.2 mm, length = 50 m) and Porapak Q column(o.d. = 1/8 inch, length = 1.8 m).

Results and discussion

Table 1 shows the activities of Cu based catalyst containing two or three components for methanol synthesis. There was a linear relationship between methanol yield and Cu surface area. However the methanol yield for the Cu/ZnO/ Ga_2O_3 was greater than that of a Cu/ZnO based catalyst even with similar surface area.

Table 1. Activities of Cu-based catalysts for CH_3OH synthesis

Catalyst	*Composition (wt %)*	*Cu surface area (m^2/g)*	*MeOH yield (wt %)*
Cu/ZnO	6:4	4.8	5.7
Cu/ZrO_2	6:4	5.4	7.2
Cu/ZnO/ZrO_2	6:3:1	7.8	10.1
Cu/ZnO/ZrO_2	6:2:2	8.0	10.5
Cu/ZnO/ZrO_2	6:1:3	8.5	11.4
Cu/ZnO/ Ga_2O_3	6:3:1	4.8	10.3

Reaction conditions : 28 atm, 250°C, W/F = 20g-cat·hr/mol, H_2/CO_2 = 3

The development of highly active Cu-based catalyst for the synthesis of methanol from CO_2 made a pilot plant scale production possible[20]. It is well known that the addition of ZrO_2 or Ga_2O_3 into the Cu/ZnO catalysts improved the catalytic performance of methanol synthesis. It was reported that ZrO_2 improved the dispersion of Cu, and that Ga_2O_3 increased the specific activity by optimizing the ratio of Cu^{1+}/Cu^0 on the surface of Cu particles[11, 19, 21, 22]. Since it is believed that Cu^{1+} species plays an important role in methanol synthesis [3,19,23], the optimization of the ratio Cu^{1+}/Cu^0 should be desirable for high yield of methanol.

The conversion of CO_2 could be increased by overcoming the thermodynamic limitation(i.e., by converting methanol to DME). Table 2 shows the results of oxygenates synthesis from CO_2 hydrogenation over different hybrid catalysts. It can be seen that the catalysts including different zeolites had different catalytic activities. The activity of hybrid catalyst with NaZSM-5 was the lowest among hybrid catalysts due to its weakest acidity. To convert methanol into DME, moderate acidity rather than weak acidity is preferable. The hybrid catalyst from Cu/ZnO/ZrO_2 and H-Ga-silicate showed the highest yield of oxygenates. Compared to the smilar catalyst system of Cu-Mo/HZSM-5 which was suggested by Qi et al.[15], the hybrid catalyst of Cu/ZnO/ZrO_2 and H-Ga-silicate showed slightly higher yield of oxygenate (12.6wt% vs 10.7 wt%) but lower oxygenate selectivity (66 wt% vs 77 wt%) at the temperature around 250℃.

Even if $Cu/ZnO/Ga_2O_3$ showed a good activity for methanol synthesis, its hybrid counterpart showed a lower yield for oxygenates and a higher selectivity for CO. It is noted that H-Ga-silicate contains Ga in its framework, but $Cu/ZnO/Ga_2O_3$ contains only Ga_2O_3 particles.

Table 2. Comparison of different types of hybrid catalysts for oxygenates synthesis

Hybrid Catalyst	*Conv. (%)*	*Selectivities (wt%)*			*Yield of Oxygenates (wt%)*
		CO	*MeOH*	*DME*	
A + NaZSM-5	15.4	50.6	19.5	29.9	7.6
A + HZSM-5	17.2	42.4	22.7	34.9	9.9
A + H-Ga-silicate	19.0	33.7	21.0	45.3	12.6
A + SAPO-34	17.5	44.0	21.1	34.9	9.8
B + NaZSM-5	18.2	73.0	10.5	16.5	4.9
B + HZSM-5	19.1	68.6	12.0	19.4	6.0
B + H-Ga-silicate	19.4	67.2	12.9	19.9	6.4
B + SAPO-34	19.6	67.9	12.7	19.4	6.3

Reaction conditions : 28 atm, 250°C, W/F = 20g-cat·hr/mol, $H_2/CO_2 = 3$
A : $Cu/ZnO/ZrO_2$(6:3:1), B : $Cu/ZnO/Ga_2O_3$(6:3:1)

As shown in Figure 1, the oxygen coverages of hybrid catalysts with $Cu/ZnO/Ga_2O_3$(data point 7,8,9 and 10) were much larger than that of original $Cu/ZnO/Ga_2O_3$(data point 6) and their oxygenate yields were smaller. It was proposed by Saito et al.[19] that there was an optimum oxygen coverage of Cu surface for the specific activity for methanol synthesis. If the optimum oxygen coverage is around 0.15, the higher oxygen coverage(0.30) of hybrid catalysts with $Cu/ZnO/Ga_2O_3$ may indicate that the hybridization made copper components less active and resulted in poor yield of oxygenates. The highest yield of oxygenates obtained by the hybrid catalyst of $Cu/ZnO/ZrO_2$ and H-Ga-silicate(data point 4) should be due to weak interaction. Its oxygen coverage(0.19) was also not much different from that(0.15) of $Cu/ZnO/Ga_2O_3$. The Ga oxide inside the pore of H-Ga-silicate seems to make Cu species in the hybrid catalyst more active by balancing the ratio Cu^{1+}/Cu^0. In our experiment, there was also an optimum oxygen coverage for DME synthesis as suggested by

Saito et al.[19]. This indicates that the methanol synthesis and DME synthesis are rate-limited by similar mechanism.

As shown in Figure 2, the reduction temperatures were shifted to higher temperature due to hybridization. The temperature at which the reduction starts (defined as reduction temperature) could be a measure of reducibility or the degree of interaction, and the reduction temperatures are listed in Table 3. Some interactions between methanol synthesis catalyst and zeolite might modify the properties of each component in the hybrid catalyst such as the reducibility of methanol synthesis catalyst and the acidity of zeolite. The reduction temperatures of hybrid catalysts were found to be higher than that of the original methanol synthesis catalysts such as $Cu/ZnO/ZrO_2$ or $Cu/ZnO/Ga_2O_3$. The increase in the reduction temperature for the hybrid catalyst containing $Cu/ZnO/ZrO_2$ was rather small(about 20°C), while that for the hybrid catalyst with $Cu/ZnO/Ga_2O_3$ was quite large(about 60°C). This difference in the reduction temperature must be related with the ability of oxygenate synthesis.

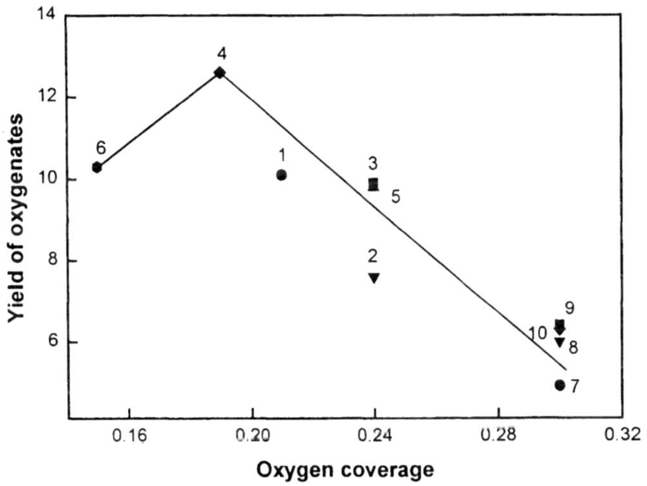

Figure 1. Yield of oxygenates vs. oxygen coverage; 1: $CuO/ZnO/ZrO_2$ (A),2:A+NaZSM-5,3:A+HZSM-5, 4:A+H-Ga-silicate,5:A+SAPO-34, 6: $CuO/ZnO/Ga_2O_3$(B),7:B+NaZSM-5,8:B+HZSM-5, 9:B+H-Ga-silicate,10:B+SAPO-34

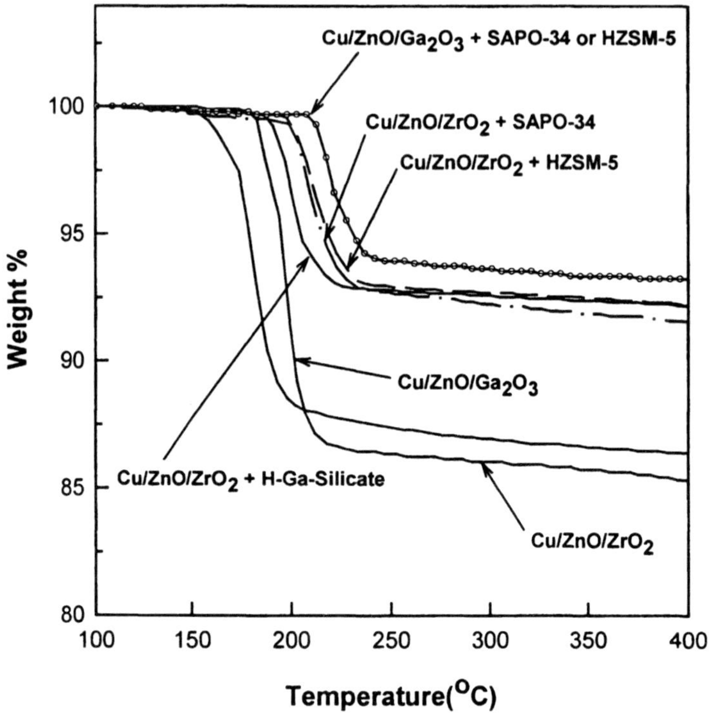

Figure 2. Temperature Programmed Reduction of Hybrid Catalysts

Table 4 shows that the increase of calcination temperature resulted in higher yield of oxygenates because the larger amount of Ga oxide particles, formed inside the pore, optimized the oxygen coverage. The gallium species in H-Ga-silicate seemed to play a better role than Ga_2O_3 of Cu/ZnO/Ga_2O_3 catalyst for improving the catalytic activity of Cu-based catalysts for the synthesis of oxygenates.

Another kind of interaction, i.e., acidity modification, seemed to be more beneficial for hydrocarbon synthesis, while it did not affect the oxygenates synthesis. It was observed that a new peak corresponding to medium acidity appeared in the NH_3 TPD spectra of hybrid catalysts and that there existed a correlation between its peak intensity and the hydrocarbon yield[14]. An example of a new peak for the hybrid catalyst(Cu/ZnO/ZrO_2 + SAPO-34) is shown in Figure 3.

Table 3. Reduction temperatures and oxygen coverages

Catalyst	*A*	*A + NaZSM-5*	*A + HZSM-5*	*A + H-Ga-silicate*	*A + SAPO-34*
Red. temp. (°C) (Shift Δ)	180 (Δ 0)	200 (Δ 20)	200 (Δ 20)	189 (Δ 9)	200 (Δ 20)
Oxygen coverage	0.21	0.24	0.24	0.19	0.24
Catalyst	*B*	*B + NaZSM-5*	*B + HZSM-5*	*B + H-Ga-silicate*	*B + SAPO-34*
Red. temp. (°C) (Shift Δ)	150 (Δ 0)	210 (Δ 60)	210 (Δ 60)	210 (Δ 60)	210 (Δ 60)
Oxygen coverage	0.15	0.30	0.30	0.30	0.30

A : $Cu/ZnO/ZrO_2$(6:3:1), B : $Cu/ZnO/Ga_2O_3$(6:3:1)

Table 4. Effect of calcination temperature of H-Ga-silicate on oxygenates yield

Calcination temperature(°C)	*550*	*600*	*650*
Oxygenates yield(wt %) (Oxygen coverage)	12.6 (0.19)	13.4 (0.18)	13.8 (0.16)

Reaction conditions : 28 atm, 250°C, W/F = 20g-cat·hr/mol, H_2/CO_2 = 3
Catalyst : $Cu/ZnO/ZrO_2$ + H-Ga-silicate

The hybrid catalyst with SAPO-34 or HZSM 5 which had higher acid amount due to solid-solid interaction showed a poor yield of oxygenates but a higher yield of hydrocarbons than that with H-Ga-silicate[12]. It is believed that the reducibility instead of acidity should play a bigger role in the synthesis of oxygenates and that the acidity should be more important for the hydrocarbon synthesis. The moderate acidity of H-Ga-silicate seems to be sufficient enough for oxygenate synthesis or methanol dehydration.

To summarize the effect of H-Ga-silicate hybridization, series of oxygenate synthesis reaction at different mixing ratio of $Cu/ZnO/ZrO_2$ and H-Ga-silicate

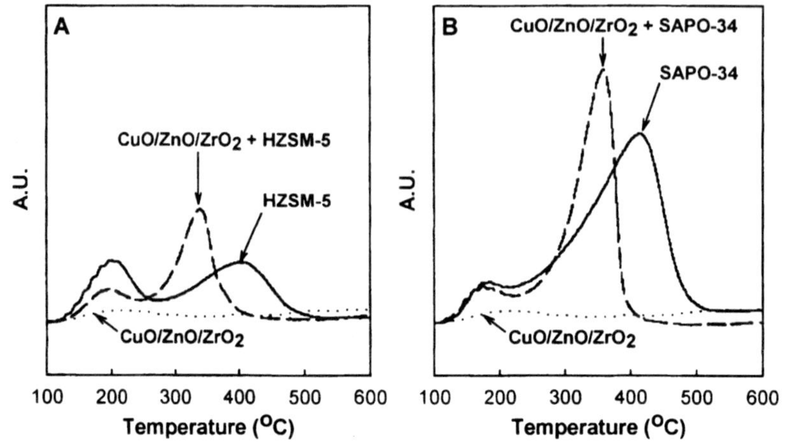

Figure 3. NH_3 TPD of (A: Cu/ZnO/ZrO_2 + HZSM-5, B: Cu/ZnO/ZrO_2 + SAPO-34) catalysts. (Appearance of a new peak in the hybrid catalyst)[14]

were carried out. Table 5 shows that even small amount of H-Ga-silicate addition significantly increased the yield of oxygenates, and the oxygenates yield depended upon the mixing ratio. The hybrid catalysts with low fraction of zeolite showed better yield for oxygenates. It was reported that the methanol formation is the rate-limiting step for DME synthesis[24], and accordingly it is desirable to increase the weight fraction of methanol synthesis catalyst.

Table 5. Enhancement of oxygenates yield due to the promotion of H-Ga-silicate

Weight ratio of hybrid catalyst [Cu/ZnO/ZrO2 : H-Ga-silicate]	*Yield of oxygenates (wt %)*
1.0 : 0.0	10.2
0.9 : 0.1	14.7
0.8 : 0.2	15.2
0.7 : 0.3	13.5
0.5 : 0.5	12.6

Reaction conditions : 28 atm, 250°C, W/F = 20g-cat·hr/mol, H_2/CO_2 = 3, catalyst loading(Cu/ZnO/ZrO_2 + H-Ga-silicate) = 1g

Conclusion

Hybrid catalysts consisting of Cu-based methanol synthesis catalyst and zeolite were investigated for the direct synthesis of oxygenates(methanol + DME) from CO_2 hydrogenation. *In situ* transformation of methanol to dimethyl ether by using hybrid catalytic system at lower temperature range was a reasonable way to improve the total oxygenates yield. The acidity of zeolites and the reducibility of Cu/ZnO-based catalysts were affected by hybridization. The modification of reducibility or oxygen coverage due to the interactions between methanol synthesis catalyst and zeolite was found influence the ability of oxygenate synthesis. To obtain higher yield of oxygenates, it was preferable to use H-Ga-silicate as a Ga source rather than that of $Cu/ZnO/Ga_2O_3$.

Acknowledgment

This work was supported by National Research Laboratory Program and Brain Korea 21 project.

References

1. Li, J.L.; Zhang, X.G.; Inui, T. *Appl. Catal. A : General,* **1997,** 164, 303.
2. Li, J.L.; Zhang. X.G.; Inui, T. *Appl. Catal. A : General,* **1996,** 147, 23.
3. Lee, J.S.; Moon, K.I.; Lee, S.H.; Kim, Y.G. *Catal. Lett.,* **1995,** 34, 93.
4. Peng, X.D.; Wang, A.W.; Toseland, B.A.; Tijm, P.J.A. *Ind. Eng. Chem. Res.,* **1999,** 38, 4381.
5. Dubois, J.L.; Sayama, K.; Arakawa, H. *Chem. Lett.,* **1992,** 1115.
6. Ng, K.L.; Chadwick, D.; Toseland, B.A. *Chem. Eng. Sci.,* **1999,** 54, 3587.
7. Machida, S.; Akitama, T.; Muramatsu, A.; YaGi, J.I. *ISIJ Tntern.,* **1997,** 37,5, 531.
8. Jeon, J.K.; Jeong, K.E.; Park,Y.K.; Ihm, S.K. *Appl. Catal. A,* **1995,** 124, 91.
9. Fujiwara, M.; Ando, H.; Tauaka, M.; Souma, Y. *Appl. Catal. A,* **1995,** 130, 105.
10. Fujiwara, M.; Kietter, R.; Ando, H.; Souma,Y. *Appl. Catal. A,* **1995,** 121, 113.
11. Ihm, S.K.; Park, Y.K.; Jeon, J.K.; Park, K.C.; Lee, D.K. *Stud. Surf. Sci. Catal.,* **1998,** 114, 505.
12. Park, Y.K.; Park, K.C.; Ihm, S.K. *Catal. Today,* **1998,** 44, 165.

13. Park, Y.K.; Baek, S.W.; Ihm, S.K. *J. Ind. Eng. Chem.*, **2001,** 7(3), 167.
14. Ihm, S.K.; Baek, S.W.; Park, Y.K.; Park, K.C. *Stud. Surf. Sci. Cat.*, **2001,** 135, 277.
15. Qi, G.X.; Fei, J.H.; Zheng, X.M.; Hou, Z.Y. *Catal. Lett.,* **2001,** 72, 121.
16. Argauer, R.J.; Landolt, G.R. *U.S. Patent* **1972,** 3 702 886.
17. Ashtekar, S.; Chilukuri, S.V.V.; Chakrabarty, D.K. *J. Phy. Chem.,* **1994,** 98, 4878.
18. Chinchen, G.C.; Hay, C.M.; Vandervell, H.D.; Waugh, K.C. *J. Catal.,* **1987,** 103, 79.
19. Saito, M.; Fujitani, T.; Takeguchi, M.; Watanabe, T. *Appl. Catal. A,* **1996,** 138, 311.
20. Ushikoshi, K.; Mori, K.; Watanabe, T.; Takeuchi, M.; Saito, M. *Stud. Surf. Sci. Catal.,* **1998,** 114, 357.
21. Toyir, J.; Piscina, P.R.; Fierro, J.L.G.; Homs, N. *Appl. Catal. B: Environmental ,* **2001,** 29, 207.
22. Toyir, J.; Piscina, P.R.; Fierro, J.L.G.; Homs, N. *Appl. Catal. B: Environmental,* **2001,** 34, 255.
23. Fujitani, T.; Saito, M.; Kanai, Y.; Kakumoto, T.; Watanabe, T.; Nakamura, J.; Uchijima, T. *Catal. Lett.,* **1994,** 25, 271.
24. Fujimoto, K.; Asami, K.; Saima, H.; Shikada, T.; Tominaga, H. *Ind. Eng. Chem. Prod. Res. Dev.,* **1986,** 25, 262

Chapter 13

Methanol Synthesis from $H_2/CO/CO_2$ over CNT-Promoted Cu–ZnO-Al_2O_3 Catalyst

Hong-Bin Zhang, Xin Dong, Guo-Dong Lin, You-Zhu Yuan, Peng Zhang, and K. R. Tsai

Department of Chemistry and State Key Laboratory of Physical Chemistry for the Solid Surfaces, Xiamen University, Xiamen 361005, China

Highly active multiwalled-carbon-nanotubes (symbolized as CNTs)-supported catalysts x%$Cu_iZn_jAl_k$/CNTs were prepared, and their catalytic performance for methanol synthesis from $H_2/CO/CO_2$ was investigated and compared with the corresponding those supported by AC and γ-Al_2O_3. The results showed that, over a 32%$Cu_6Zn_3Al_1$/CNTs catalyst, the intrinsic reaction rate of methanol synthesis at 508 K was 1.38 and 1.42 times as high as those of the AC- and γ-Al_2O_3-supported catalysts at the respective optimum operating temperatures, 523 and 543 K. The nature of promoter action by the CNTs was discussed together with the results of spectroscopic characterization of the CNT-supported catalysts. The present work indicated that the unique feature of the CNTs in hydrogen-adsorption might play an important role in effectively promoting the methanol synthesis.

Introduction

Among the C_1 chemicals, methanol is the species most widely used in various chemical applications. More recently, it has been used as a clean synthetic fuel additive. Serious considerations have been given to it as an alternative fuel source, including as a convenient hydrogen carrier for PEM fuel cells (*1*). More active catalysts and lower-temperature processes with high syngas single-pass conversion have been the objectives for many research and development efforts.

Kinetics and mechanism of the methanol synthesis on Cu-ZnO-Al_2O_3 catalysts have been extensively studied since the late 1970s. A number of study and excellent review articles have been published on this subject, including ones by Klier (*2*), Fleisch et al. (*3*), Chinchen et al. (*4*), Herman (*5*). Progress in this field has considerably contributed to our growing understanding of the nature of this catalytic reaction system.

Carbon nanotubes are drawing increasing attention recently (*6*). This type of new carbon material possesses a series of unique features, such as, its nanosize channel, the highly conductive graphite-like tube-wall, the sp^2-carbons constructed surface, and its excellent performance of hydrogen adsorption. These features make the CNTs full of promise to be a novel catalyst carrier (*7-10*) or even promoter. In the present work, a series of CNT-supported/promoted Cu-ZnO-Al_2O_3 catalysts were prepared. Their catalytic performance for CO/CO_2 hydrogenation to methanol was studied and compared with those of the corresponding systems supported by AC (activated carbon) and γ-Al_2O_3. The results shed some light on understanding of the nature of promoter action by the CNTs and the prospect of developing highly active catalysts.

Experimental

Catalyst Preparation

The CNTs were synthesized in-house by the catalytic method reported previously (*11*). The prepared CNTs were a type of multi-walled carbon nanotubes. The freshly prepared CNTs were treated with boiling nitric acid for 4 h, followed by rinsing with de-ionized water, then drying at 473 K under dry N_2. Open-end CNTs with somewhat hydrophilic surface were thus obtained. A series of Cu-ZnO-Al_2O_3 catalysts supported by CNTs (or AC or γ-Al_2O_3), symbolized as x%(mass percentage)$Cu_iZn_jAl_k$/CNTs (or AC, or γ-Al_2O_3), was prepared by impregnating a certain amount of CNTs (or AC, γ-Al_2O_3) with aqueous or ethanol solution containing a calculated amount of nitrates of Cu, Zn, and Al, followed by drying at 373 K for 6 h, and calcining at 473 K for 4 h. All catalyst samples were pressed, crushed, and sieved to a size of 40~80 mesh.

Catalyst Evaluation

The catalyst was tested for methanol synthesis in a continuous-flow micro-reactor-GC combination system. 0.5 g of catalyst sample (i.e., equivalent to ~1.0 ml of sample) was used for each test. Prior to the reaction, catalyst sample was pre-reduced by 5%H_2+N_2 for 14 h. Methanol synthesis from $H_2/CO/CO_2$ over the catalysts was conducted at a stationary state with feedgas of $H_2/CO/CO_2/N_2$ = 62/30/5/3 (v/v) at 453~563 K and 2.0 MPa. The reactants and products were determined by an on-line GC (Model 102-GD) equipped with a TC detector and dual columns filled with 5A zeolite molecular sieve and 401-porous polymer, respectively. The former column was used for analysis of CO and N_2 (as internal standard), and the latter for CH_3OH, CO/N_2, CO_2, MF, and DMC. CO-conversion and methanol-STY were evaluated by an internal standard method and an external standard (i.e., working curve) method, respectively.

Catalyst Characterization

Specific surface area (SSA) was measured by N_2 adsorption using a Sorptomatic-1900 (Carbo Erba) system, and Cu surface area was determined referring to the method improved by Bond et al. (*12*). TEM observations were performed by a JEOL JEM-100CX TE microscope. XRD measurements were carried out on a Rigaku D/Max-C X-ray Diffractometer with Cu-Kα radiation. Raman spectra was taken by a Renishaw UV-Vis Raman System 1000R spectrometer, with 325 nm line from a He-Cd laser used for excitation. XPS measurements were done on a VG ESCA LAB MK-2 system with Mg-Kα radiation. H_2-TPR/TPD test of catalyst was conducted on a fixed-bed continuous-flow micro-reactor or adsorption-desorption system.

Results and Discussion

Reactivity of Methanol Synthesis over the CNT- Supported Catalysts

Investigation of the effect of feedgas GHSV on the reactivity of methanol synthesis over the 40%$Cu_6Zn_3Al_1$/CNTs catalyst showed that, at a certain temperature (e.g. 503 K), CO conversion increased with increasing GHSV from zero to ~3000 ml (STP) h^{-1} (g-catal.)$^{-1}$, and reached a maximum at GHSV close to 3000 ml (STP) h^{-1} (g-catal.)$^{-1}$. It then started to decrease with further increasing GHSV. It was believed that, as the GHSV approached 3000 ml (STP) h^{-1} (g-catal.)$^{-1}$, the reaction of methanol formation was in the kinetics-controlled region. In the present work, all reactivity tests of methanol synthesis were performed under the reaction condition of GHSV= 3000 ml (STP) h^{-1} (g-catal.)$^{-1}$, unless otherwise specified.

It is well known that CO_2 in the feedgas is a participating component indispensable to continuously fast conversion of the syngas to methanol. However, the net-conversion of CO_2 indicated by GC analysis of the inlet and exit gases of the reactor was quite low, and all assay results of the catalyst activity in the present work are shown with CO conversion, methanol STY, and intrinsic methanol-formation rate.

Figure 1 shows the assay results of reactivity of methanol synthesis in 463~563 K over a series of 40%$Cu_iZn_jAl_k$/CNTs catalysts with different compositions of Cu/Zn/Al (molar ratio). On all five catalysts, CO conversion and methanol STY both went up first with increasing temperature, reached a maximum at their respective optimum operating temperature, and then went down as the temperature was further increased. This could be understood considering that at the lower temperature regions before the maximum, the methanol synthesis reaction was kinetically controlled, and after the maximum, the reaction turned to be limited by thermodynamics equilibrium due to the high reaction temperature.

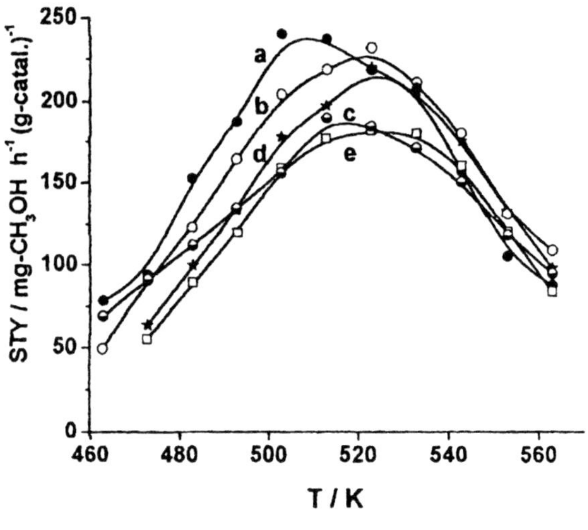

Figure 1. Space-time-yield of CH_3OH synthesis at different temperatures over: a) 40%$Cu_6Zn_3Al_1$/CNTs; b) 40%$Cu_{10}Zn_{10}Al_1$/CNTs; c) 40%Cu_2Zn_1/CNTs; d) 40%$Cu_{4.5}Zn_{4.5}Al_1$/CNTs; e) 40%Cu_1Zn_1/CNTs.

It can be seen from Figure 1 that the observed activity sequence of those catalysts for methanol synthesis was: a) 40%$Cu_6Zn_3Al_1$/CNTs > b) 40%$Cu_{10}Zn_{10}Al_1$/CNTs > d) 40%$Cu_{4.5}Zn_{4.5}Al_1$/CNTs > c) 40%Cu_2Zn_1/CNTs > e) 40%Cu_1Zn_1/CNTs. Over the catalyst of 40%$Cu_6Zn_3Al_1$/CNTs with optimal molar ratio of Cu/Zn/Al= 6/3/1, methanol-STY reached 240 mg-CH_3OH h^{-1} (g-catal.)$^{-1}$ with corresponding CO-conversion at 19.6% under the reaction conditions of 2.0 MPa, 503 K, $H_2/CO/CO_2/N_2$=62/30/5/3 (v/v), GHSV=3000 ml (STP) h^{-1} (g-catal.)$^{-1}$.

Figure 2 shows the assay results of reactivity of methanol synthesis over the CNT-supported catalysts with different loading amounts, x%$Cu_6Zn_3Al_1$/CNTs. On all five catalysts, methanol-formation rate reached a maximum in the region of 503~523 K. Their activity sequence for methanol synthesis is: a) 32%$Cu_6Zn_3Al_1$/CNTs > b) 38%$Cu_6Zn_3Al_1$/CNTs > c) 40%$Cu_6Zn_3Al_1$/CNTs > d) 42%$Cu_6Zn_3Al_1$/CNTs ≈ e) 24%$Cu_6Zn_3Al_1$/CNTs. The optimal loading amount of the $Cu_6Zn_3Al_1$ seemed to be ~32%. Over this catalyst under the reaction conditions of 508 K, 2.0 MPa, feed-gas $H_2/CO/CO_2/N_2$=62/30/5/3 (v/v), GHSV=3000 ml (STP) h^{-1}(g-catal.)$^{-1}$, methanol-formation rate reached 0.109 μmol-CH_3OH s^{-1}(m^2-surf. Cu)$^{-1}$, with the corresponding CO-conversion and methanol-STY at 21.6% and 287 mg-CH_3OH h^{-1}(g-catal.)$^{-1}$, respectively, and the CH_3OH-selectivity in the products at ≥ 98.0%.

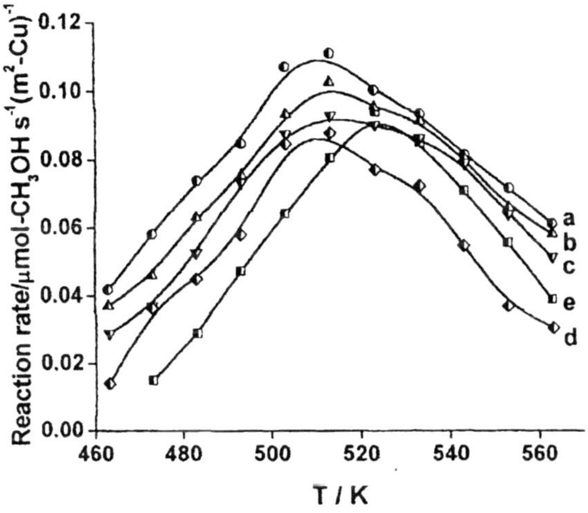

Figure 2. Reaction rate of CH_3OH synthesis at different temperatures over: a) 32%$Cu_6Zn_3Al_1$/CNTs; b) 38%$Cu_6Zn_3Al_1$/CNTs; c) 40%$Cu_6Zn_3Al_1$/CNTs; d) 42%$Cu_6Zn_3Al_1$/CNTs; e) 24%$Cu_6Zn_3Al_1$/CNTs.

It is conceivable that the dispersion of Cu-Zn-Al components on the surface of CNTs constructed by sp^2-C could be improved by using proper organic solvent (e.g., ethanol) in place of water for the preparation of the catalyst precursor solution, which would be conducive to enhancing the catalyst activity. That turned out to be the case. Over a 32%$Cu_6Zn_3Al_1$/CNTs catalyst prepared from the corresponding ethanol solution, the highest methanol-formation rate attained 0.114 μmol-CH_3OH s^{-1} (m^2-surf. Cu)$^{-1}$, with the corresponding CO-conversion and methanol-STY at 23.1% and 307 mg h^{-1} (g-catal.)$^{-1}$, respectively, ~7% higher as compared to that prepared from the aqueous solution.

It was experimentally found that the support could significantly affect the activity of methanol synthesis. Figure 3 shows the results of activity evaluation of methanol synthesis over the catalysts supported by different carriers. Over the 32%$Cu_6Zn_3Al_1$/CNTs catalyst, the methanol synthesis exhibited considerable reactivity at operating temperature as low as 463 K, and reached the maximum of reaction rate, 0.109 μmol-CH_3OH s^{-1} (m^2-surf. Cu)$^{-1}$, at 508 K, which was 1.38 and 1.42 times of those over the catalysts supported by AC and γ-Al_2O_3 (i.e., 0.079 and 0.077 μmol-CH_3OH s^{-1} (m^2-surf. Cu)$^{-1}$) at the respective optimal operating temperatures, 523 and 543 K, respectively; correspondingly, the methanol-STY (287 mg-CH_3OH h^{-1} (g-catal.)$^{-1}$) was 1.9 and 2.5 times of those over the catalysts supported by AC and γ-Al_2O_3 (i. e., 150 and 114 mg-CH_3OH h^{-1}(g-catal.)$^{-1}$) at 523 and 543 K, respectively. Moreover, on the CNT-supported catalyst, the optimal working temperature for methanol synthesis could be lower than that on the AC or γ-Al_2O_3-supported catalyst by 20 or 40 K.

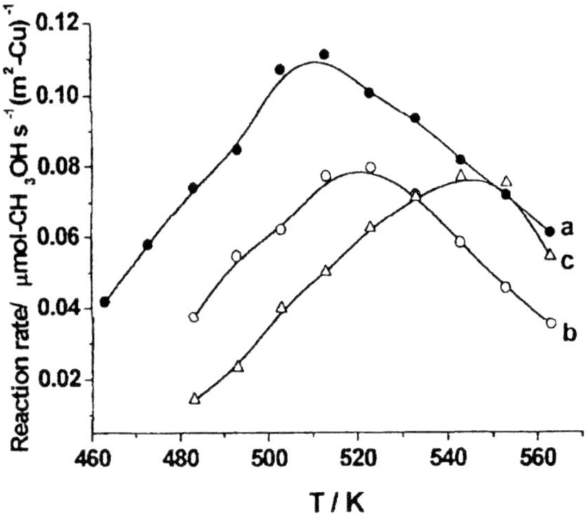

Figure 3. Reaction rate of CH_3OH synthesis at different temperatures over: a) 32%$Cu_6Zn_3Al_1$/CNTs; b) 32%$Cu_6Zn_3Al_1$/AC; c) 32%$Cu_6Zn_3Al_1$/γ-Al_2O_3.

Physico-Chemical Properties of the CNTs as Carrier/Promoter

The results of reactivity assay of the methanol synthesis over a series of the x%$Cu_iZn_jAl_k$/CNTs (or AC, or γ-Al_2O_3) catalysts are summarized in Table I. It is quite evident that the considerably better performance of the CNT-supported catalysts for the methanol synthesis is closely related to the unique structures and properties of the CNTs as carrier. TEM observation showed that CNTs were almost the only species in the purified CNTs products. In elemental analysis, there were no elements detected other than carbon. Figure 4 displayed the TEM image of these CNTs, which were pretty uniform in diameter along the tube-length with the O. D. in the range of 10~50 nm. The HRTEM observation demonstrated that those CNTs derived from CH_4 were constructed by a superposition of many graphene layer facets which were tilted at a certain angle with respect to the axis of central hollow nanofibre, as if a number of cones were placed one on top of another (*9-11*). From the HRTEM image, the I. D. of those CNTs was estimated at ~3 nm. XRD pattern of those CNTs (Fig. 5b) was close to that of graphite (Fig. 5a), but the strongest peak was weakened and somewhat broadened. The position of this peak shifted from $2\theta = 26.5°$ for graphite to $2\theta = 26.1°$ for the CNTs, which corresponded to an increase in the spacing between the graphene layers from 0.336 nm for graphite to 0.341 nm for the CNTs; while the half-peak width was enlarged from ~0.50° for graphite to ~1.43° for the CNTs, indicating that the degree of long-range order of these nanostructures was lower than that of graphite.

Table I. Reactivity of CH_3OH synthesis from $H_2/CO/CO_2$ over a series of the supported catalysts, x%$Cu_iZn_jAl_k$/CNTs (or AC or γ-Al_2O_3)*

Catalyst sample	*SSA / Cu Surf. area (m^2 (g-catal.)$^{-1}$)*	*Reaction temp.(K)*	*STY (mg-CH_3OH h^{-1} (g-catal.)$^{-1}$)*	*Reaction rate (μmol-CH_3OH s^{-1} (m^2-surf. Cu)$^{-1}$*
42%$Cu_6Zn_3Al_1$ /CNTs	69.5 / 21.1	473	73.0	0.030
		523	238	0.098
40%$Cu_6Zn_3Al_1$ /CNTs	83.0 / 21.6	463	79.4	0.032
		508	240	0.097
38%$Cu_6Zn_3Al_1$ /CNTs	92.2 / 22.0	463	101	0.040
		508	261	0.103
32%$Cu_6Zn_3Al_1$ /CNTs	94.5 / 22.9	463	110	0.042
		508	287	0.109
24%$Cu_6Zn_3Al_1$ /CNTs	104 / 20.5	463	78.4	0.033
		508	239	0.101
32%$Cu_6Zn_3Al_1$ /CNTs**	122 / 23.3	463	100	0.037
		513	307	0.114
32%$Cu_6Zn_3Al_1$ /AC	195 / 16.4	483	70.8	0.037
		523	150	0.079
32%$Cu_6Zn_3Al_1$ /γ-Al_2O_3	132 / 13.0	483	20.9	0.014
		543	114	0.077

*Reaction conditions: 2.0MPa, $H_2/CO/CO_2/N_2$= 60/30/5/5(v/v), GHSV= 3000 ml(STP) h^{-1} (g-catal.)$^{-1}$. **Prepared from ethanol solution of the precursor.

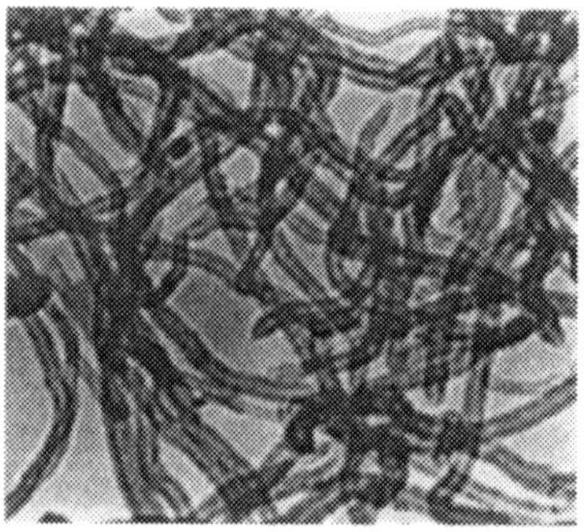

Figure 4. TEM image ($\times 10^5$) of the CNTs grown catalytically from CH_4.

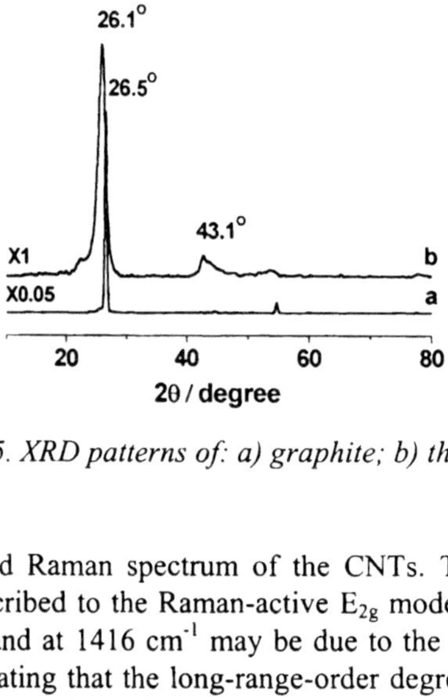

Figure 5. XRD patterns of: a) graphite; b) the CNTs.

Figure 6a showed Raman spectrum of the CNTs. The strongest peak at 1580 cm^{-1} may be ascribed to the Raman-active E_{2g} mode analogous to that of graphite, while the band at 1416 cm^{-1} may be due to the "D-band" induced by defects (*13-14*), indicating that the long-range-order degree of the arrangement of carbon atoms at the surface of CNTs is not as high as that of graphite; the bands at 2832, 2996, 3160 cm^{-1} and the 4412 cm^{-1} region may be due to the second- and third-order combination frequencies: 2D, D+E_{2g}, 2E_{2g}, and 2D+E_{2g}, respectively (*14*).

Raman spectra also provide significant information of hydrogen adsorption on the CNTs. On the H_2/CNTs adsorption system, the observed Raman peaks at 2856, 2967 and 3950 cm^{-1} (Fig. 6b) could be assigned to symmetric C-H stretch of surface $\underline{C}H_2$, asymmetric C-H stretch of surface $\underline{C}H_3$, and H-H stretch of molecularly adsorbed H_2(a), respectively, implying that adsorption of H_2 on the CNTs could occur in associative and dissociative forms (*14*).

The spectra of O_2-TPO (temperature-programmed oxidation with a 5v%O_2-He gaseous mixture as oxidant) showed that the carbon nanostructures in the CNT sample were predominantly graphite-like, with the corresponding main

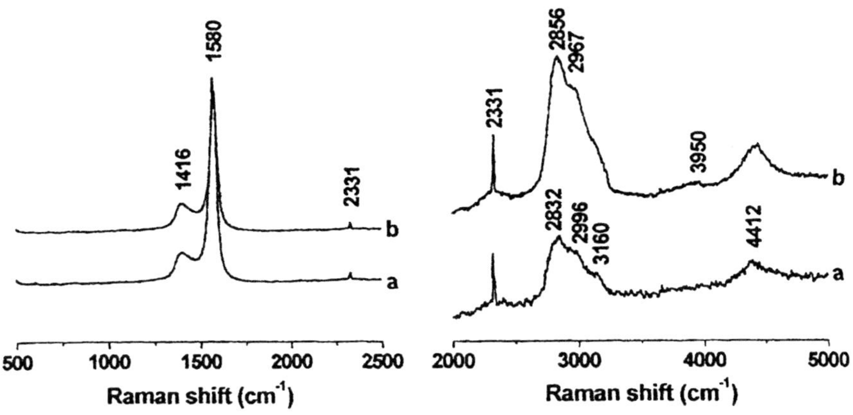

Figure 6. Raman spectra of: a) the CNTs; b) the CNTs adsorbing H_2.

oxidation peak at 960-1000 K. The content of amorphous carbon (the oxidation peak at ~700K) was very low (less than 10% estimated).

It was indicated by the N_2-BET measurements that the specific surface area (SSA) of this type of CNTs was ~135 m^2g^{-1} (by comparison, the SSA of the AC and γ-Al_2O_3 was 592 and 187 m^2g^{-1}, respectively). The size of their inner-diameter was in the range of 2.6-3.4 nm, which is consistent with that estimated from its HRTEM image (*11*).

TPR, XRD and XPS Characterization of the Catalysts

The H_2-TPR spectra of the oxidation precursors of the three supported catalysts were shown in Figure 7. The H_2-reduction of 32%$Cu_6Zn_3Al_1$/CNTs sample started at ~463 K, with the main H_2-TPR peak centered at around 530 K. The two samples supported by AC and γ-Al_2O_3 were not reduced by H_2 until ~500 K, and each reached the reduction-peak round 578 K, which was ~48 degrees (K) higher than that of the CNT-supported system. The observed reducibility sequence of the three samples was as follows: $Cu_6Zn_3Al_1$/CNTs >> $Cu_6Zn_3Al_1$/AC ≥ $Cu_6Zn_3Al_1$/γ-Al_2O_3, in line with the sequence of their catalytic activity for the methanol synthesis.

Figure 8 shows the XRD patterns of the 32%$Cu_6Zn_3Al_1$/CNTs system in three different states, in which a strong graphite-like feature due to the CNTs appeared at 2θ = 26.1°. The features at 2θ = 35.8° and 39.5° ascribed to CuO-phase could be easily identified for the oxidized sample. With the oxidized sample being reduced, a weak but distinguishable peak ascribed to Cu_2O-phase appeared at 2θ = 36.4° in addition to the two peaks at 2θ = 43.2° and 31.7° due to Cu^0 and ZnO phases, respectively. On the functioning catalyst, strong Cu^0 peaks (2θ = 43.2°, 50.4°, and 74.4°) and a weak Cu_2O peak (2θ = 36.4°) were

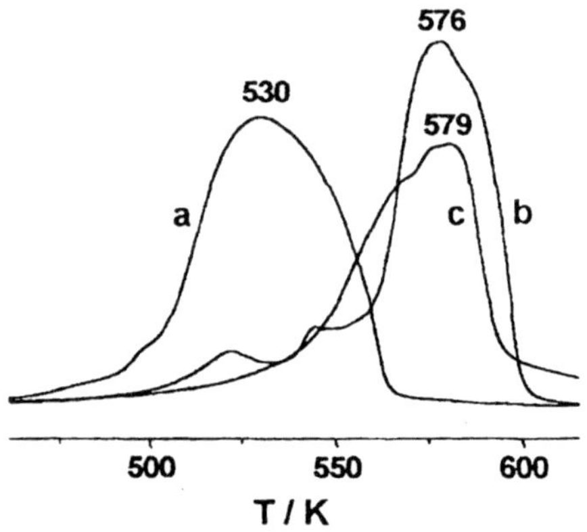

Figure 7. H_2-TPR spectra of the supported catalyst precursors: a) 32%$Cu_6Zn_3Al_1$/CNTs; b) 32%$Cu_6Zn_3Al_1$/AC; c) 32%$Cu_6Zn_3Al_1/\gamma$-Al_2O_3.

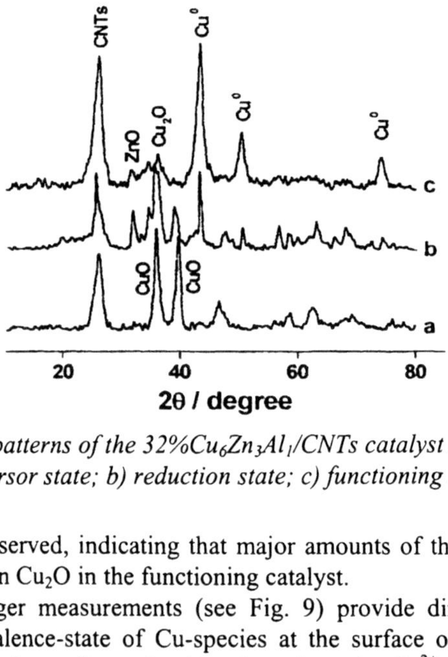

Figure 8. XRD patterns of the 32%$Cu_6Zn_3Al_1$/CNTs catalyst at: a) oxidation precursor state; b) reduction state; c) functioning state.

simultaneously observed, indicating that major amounts of the Cu-species was in Cu^0 and minor in Cu_2O in the functioning catalyst.

The XPS-Auger measurements (see Fig. 9) provide direct experimental evidence about valence-state of Cu-species at the surface of 32%$Cu_6Zn_3Al_1$/ CNTs catalyst. At the surface of oxidized catalyst, the Cu^{2+}-species existed in two different microenvironments, with the corresponding $Cu^{2+}(2P_{3/2}/2P_{1/2})$= 933.6/953.6 and 943.9/963.0 eV(B.E.), respectively; on the catalyst prereduced by 5%H-N_2 at 503 K, there existed the surface Cu-species in mixed valence-states, relatively large amounts in Cu^0 ($Cu^0(2P_{3/2}/2P_{1/2})$= 932.6/952.5 eV(B.E.)) and minor in Cu^+ ($Cu^+(2P_{3/2}/2P_{1/2})$= 943.0/962.2 eV(B.E.)), with the latter probably located at the Cu/Zn interphasial sites; while on the functioning catalyst, the concentration of surface Cu^+ was under the XPS detection limit, and Cu^0 was the dominant surface Cu-species, with $Cu^0(2P_{3/2}/2P_{1/2})$= 932.7/952.7 eV(B.E.) and Cu^0(LMM)= 568.3 eV(B.E.) (*15*).

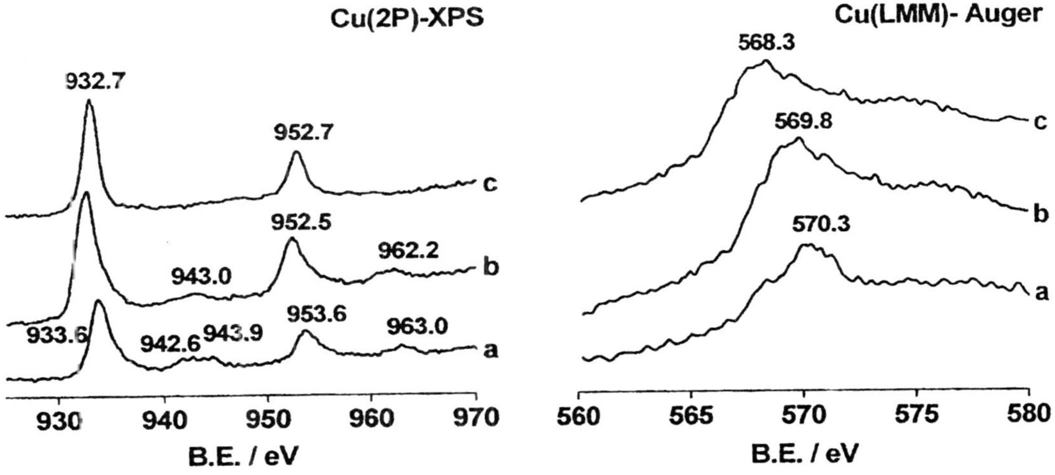

Figure 9. Cu(2P)-XPS-Auger spectra of the 32%$Cu_6Zn_3Al_1$/CNTs catalyst at: a) oxidaiion precursor state; b) reduction state; c) functioning state.

TPD Characterization of Hydrogen Adsorption on the CNTs and Catalysts

Figure 10 (A) showed the TPD spectrum of the CNT sample adsorbing H_2 (99.999% purity) at room temperature. With increasing temperature, the first main TPD peak was present at ~400 K, followed by a shoulder-peak at ~485 K and the second main peak at 990 K. Further analysis by GC and MS indicated that the desorbed product was almost exclusively gaseous hydrogen at temperatures lower than ~693 K. At temperatures higher than ~693 K, the product contained CH_4, C_2H_4 and C_2H_2, in addition to a considerable amount of H_2. This result suggested that H_2 adsorption on this type of CNTs may be in the two forms: associative (molecular state) and dissociative (atomic state), in line with the results of Raman characterization of the H_2/CNTs adsorption systems described above and Ref. (*14*).

Figure 10 (B) showed the TPD spectra of hydrogen adsorbed on the pre-reduced catalysts supported by the CNTs, AC and γ-Al_2O_3, respectively. Overall, each spectrum involved a lower-temperature peak (peak-I) centered round 383 K and a higher-temperature peak (peak-II) spanning from 453 K to 773 K or higher. The low-temperature peaks resulted from the desorption of weakly adsorbed species, most probably molecularly adsorbed hydrogen H_2(a), and the high-temperature peaks were attributed to strongly adsorbed species, perhaps dissociatively adsorbed hydrogen H(a). The relative area-intensities of the peak-I and II for these catalyst samples were estimated, and the results were shown in Table II. It is conceivable that, at the temperatures for methanol synthesis (453~563 K for the present work), the concentration of surface hydrogen ad-species associated with peak-I was expected to be very low, and most of hydrogen adspecies at the surface of functioning catalysts was those corresponding to peak-II. It was probably those strongly adsorbed hydrogen species that was closely associated with the reaction activity of methanol synthesis. The ratio of relative area-intensities of the peak-II for these catalyst samples was (32%$Cu_6Zn_3Al_1$/CNTs) / (32%$Cu_6Zn_3Al_1$/AC) / (32%$Cu_6Zn_3Al_1$/γ-Al_2O_3)=100/63/35 (see Table II). This suggested that the sequence of increasing concentration of hydrogen-adspecies at the surface of functioning catalysts was: 32%$Cu_6Zn_3Al_1$/CNTs > 32%$Cu_6Zn_3Al_1$/AC > 32%$Cu_6Zn_3Al_1$/γ-Al_2O_3, in line with the observed sequence of catalytic activity of these catalysts for the methanol synthesis.

Nature of the CNT Promoter Action

The results of the surface area measurements (see Table I) showed that Cu exposed area of the 32%$Cu_6Zn_3Al_1$/CNTs catalyst reached 22.9 m^2g^{-1}, which was 40% and 76% higher than those (i.e., 16.4 and 13.0 m^2g^{-1}) of the corresponding AC- and γ-Al_2O_3-supported catalysts, respectively, though its SSA was only 48% and 72% of those of the corresponding systems supported by AC and γ-Al_2O_3, respectively. To be sure, the increment of Cu surface area was undoubtedly in favor of enhancing the specific activity of the catalyst.

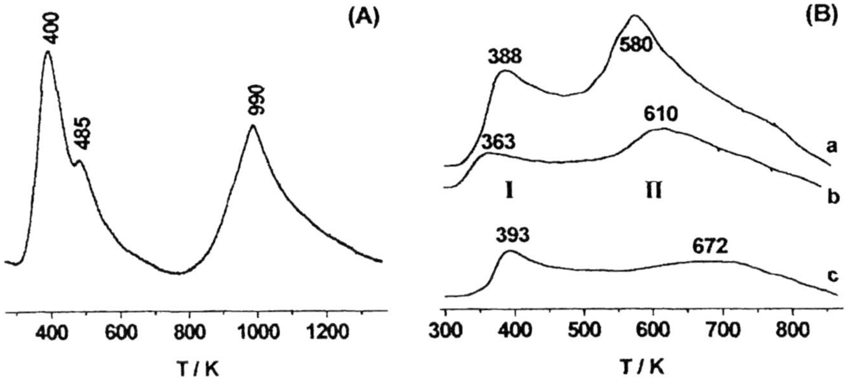

Figure 10. TPD spectra of hydrogen adsorbed on: A) the CNTs; and B) the pre-reduced supported catalysts: a) 32%$Cu_6Zn_3Al_1$/CNTs; b) 32%$Cu_6Zn_3Al_1$/AC; c) 32%$Cu_6Zn_3Al_1$/γ-Al_2O_3.

Table II. Relative area-intensities of the TPD peaks I and II for the 32%$Cu_6Zn_3Al_1$/CNTs (or AC or γ-Al_2O_3) catalysts

Catalyst sample	*Relative area-intensity**	
	Peak-I	*Peak-II*
32%$Cu_6Zn_3Al_1$/CNTs	55	100
32%$Cu_6Zn_3Al_1$/AC	16	63
32%$Cu_6Zn_3Al_1$/γ-Al_2O_3	19	35

*With area-intensity of the strongest Peak-II as 100.

Nevertheless, it would be difficult to be convinced that the 91% or 152% increase in methanol-STY (i.e., 287 vs. 150 mg h^{-1}(g-catal.)$^{-1}$ for the 32%$Cu_6Zn_3Al_1$/CNTs vs. 32%$Cu_6Zn_3Al_1$/AC, or 287 vs. 114 mg h^{-1}(g-catal.)$^{-1}$ for 32%$Cu_6Zn_3Al_1$/CNTs vs. 32%$Cu_6Zn_3Al_1$/γ-Al_2O_3) was solely attributed to the difference in their Cu surface area. Besides, the difference in the surface areas could hardly rationalize the 38% and 42% increase of the intrinsic methanol-formation rate, i.e., from 0.079 and 0.077 μmol-CH_3OH s^{-1}(m^2-surf. Cu)$^{-1}$ for AC- and Al_2O_3-supported systems, respectively, going up to 0.109 μmol-CH_3OH s^{-1}(m^2-surf. Cu)$^{-1}$ for the CNT-supported system (see Table I).

Thus, it appears that the high activity of the CNT-supported catalyst for methanol synthesis is also closely associated with the peculiar structure and properties of the CNTs as carrier. In view of chemical catalysis, in addition to its high mechanical strength, nanosize channel, sp^2-C constructed surface and graphite-like tube-wall, the excellent performance of the CNTs in hydrogen-adsorption/storage is also very attractive. It could be inferred from the above TPD investigations that there would exist a considerably greater amount of hydrogen-adspecies on the CNT carrier under the condition of methanol synthesis used in the present study. This would lead to higher stationary-state concentration of active hydrogen-adspecies on the surface of the functioning catalyst, and thus favor a series of surface hydrogenation reactions in the process of CO/CO_2 hydrogenation to methanol. On the other hand, the operation temperature of the catalyst supported by CNTs for methanol synthesis can be 20 or 40 K lower than that of the corresponding system supported by AC or γ-Al_2O_3. This would contribute considerably to an increase in equilibrium conversion of CO and space-time-yield of CH_3OH. The results of the present study indicated that the CNTs served not only as carrier, but also as an excellent promoter, and that its unique feature of adsorbing H_2 may play an important role in effectively promoting the methanol synthesis. The probable effect of promoting hydrogen spillover is currently under investigation.

Preparation of Highly Active CNT-promoted Coprecipitated Catalyst

Making use of the promoter action by the CNTs, highly active and practical CNT-promoted catalysts, $Cu_iZn_jAl_k$-y%CNTs, have been designed and prepared by coprecipitation method, and their catalytic performance for methanol synthesis from $H_2/CO/CO_2$ was evaluated and compared with that of the corresponding CNT-free coprecipitated $Cu_iZn_jAl_k$ catalyst. It is experimentally found that appropriate incorporation of a minor amount of the CNTs into the $Cu_iZn_jAl_k$ can significantly advance the catalyst activity for methanol synthesis. Over a coprecipitated $Cu_6Zn_3Al_1$-12.5%CNTs catalyst under the reaction conditions of 493 K, 5.0 MPa, $H_2/CO/CO_2/N_2$=62/30/5/3 (v/v), GHSV= 8000 ml (STP) h^{-1} (g-catal.)$^{-1}$, the observed intrinsic reaction rate and STY of methanol synthesis reached 0.291μmol-CH_3OH s^{-1} (m^2-surf. Cu)$^{-1}$ and 1320 mg-CH_3OH h^{-1} (g-catal.)$^{-1}$, respectively, which were ~25% and ~ 40% higher than those (i. e., 0.233 μmol-CH_3OH s^{-1} (m^2-surf. Cu)$^{-1}$ and 916 mg h^{-1} (g-catal.)$^{-1}$) over the corresponding CNT-free coprecipitated $Cu_6Zn_3Al_1$ catalyst, respectively. Meanwhile, the operation temperature for methanol synthesis over the CNT-promoted coprecipitated catalyst can be 15~25 degrees (K) lower than that over the corresponding CNT-free catalyst. The work is in progress.

Acknowledgments

The authors are grateful for the financial supports from the National Natural Science Foundation (Project No. 50072021) and Fujian Provincial Natural Science Foundation (Project No. 2001H017) of China.

References

1. MacDougall, L. V. *Catal. Today* **1991**, *8*, 337.
2. Klier, K. *Adv. Catal.* **1982**, *31*, 243.
3. Fleisch, T. H.; Mieville, R. L. *J. Catal.* **1984**, *90*, 165.
4. Chinchen, G. C.; Waugh, K. C.; Whan, D. A. *App. Catal.* **1986**, *25*, 101.
5. Herman, R. H. *Stud. Surf. Sci. Catal.* (Ed. Guczi, L.) Amsterdam, Elsevier, **1991**, *64*, 265.
6. De Jong, K. P.; Geus, J. W. *Catal. Rev.-Sci. Eng.* **2000**, *42*, 481.
7. Planeix, J. M.; Coustel N.; Coq B.; Brotons, V.; Kumbhar, P. S.; Dutartre, R.; Geneste, P.; Bernier, P.; Ajiayan, P. M. *J. Am. Chem. Soc.* **1994**, *116*, 7935.
8. Hoogenraad, M. S.; Onwezen, M. F.; van Dillen, A. J.; Geus, J. W. *Stud. Surf. Sci. Catal.* (Eds. Hightower, J. W., Delgass, W. N., Iglesia, E., and Bell, A. T.) Amsterdam, Elsevier, **1996**, *101*, 1331.
9. Zhang, Y.; Zhang, H. B.; Lin G. D.; Chen, P.; Yuan, Y. Z.; Tsai, K. R. *Appl. Catal. A: General* **1999**, *187*, 213.
10. Zhang, H. B.; Zhang, Y.; Lin, G. D.; Yuan, Y. Z.; Tsai, K. R. *Stud. Surf. Sci. Catal.* (Eds. Corma, A., Melo, F. V., Mendioroz, S., and Fierro, J. L. G.) Amsterdam, Elsevier, **2000**, *130*, 3885.
11. Chen, P.; Zhang, H. B.; Lin, G. D.; Hong, Q.; Tsai, K. R. *Carbon* **1997**, *35*, 1495.
12. Bond, G. C.; Namuo, S. N. *J. Catal.* **1989**, *118*, 507.
13. Eklund, P. C.; Holden, J. M.; Jishi, R. A. *Carbon* **1995**, *33*, 959.
14. Zhang, H. B.; Lin, G. D.; Zhou, Z. H.; Dong, X.; Chen, T. *Carbon* **2002**, *40*, 2429.
15. Marisa, S. *Chem. Phys. Lett.* **1979**, *63*, 52.

Biochemical Fixation of CO_2

Chapter 14

Enzymatic Conversion of Carbon Dioxide to Methanol by Dehydrogenases Encapsulated in Sol–Gel Matrix

Zhongyi Jiang, Hong Wu, Songwei Xu, and Shufang Huang

State Key Laboratory of C_1 Chemistry and Chemical Technology and School of Chemical Engineering and Technology, Tianjin University, Tianjin 300072, China

The effective generation of methanol directly from gaseous carbon dioxide has received considerable attention in recent years since it can recycle the greenhouse gas and produce a clean fuel. Herein, we report an enzymatic approach for carbon dioxide fixation using formate dehydrogenase ($F_{ate}DH$), formaldehyde dehydrogenase ($F_{ald}DH$) and alcohol dehydrogenase (ADH) co-encapsulated in a silica gel as the catalysts. The gels were prepared by a modified sol-gel process that uses tetramethoxysilane (TMOS) as the precursor and nicotinamide adenine dinucleotide (NADH) as an electron donor. The enzymatic conversion of CO_2 to methanol was carried out at low temperatures and low pressures. The effects of the reaction temperature, pH value, the amount of enzyme and the amount of NADH on the yields of methanol have been investigated. The highest yield of methanol was 92.1%. The activity of immobilized enzymes was a little lower than that of the free enzymes due to the minor conformation change of the enzyme and the existence of additional diffusion hindrance.

Introduction

Effective generation of methanol directly from carbon dioxide has received considerable attention in recent years since it can recycle the greenhouse gas and produce a clean fuel. Hydrogenation of carbon dioxide through heterogeneous catalysis, electrocatalysis and photocatalysis has been the predominant methods investigated (*1-4*).

Herein, we report a novel and promising approach to convert carbon dioxide into methanol through consecutive reduction reactions catalyzed by three different dehydrogenases, which were encapsulated in silica gel matrices (*5,6*). The whole process consists of three steps (*7-9*): reduction of CO_2 to formate catalyzed by formate dehydrogenase ($F_{ate}DH$), reduction of formate to formaldehyde by formaldehyde dehydrogenase ($F_{ald}DH$), and reduction of formaldehyde to methanol by alcohol dehydrogenase (ADH). Reduced nicotinamide adenine dinucleotide (NADH) acts as a terminal electron donor for each of the dehydrogenase-catalyzed reductions:

$$CO_2 \xrightarrow[NADH]{F_{ate}DH} HCOOH \xrightarrow[NADH]{F_{ald}DH} HCHO \xrightarrow[NADH]{ADH} CH_3OH$$

The enzymatic conversion of CO_2 to methanol was carried out at low temperatures and low pressures. The suitable procedures for the efficient sol-gel co-encapsulation of the three dehydrogenases have been studied. The effects of the reaction temperature, pH, the amount of enzyme and the amount of NADH on the yield of methanol have been investigated. The highest yield of methanol was 92.1%. The activity of the immobilized enzymes was a little lower than that of the free enzymes. This is due to the minor conformation change of the enzymes and the existence of additional diffusion hindrance.

Experimental

Preparation of Immobilized Enzymes

The sol-gel encapsulated enzymes were prepared as follows: Tetramethoxysilane (TMOS) was used as the precursor for making the silica sol-gel. The initial sol was prepared by mixing 1.94 g of TMOS and 1.10 g of 3%(v/v) HCl solution. The mixture was then vigorously mixed for 10 min to form a sol.

The $1^{\#}$ gels were prepared by adding 1.0 mL of the stock enzyme solution, which was comprised of 7.0 mg of $F_{ate}DH$, 2.0 mg of $F_{ald}DH$ and 2.0 mg of ADH, to 1.0 mL of the sol in a polystyrene cuvette. Typical gelation times were on the order of 50-60 s. The cuvette was then covered with parafilm and the gel was allowed to age at 4 ℃ for 24-48 h.

The $2^{\#}$ gels were prepared by adding 1.0 mL of the stock enzyme solution, which was comprised of 14.0 mg of $F_{ate}DH$, 4.0 mg of $F_{ald}DH$ and 4.0 mg of

ADH, to 1.0 mL of the sol in a polystyrene cuvette. Typical gelation times were on the order of 5-10 s. The cuvette was then covered with parafilm and the gel was allowed to age at 4 ℃ for 24-48 h.

Characterization

The size of gel particles was determined using a Materizer 2000 instrument (Malvern Instruments Ltd., UK).

All adsorption measurements of the dehydrogenases encapsulated in the aqueous TMOS-based silica gels were made with an UV-3010 Spectrometer (Hitachi, Japan). The pore size and surface areas were obtained using a CHEMBET-3000.

Enzymatic Reactions

The aged gels were put into a dialysis membrane using 250 mL of 0.1 M phosphate buffer at pH of 7 as the dialysis solution, and then placed in a refrigerator at 4 ℃ with frequent change of the 0.1 M phosphate buffer. The dialysis lasted 24-48 h in order to completely remove the methanol or ethanol generated in the sol-gel process. Through dialysis 0.1 mL of NADH solution diffused into the gel (the final concentration of NADH varied from 0.025 to 0.1 M). The sample containing the gel and the NADH solution was left undisturbed for 48 h. CO_2 was then bubbled for 8 h through this mixture in order to produce methanol. The concentration of methanol produced was determined by gas chromatography (HP 6890). A calibration curve was established for aqueous methanol solutions at known concentrations of methanol.

Results and Discussions

Characterization of 1# Gel and 2# Gel

Table 1 gives the values of the relative absorbance (k) of the spectra and the corresponding spectral shift ($\Delta\lambda$) for the 1# gel and the 2# gel. Both k and $\Delta\lambda$ were determined based on the characteristic absorption peak of the corresponding dehydrogenases in buffer solution, relative to those in the gels. Dehydrogenases encapsulated in TMOS-based gels exhibited no or little spectral shift, indicating that the enzyme conformation was preserved to a considerable extent after encapsulation (*10*). The change in spectra intensity, nevertheless, suggested an effect on the enzyme encapsulation by the sol-gel process (*11,12*).

The surface areas of the 1# gel and the 2# gel were 260 m^2/g and 200 m^2/g, respectively, whereas the surface areas of the plain TMOS gel that contained no dehydrogenases was 280 m^2/g. The small decrease in surface area was probably due to the difference in the microenvironment created by the interactions of the enzymes with the matrices, which involved van der Waals forces and hydrogen bonding. The average pore size of the 1# gel and the 2# gel were 3.4 nm and 2.6 nm, respectively, whereas the average pore size of the plain TMOS gel was 2.4 nm. This shows that dehydrogenases acted as the template molecules to modify the sol-gel process (*13,14*).

Table 1. Comparison of the spectral shift ($\Delta\lambda$, nm) and the relative change of adsorbance intensity (k,%) of dehydrogenases encapsulated in the TMOS-based silica gels

Gel	$\Delta\lambda$	*k*
1#	0	26.5
2#	-0.5	40.8

Enzymatic Reactions

The yields of methanol were calculated based on the amount of NADH. As seen from the reaction scheme mentioned above, 3 mol of NADH are consumed per mol of methanol produced. As such, for a 100% methanol yield, the moles of methanol produced should be 1/3 of the amount of NADG added. Therefore:

Yield of methanol (%)=3×moles of methanol product/initial moles of NADH ×100%

From the preliminary experimental results, the feasibility of the enzymatic conversion of carbon dioxide to methanol has been successfully proved. The production of methanol is due to the enzymatic reactions and not due to the possible hydrolysis of the residual methoxides present in the TMOS sol-gels. This was confirmed by a control experiments performed with plain TMOS sol-gels. The plain TMOS sol-gels, without the encapsulated hydrogenases, did not produce any methanol under the identical reaction conditions. In addition, it was also found that all four species (i.e. F_{ate}DH, F_{ald} DH, ADH, NADH) must be present to generate methanol. This was established by preparing several sol-gel with systematic exclusion of one or more of the four components. It was observed that the sol-gels prepared without any one of the four components failed to produce methanol. The highest yield of methanol using the 1# gel as the catalyst was 92.1%. Under the same reaction conditions, the yield of methanol using dehydrogenases in the free form as the catalyst was 98.1%. The decrease in the activity of the immobilized enzymes was mainly due to the minor conformation change of the enzymes and the extra diffusion hindrance for the substrate and product.

Effects of the Enzyme Amount Encapsulated in the Gels on the Methanol Yields

Table 2 gives the methanol yield comparison for the 1# gel and 2# gel. The reason of the yield of methanol for the 2# gel is lower than that for the 1# gel is explained as follows: for the 2# gel, the gelation time was 5-10 s; for the 1# gel, the gelation time was 50-60 s. The faster gelation induces a more uneven distribution of the hydrogenases in the gel and also an un-favored mass transfer due to the decrease in pore size and surface area.

Table 2. Effects of enzyme amount encapsulated in the 1# gel and the 2# gel on the enzymatic conversions of CO_2

Gel	*pH*	*Temperature/℃*	*NADH amount Mmol*	*Pressure /Mpa*	*Yield /%*
1#	7.0	37	100	0.3	92.1
2#	7.0	37	100	0.3	52.3
1#	7.0	37	150	0.3	42.2
2#	7.0	37	150	0.3	35.1

Effects of Reaction Temperatures on the Methanol Yields

The three dehydrogenases have different optimum reaction temperatures. The optimum temperatures were 37 ℃ for $F_{ate}DH$ and $F_{ald}DH$, whereas the optimum temperatures for ADH was 25 ℃. The experiments were carried out at 25 ℃ and 37 ℃ in order to find the optimum temperature for the three enzyme co-immobilized in the same gel. The experimental results are shown in Table 3. They showed that the better reaction temperature is 37 ℃.

Table 3. Effects of temperature on the enzymatic conversions of CO_2

Temperature /℃	*PH*	*NADH amount Mmol*	*Pressure /Mpa*	*Yield /%*
25	7.0	100	0.3	30
37	7.0	100	0.3	92.1
25	7.0	150	0.5	27
37	7.0	150	0.3	42.2

Effects of pH Values on the Methanol Yields

Table 4 shows the effect of pH on the methanol yield. The highest methanol yield was obtained at pH=7.0. According to the tentative analysis of the reaction mechanism for dehydrogenase, the reduction uses NADH as the electron donor and needs the existence of a proton. The lower pH was more favorable for protonation and thus more favorable for the increase in methanol yield.

Table 4. Effects of pH on the enzymatic conversions of CO_2

Temperature /℃	*PH*	*NADH amount Mmol*	*Pressure /Mpa*	*Yield /%*
37	7.0	100	0.3	92.1
37	7.5	100	0.3	66.9
37	8.0	100	0.3	49.5

Effect of the Amount of NADH on the Methanol Yield

The effects of the amount of NADH on the methanol yields are shown in Table 5. It is important to note that the overall yield of the enzymatic reactions was lower at higher concentrations of NADH. This is presumably due to an increased tendency of this system to undergo the reverse reaction (i.e., conversion of methanol to carbon dioxide)(*15,16*).

Table 5. Effects of the amount of NADH on the enzymatic conversions

NADH amount Mmol	*Temperature /℃*	*pH*	*Pressure /MPa*	*Yield /%*
50	25	8.0	0.3	68.8
100	37	8.0	0.5	49.5
150	37	8.0	0.5	32.0

Conclusions

The feasibility of enzymatic conversion of carbon dioxide to methanol is tentatively explored. The consecutive reduction of carbon dioxide by three different dehydrogenases encapsulated in a TMOS sol-gel matrix results in a high yield of methanol. This will open up a new avenue not only for on-site production of methanol from cheap raw materials but also for the efficient fixation of the greenhouse gas carbon dioxide.

Acknowledgement

The financial supports from the National Natural Science Foundation of China (under the contract 20176039) and from the 985 Project of State Key Laboratory of C1 Chemistry and Technology of Tianjin University are greatly appreciated.

References

1. Wieser, M.; Yoshida, T.; Nagasawa, T. *J. Mol Cata. B: Enzymatic.* **2001**, *11*, 179-184.
2. Kuwabata, S.; Tsuda, R., Yoneyama, H. *J. Am Chem. Soc.* **1994**, *116*, 5437-5443.
3. Heleg, V.; Willner, I., *Chem. Commun.* **1994**, *18*, 2113-2114.
4. Aresta, M.; Quaranta, E.; Liberio, R. *Trtrahedron* **1998**, *54*, 8841-8846.
5. Obert, R.; Dave, B.C. *J. Am. Chem. Soc.* **1999**, *121*, 12192-12193.
6. Kumar, A.; Malhotra, R.; Malhotra, B.D. *Anayltica Chimica Acta* **2002**, *414*, 43-50.
7. Gill, I.; Ballesteros, A. *Trends in Biotechnology* ,**2000**,*18*, 282-296.
8. Kutzenko, A. S.; Lamzin, V. S.; Popov, V. O. *FEBS Letters* **1998**, *423*, 105-109.
9. Palmore, G.T. R.; Bertschy, H.; Bergens, S. H. *J. Electroanalytical Chem.* **1998**, *443*, 155-161.
10. Pocker, Y.; Page, J.D.; Li, H. *Chemical-Biological Interactions* **2001**, *130-132*, 371-381.
11. Dunn, B.; Miller, J. M.; Dave, B.C. *Acat mater.* **1998**, *46*, 737-741.
12. Bhatia, R.B.; Brinker, C.J. *Chem.Mater.* **2000**,*12*, 2434-2441.
13. Liu, D. M.; Chen, I-W. *Acta mater.* **1999**, *47*, 4535-4544.
14. Wei, Y.; Xu, J-G.; Feng, Q-W. *Mater. Lett.* **2000**, *44*, 6-11.
15. Li, C-I.; Lin, Y-H; Shih, C-L. *Biosensors and Bioelectronics* **2002**,*17*, 323-330.
16. Pocker, Y. *Chemical-Biological Interactions* **2001**, *130-132*, 383-393.

Chapter 15

Toward Solar Energy Conversion into Fuels: Design and Synthesis of Ruthenium–Manganese Supramolecular Complexes to Mimic the Function of Photosystem II

Licheng Sun[1,*], Björn Åkermark[1], Leif Hammarström[2], and Stenbjörn Styring[3]

[1]Department of Organic Chemistry, Arrhenius Laboratory, Stockholm University, 106 91 Stockholm, Sweden
[2]Department of Physical Chemistry, Uppsala University, P.O. Box 532, 751 21 Uppsala, Sweden
[3]Department of Biochemistry, Centre for Chemistry and Chemical Engineering, University of Lund, P.O. Box 124, S–221 00 Lund, Sweden
*Corresponding author: email: licheng.sun@organ.su.se

To mimic the electron donor side of Photosystem II (PSII), a number of supramolecular model complexes have been designed and synthesized. Ruthenium(II) *tris*-bipyridyl complexes have been used in most cases as photosensitizers, mimicking the function of P_{680} in PSII. As electron donors, monomeric and dimeric manganese complexes and tyrosine have been introduced into the supramolecular systems, modelling the Mn cluster and $Tyrosine_Z$ respectively in PSII. For monomeric manganese complexes, di-, tri- and tetra-dentate ligands have been linked to a Ruthenium(II) *tris*-bipyridyl type complex; and for dimeric manganese complexes hepta-dentate ligands containing pyridines have been used. Some related ligands, where two pyridines have been replaced by phenolate groups have also been synthesized, in order to get ligands that can stabilize manganese complex in high valence states. Photophysical and photochemical studies showed that the electron transfer rate from monomeric Mn complex to photo-generated Ru(III) was low when Mn-Ru distance was long, while the electron transfer rate was enhanced when the Mn-Ru distance was short. However, the excited state of ruthenium complex was quenched if Mn got close to the Ru, leading to a short lifetime. By synthetically inserting a tyrosine unit between Ru and Mn moieties, quenching of the excited state lifetime of Ru was reduced, and the electron transfer from Mn to Ru(III) was very fast although the Mn-Ru distance was long. These supramolecular Ru-Mn systems are closely modelling the electron donor side of PSII both functionally and structurally. The design and synthesis of these model systems are summarized and discussed in this chapter.

1. Introduction

Photosynthesis is one of the most complicated biosystems in Nature. Solar energy is converted via photosynthesis into chemical energy. This energy is then used by all other organisms including human beings to develop and sustain life. However, the total amount of solar energy that is stored on the earth after several billion years photosynthesis is limited. This kind of energy in forms of fossil fuels such as coal, oil, natural gas, etc, is rapidly being consumed, and in the process, gives rise to environmental problems. In the long run, the best way to solve the problem of future energy is to use the "clean fuels" such as hydrogen or methanol. There are many ways to produce hydrogen or methanol. One way is to use solar energy to split water or to reduce CO_2 photochemically.[1] Research on photochemical water splitting or CO_2 reduction has been conducted for many years in many countries. Some achievements have been made, but they are mainly based on using UV light. The technical obstacle is, among others, to draw electrons from water, and to use the electrons to reduce protons in water to hydrogen or to reduce the CO_2 in the air(green house gases) to methanol. The whole process should ultimately be driven by visible light. To overcome the difficulties, a new concept to oxidize water was introduced recently. This concept was based on the principles of natural Photosystem II(PSII).

PSII is the key enzyme in photosynthesis in plants, algae and cyanobacteria.[2] It is the only photosynthetic reaction centre (RC) that can use water as an electron source to provide the plants with an unlimited source of electrons. In addition to common reactions of all photosynthetic reaction centers, PSII has the ability to use visible light to drive electron transport from water to a quinone acceptor (plastoquinone). The reduced plastoquinone subsequently carries the reducing equivalents further in the photosynthetic electron transfer chain. This provides reducing power to many biosynthetic reactions in the plants. CO_2 in the air is reduced to biomass by the electrons derived from PSII. As a waste product, water oxidation in PSII forms molecular oxygen which is the prerequisite for the existence of most higher life forms, placing the PSII reaction centre in a central position in the biosphere.

The structure of PSII has very recently been solved at a resolution of 3.8 Å.[2] A schematic presentation of its function is shown in Figure 1. On the electron donor side of PSII, the key components are chlorophylls P_{680}, a redox active amino-acid $Tyrosine_Z$, and a Mn-cluster composed of four Mn ions of high valence (the oxygen evolving complex, OEC),[3,4] see Figure 1. The structure of OEC is far from clear, it was believed that a Cl^- and a Ca^{2+} ion are also involved and tightly coupled to the Mn ions.

When PSII is irradiated with solar light, P_{680} is excited, and transfers an electron to the acceptors (pheophytin and two quinones Q_A and Q_B).[5] The photooxidized primary electron donor, P_{680}^+, one of nature's most oxidizing species with a potential of ~+1.2V vs NHE, is rapidly reduced by $Tyrosine_Z(Tyr_Z)$. Tyr_Z is a very fast electron donor to P_{680}^+, so it can prevent the back reaction from the quinone radical after the charge separation on the acceptor side. The oxidised Tyr_Z radical is then reduced by an electron from the Mn-cluster. After four photo processes, four electrons are drawn from the Mn-cluster, and two water molecules are subsequently oxidised to molecular oxygen with the release of four protons.

In recent years, much attention has been focused on the molecular catalysts for water oxidation.[6] There are two main reasons for this interest. The first is to get a clear understanding of the mechanism in water oxidation in PSII.[3] The second is to synthesize artificial molecular catalysts for water oxidation.[7] The latter can be used to drawn electrons from water and further utilize the electrons for fuel production, such as the reduction of water to hydrogen or the reduction of carbon dioxide to methanol. The best energy source to drive this process is the solar light. An idealized system for the light induced water splitting was proposed by V. Balzani et al.[8] A similar system for photochemically CO_2 reduction is shown in Figure 2.

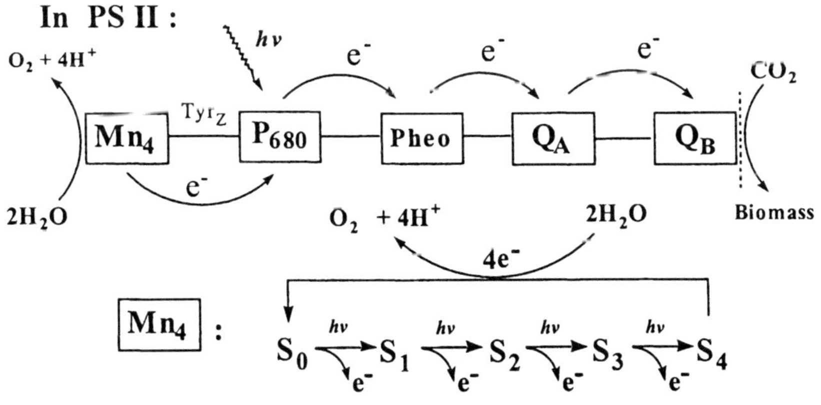

Figure 1. Schematic presentation of the main components in functional blocks and the electron transfer processes involved in the natural PSII reaction centre(top), and the five oxidation states involved in the water oxidation cycle(bottom).

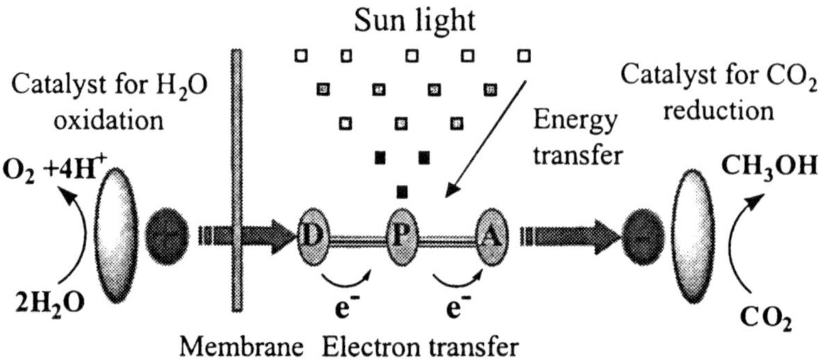

Figure 2. Schematic presentation of the energy and electron transfer processes in an artificial photosynthetic system. Photo-chemical reduction of carbon dioxide is illustrated. P: photosensitizer; D: electron donor; A: electron acceptor.

An ideal artificial system for photochemically CO_2 reduction

In Figure 2, a photosensitizer(P) is connected to an electron donor(D) and an electron acceptor(A). P is also coupled to an attenna system which can harvest the solar light and converts to the reaction center P. After photoinduced electron transfer processes, the oxidized electron donor transfers the charge to a catalyst which can oxidize water into molecular oxygen, while the reduced electron acceptor transfers the electron to another catalyst which can reduce CO_2 to methanol. The construction of such a complicated system is a great challenge for both chemists and material scientists. So far, few examples have been reported.[1]

In the past twenty years much effort has been devoted to studies of photoinduced electron transfer (ET) reactions from chlorophyll and analogues to acceptors, e.g. to mimic **the electron acceptor side** of Photosystem II, for example, porphyrin(s)-quinone(s), porphyrin-fullerence or transition metal complex systems[5,9,10]. In contrast, the regeneration of chlorophyll and analogues from their photo-oxidised forms by electron transfers from manganese or tyrosine, e.g. to mimic **the electron donor side** of Photosystem II, has received very limited attention. A few studies have been presented,[11-13] which show that manganese complexes can catalyse the conversion of water into oxygen. However, in these studies the manganese oxidation has not been coupled to photochemical charge separation. In 1994, we launched a project with the major goal of developing an artificial model system for PSII. More specifically, the aim was to construct a supramolecular system by using Nature's principles. Our hope is that such a system can be used to produce a sustainable and renewable fuel, e.g. hydrogen from water, or methanol from CO_2, in a similar way as nature produces biomass.

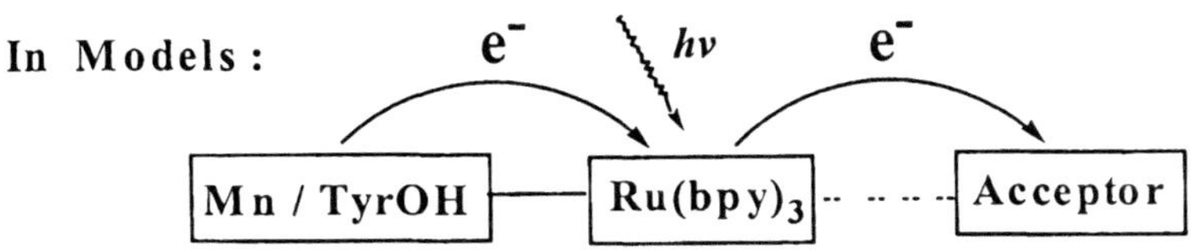

*Figure 3. Building blocks and photo-induced electron transfer processes on model systems. Ruthenium-*tris*-bipyridine is used as photosensitizer to mimic the P_{680}, Mn and/or tyrosine are used as internal electron donors, an external electron acceptor is used in the models.*

The primary charge separation in photosynthesis is the key reaction for life on earth in witch it is the only reaction that converts solar energy into chemical energy. It is also remarkably efficient, and almost all the energy in a photon of 680 nm is conserved in the first charge separated state of Photosystem II. However, the primary radical pair is short lived and will recombine within nanoseconds if the separated charges are not stabilised. This is accomplished by a series of charge stabilizing reactions (Figure 1) where the electron and the electron hole are moved away from each other by a series of highly controlled reactions within the protein.

In natural oxygen producing photosynthesis, the solar energy is converted into chemical energy, by photo-induced reduction of CO_2 using water as electron source. The efficiency in the crucial steps is remarkably high.[3,14] The construction of an artificial system that exactly mimics these reactions remains difficult for chemists.[15-17] In this review we will describe our attempts to synthesize artificial systems that mimic the electron donor side of PSII, one of the most vital parts of natural photosynthesis, the light-driven oxidation of water. We use ruthenium *tris*-bipyridyl complexes as photosensitizers to mimic the function of P_{680} in PSII, since they are stable, relatively easy to functionalize, and have favourable photophysical properties. In addition, their oxidation potentials($Ru^{III/II}$) are very close to P_{680}^{+} in PSII.

2. Synthesis of ruthenium *tris*-bipyridine and manganese dinuclear complexes.

At the beginning of our project, the first question was if it was at all possible to demonstrate an intramolecular electron transfer from a manganese complex to a photo-oxidized ruthenium *tris*-bipyridyl complex, as shown in Figure 4. In this figure, the Ru *tris*-bipyridine moiety in the Ru-Mn dinuclear complex is first irradiated by light, then an electron is transfered from the excited state of Ru complex to an external electron acceptor, such as methylviologen MV^{2+}, forming Ru^{III} and methylviologen radical $MV^{\bullet+}$. The photo-generated Ru^{III} is then reduced back to Ru^{II} by an intramolecular electron transfer from the covalently

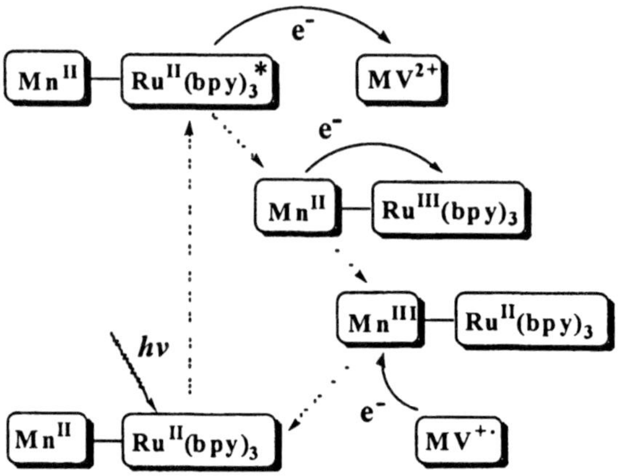

Figure 4. A general presentation of photo-induced electron transfer reactions in supramolecular dinuclear Ru-Mn model complexes. $Ru^{II}(bpy)_3$ is used as photosensitizer, covalently linked Mn^{II} works as electron donor, MV^{2+} works as an external electron acceptor.

linked Mn^{II} complex moiety, before the Ru^{III} and $MV^{\bullet+}$ recombine. During this process, the Mn^{II} complex is photo-oxidized to Mn^{III}. To test the possibility of the above mechanism, a series of Ru-Mn complexes was designed and synthesised.[18-21]

The first Ru-Mn dinuclear complex (**1**)[19] was prepared as shown in Scheme 1. The bis-bipyridine ligand **2** was made by litiation of 4,4'-dimethyl-2,2'-bipyridine with LDA(lithium di-isopropane amine) followed by a reaction with dibromoethane. Reaction of this bis-chelating ligand with $Ru(bpy)_2Cl_2$ in ethanol gave a mixture of the mono Ru complex with a free ligand **3** and a dinuclear Ru complex. Since the dinuclear complex was not the desired product, excess of ligand **2** (4-5 equivalents) was used in the reaction to decrease its formation. The free ligand in complex **3** was used to coordinate a Mn^{II} ion by reaction with $MnCl_2$ in methanol. The coordination of the Mn^{II} ion to the bipyridine ligand is not so strong, therefore an equilibrium between **1** and **3** was found in methanol solution.[22] In general, when a Mn^{II} ion is introduced in a supramolecular complex, little information about the structure from the NMR spectrum can be obtained because Mn^{II} ion has d^5 electron configuration, and is paramagnetic. Since in most cases, the dinuclear complex has at least two counter ions from the Ru moiety, electrospay ionization mass spectrometry (ESI-MS) together with EPR was found to be the best way to characterize the complexes.

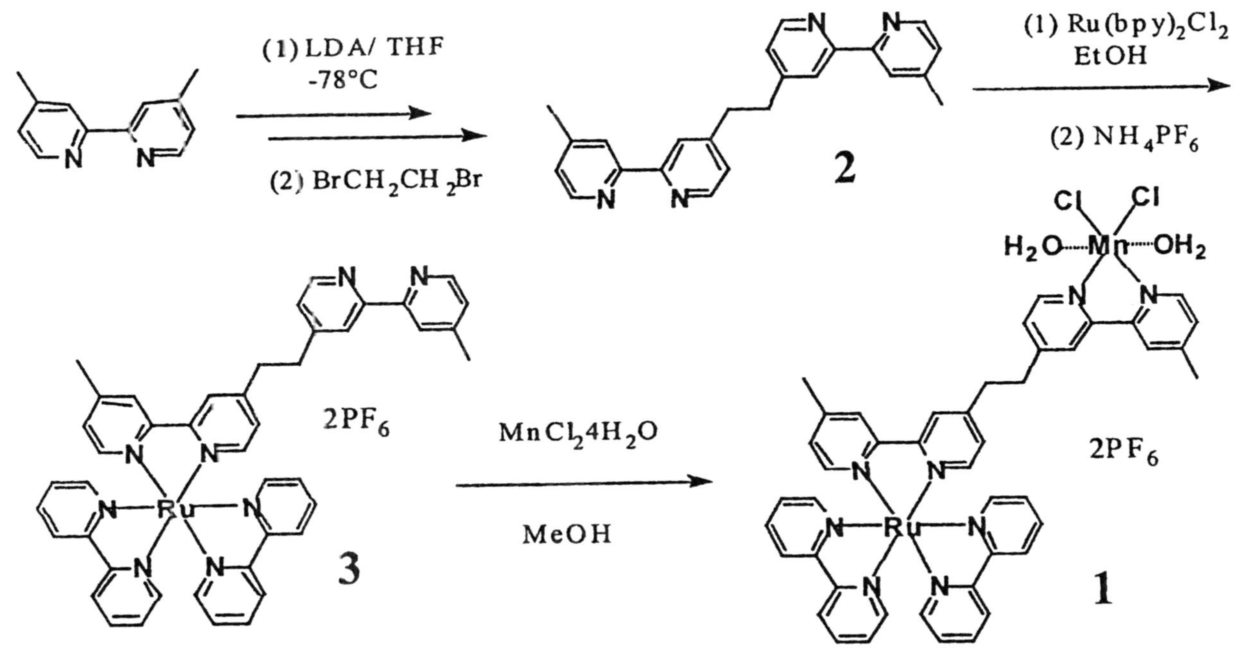

Scheme-1. Synthetic route of complex 1.

The intramolecular electron transfer from the coordinated Mn^{II} to the photo-generated Ru^{III} in acetonitrile solutions of the dinuclear complexes **1** was studied with laser flash photolysis.[22] It was demonstrated that the photo-generated Ru^{III} in the dinuclear complex **1** can indeed oxidize the Mn^{II} in the complex. The first order rate constant for electron transfer from the Mn^{II} was $1.8 \times 10^5\ s^{-1}$.[22] This was the first time a manganese complex was used as an intramolecular electron donor to a photo-oxidised photosensitizer. [19]

Complex **4**[19] was designed with a shorter link between the Ru and Mn ions, and also with dipycolylamine(dpa) as a tri-dentate ligand for Mn ion instead of bipyridine.[19] The synthesis of **4** was started by oxidation of 4,4'-dimethyl-2,2'-bipyridine with selenium dioxide(Scheme-2). The product 4-methyl-4'-formyl-2,2'-bipyridine was reduced to 4-methyl-4'-hydroxymethyl-2,2'-bipyridine by $NaBH_4$ in methanol in quantitative yield. Replacement of hydroxy group with bromide was conducted in concentrated HBr solution (48%), see Scheme-2, giving the ligand 4-methyl-4'-bromo-2,2'-bipyridine **5**. The ligand **5** was also synthesized by the radical substitution reaction of 4,4'-dimethyl-2,2'-bipyridine with NBS. However, the yield was low (ca 30%) and the product mixture was difficult to separate by column chromatography from the other byproducts. There are two possible routes to synthesize the Ru complex covalently linked to dpa ligand from **5**. One is to link the dpa ligand first to **5**, then introduce the Ru complex. This route could give a complicated mixture of products due to the two different coordination sites of the ligands. Therefore, the second route was applied: remove the Cl^- from $Ru(bpy)_2Cl_2$ by reaction with AgOTf, then coordination of the Ru complex with ligand **5** in acetone at r.t., giving 92% yield of the Ru tris-bpy complex **6** with a bromomethyl functional group. Reaction with the nucleophile dpa in acetonitrile in the presence of triethylamine as base gave **7**. It was found that the nucleophilic substitution of **6** with dpa was enhanced in comparison with **5** due to the coordination of Ru complex. This could be explained by the increasing of positive charge on the benzylic carbon. The final product, the dinuclear Ru-Mn complex **4**, was more stable than complex **1** in acetonitrile solution, probably due to the stronger coordination of the tri-dentate ligand dpa than the bi-dentate ligand bpy.

A similar dinuclear complex **8**[21] with a longer Ru-Mn distance was synthesized by insertion of a phenyl group as a bridge between Ru and Mn complex moieties. This was achieved by a multi-step synthesis of the ligand **9**, see Scheme-3. The bromo-substituted ligand **10** was prepared by conventional radical substitution reaction of **9** with NBS. The coordination of Ru to syntheisze complex **11** and the nucleophilic substitution with dpa to form the complex **12** were conducted in the similar way to the complexes **6** and **7**, respectively. In methanol solution, the dpa ligand in Ru complex **12** can coordinate to a Mn(II) ion, forming a Ru-Mn dinuclear complex in satisfactory yield.

Although dpa is a better ligand than 2,2'-bipyridine in terms of binding the Mn(II) ion, dissociation of Mn(II) ion in complexes **4** and **8** was still observed. It has been reported that tetra-dentate ligand bispicen was able to coordinate Mn(II) ion relatively strongly. We therefore designed Ru-Mn dinuclear complexes where bispicen was used as ligand for Mn^{II} ion, see the structures of complexes **13** and **16** in Scheme-4 and Scheme-5, respectively.[19]

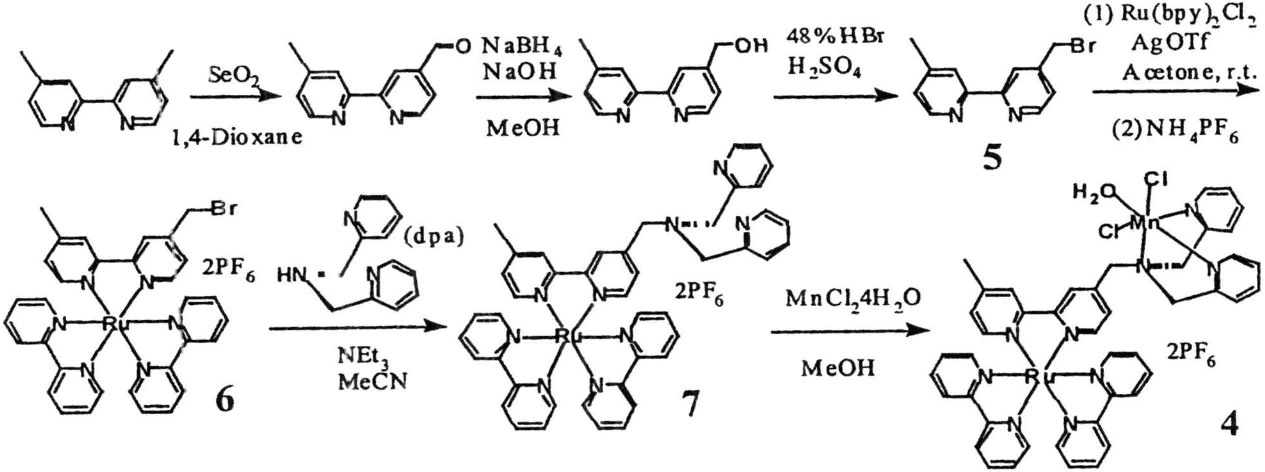

Scheme-2. Synthetic route of complex 4.

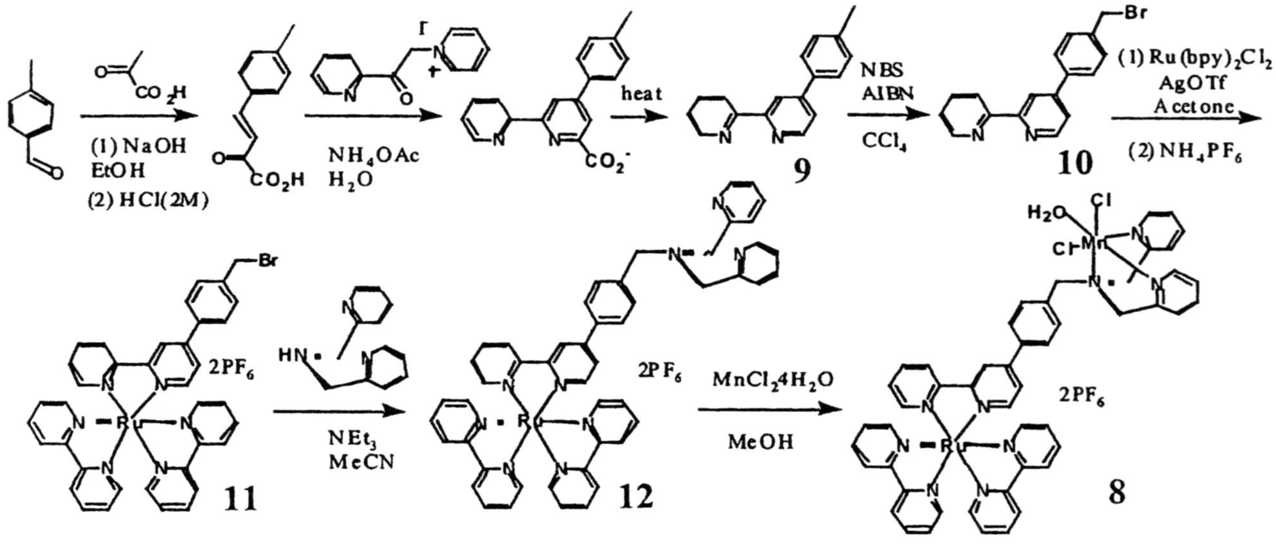

*Scheme-3. Synthetic route of complex **8**.*

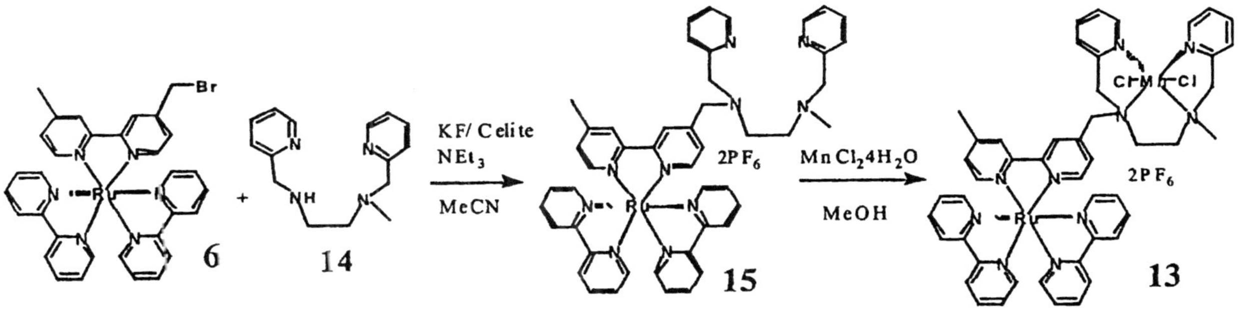

Scheme-4. Synthetic route of complex 13.

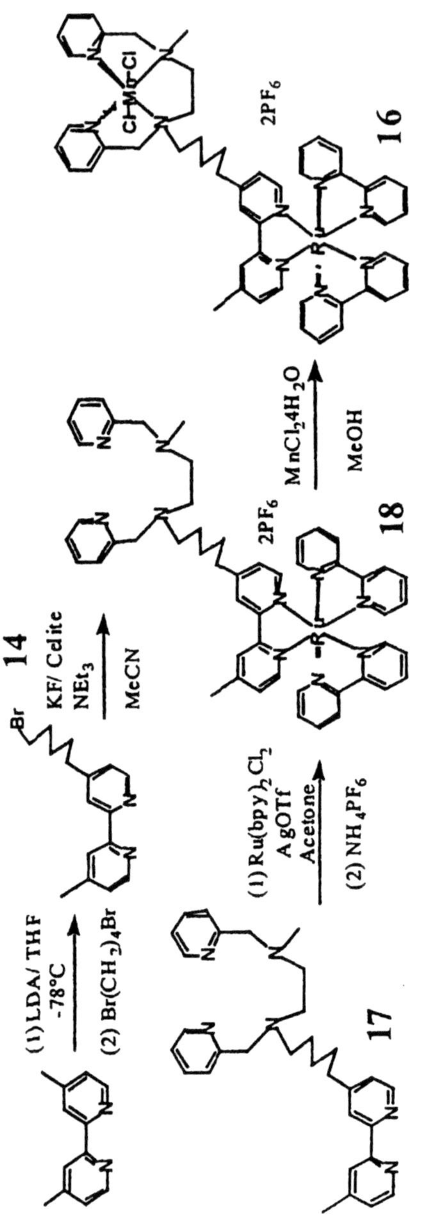

Scheme-5. Synthetic route of complex 16.

For the synthesis of both complexes, a precursor mono-methyl-substituted bispicen **14** was prepared first by a reaction of N-methyl-ethylenediamine and picolylamine in acetonitrile in the presence of bases(KF and NEt_3). The yield of this reaction was low(ca 30%) and the product must be separated and purified by column chromatography. Covalent linking of bispicen to the Ru complex was achieved by reaction of **14** and **6** in the presence of bases. In contrast to the formation of complexes **1**, **4** and **8**, the complex **13** was rapidly formed as red coloured precipitate on reaction between the ligand **15** and the manganese(II) ion in methanol solution. To make complex **16**, the bispicen ligand **14** was linked to the bipyridine ligand by nucleophilic substitution reaction, giving the ligand **17** in a reasonable yield. There are many possible coordination sites for $Ru(bpy)_2Cl_2$ in ligand **17**. In order to increase the chance of coordination of the Ru to the bipyridine moiety, and to decrease the undesired coordination of Ru with bispicen moiety, an acidic (pH 4-5) condition was used. Under the conditions, the bispicen moiety was protonated, preventing coordination of the $Ru(bpy)_2Cl_2$ complex. Some by-products were still formed and the yield of **18** was only ca 25%. Coordination of Mn^{II} ion to the bispicen ligand in **18** was found to be almost quantitative, but no direct precipitation of the product **16** was observed after adding the Mn salt into the methanol solution of **18**.

A similar tetra-dentate ligand, tripicolylamine(tpa), was also linked to Ru-*tris*-bipyridine complex, see structures in Scheme-6.[21] The synthesis of tpa with a hydroxyl functional group **20** was performed by a reaction of dpa ligand with 2-bromomethyl-3-hydroxy-pyridine under basic condition, giving the product **20** in a good yield. Deprotonation of the hydroxyl group in ligand **20** with sodium methoxide and subsequent reaction with the Ru complex precursor **6** in DMSO for four days gave the complex **21** in 46% yield after purification by column chromatography. The dinuclear Ru-Mn complex was obtained quantitatively as a red precipitate by reaction of Mn salt with **21** in methanol. Among all Ru-Mn dinuclear complexes mentioned above, complex **19**[21] is the most stable. Measurements of the Ru-emission quenching showed that no obvious dissociation of Mn^{II} ion from the ligand occurred in dilute acetonitrile solution.

By time-resolved emission and flash photolysis measurements, the excited state lifetimes and intromolecular electron transfers of complex **1, 4, 8, 13, 16** and **19** were studied in acetonitrile solution. It was found that the coordination of Mn^{II} ion with the ligand of complex **3, 7, 12, 15, 18** or **21** would quench the excited state lifetime of Ru complex.[20,23] The quenching effects are dependant on the distance between the Ru and Mn metal centers, and the excited state of Ru complex is strongly quenched when the distance is short. In case of **13**, for instance, the excited state lifetime of Ru complex was 7 ns. When the distance was increased from 9 Å (**13**) to 14 Å (**16**), the excited state lifetime of Ru complex was increased to 300 ns. It was also found that the rate constant of intramolecular electron transfer from Mn^{II} to the photo-generated Ru^{III} was decreased when the distance between manganese and ruthenium was increased. It is thus very difficult to have both a long lived excited state of Ru complex, and fast intramolecular electron transfer from Mn^{II} to the photo-generated Ru^{III}.

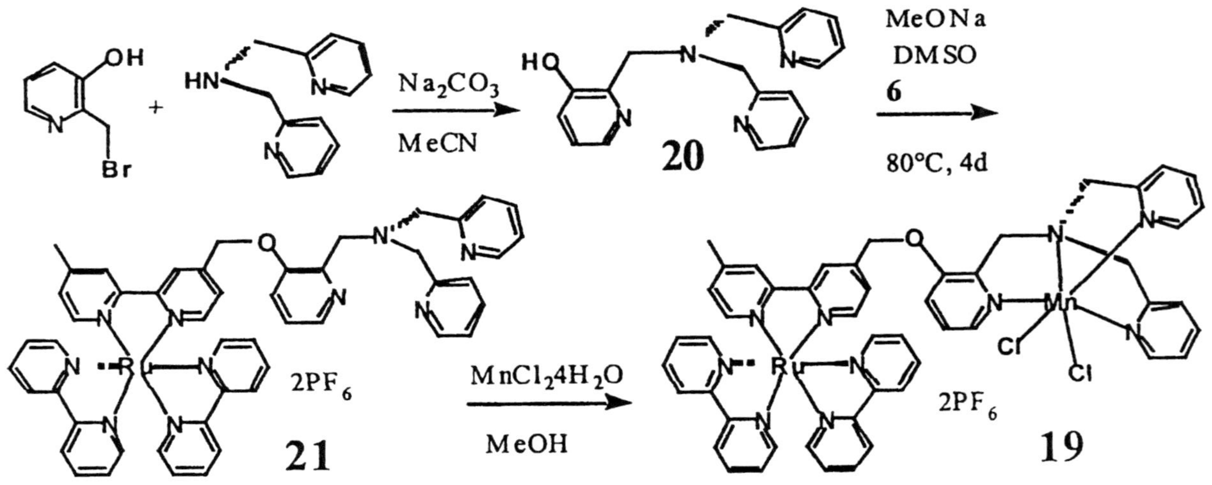

Scheme-6. Synthetic route of complex **19**.

It is interesting to note by fluorescence measurement that the singlet excited state of porphyrin was also rapidly quenched by the Mn^{II}-complex in a porphyrin-Mn(dpa) complex. Thus, we could guess that the excited state quenching of P_{680} chlorophylls in PSII by the manganese cluster could also be a potential problem, since Mn^{II} is involved at least in the fully reduced S_0 state of the OEC. It seems that Nature has solved the problem by keeping the manganese cluster at a large distance from P_{680}. To maintain the rapid electron transfer reaction, Nature invented $Tyrosine_Z$ as an electron transfer relay on the electron donor side of Photosystem II.

3. Synthesis of ruthenium *tris*-bipyridyl complex covalently linked to *L*-tyrosine.

Having the principles of Photosystem II in mind, we designed a system in which tyrosine unit was introduced into the model complex as shown in Figure 5.[24] This was to avoid the problem of quenching by manganese and also to test if tyrosine could act as an intromolecular electron donor relay. By excitation with visible light, the Ru complex moiety was found to transfer an electron to an external acceptor, MV^{2+} for example, forming Ru^{III} and methylviologen radical $MV^{\bullet+}$. Then the tyrosine moiety transfered an electron to the photo-generated Ru^{III}, forming a tyrosyl radical and regenerating the Ru^{II} photosensitizer.

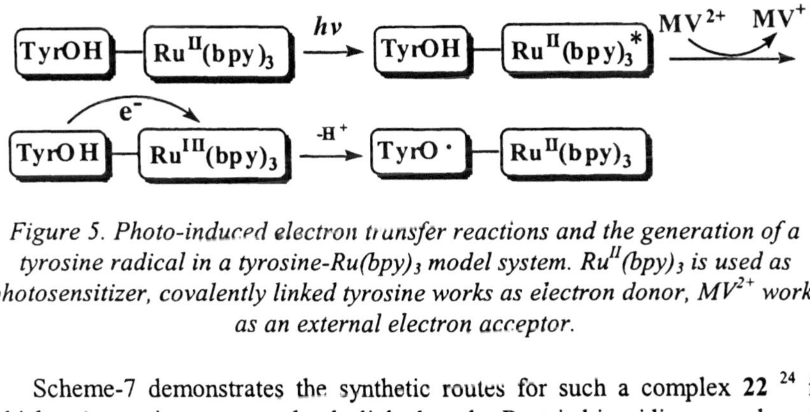

Figure 5. Photo-induced electron transfer reactions and the generation of a tyrosine radical in a tyrosine-Ru(bpy)₃ model system. $Ru^{II}(bpy)_3$ is used as photosensitizer, covalently linked tyrosine works as electron donor, MV^{2+} works as an external electron acceptor.

Scheme-7 demonstrates the synthetic routes for such a complex **22** [24] in which a *L*-tyrosine was covalently linked to the Ru tris-bipyridine complex via an amide bond. First, the mono carboxylate substituted bipyridine **23** was made through two step oxidations with selenium dioxide and silver oxide. Compound **23** was converted to its acid chloride quantitatively by reaction with thionyl chloride, and then reacted with *L*-tyrosine ethyl ester in the presence of triethylamine to give tyrosine-bpy ligand **24** in 70% yield after purification by column chromatography on silica gel. Coordination of $Ru(bpy)_2Cl_2$ with ligand **24** in ethanol gave the final product **22** in almost 94% yield. ESI-MS showed the clean mono-charged and double-charged peaks of complex **22**. Another route to synthesize the complex **22** was to make complex **25** first, then to convert the acid to acid chloride **26**. This was achieved by directly refluxing Ru complex **25** in thionyl chloride for 2 hours. This experiment showed that the Ru-*tris*-bipyridine complex is stable even in the presence of reagents, such as thionyl chloride. The sturdiness of the Ru(II)-*tris*-bipyridine system **25** suggests that it can be used as a convenient synthetic building block in organic synthesis. The reaction of Ru complex acid chloride **26** with *L*-tyrosine ethyl ester in acetonitrile solution with triethylamine as base gave the complex **22** in ca 90% yield.

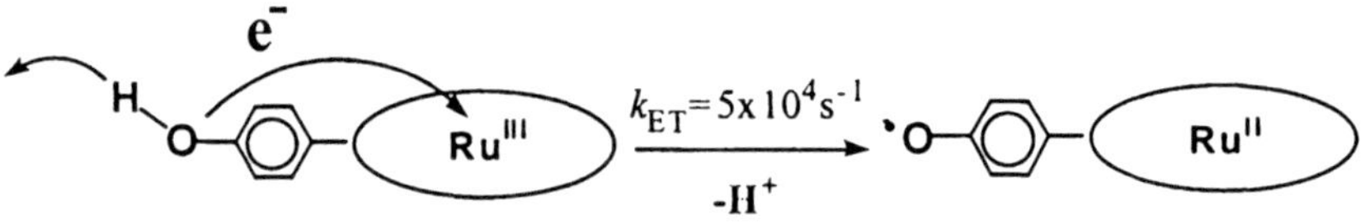

Figure 6. Mechanism of the tyrosine radical formation in an artificial tyrosine-ruthenium complex. The electron transfer from the tyrosine to the photo-generated Ru^{III} and the de-protonation is a concerted reaction when the pH of solution is below the pKa of tyrosine.

Time-resolved emission measurements showed that the excited state of the Ru moiety in **22** was indeed not affected by the tyrosine unit at pH<10. In de-oxygenated aqueous solution at pH 7, the lifetime of the excited state of **22** was 370 ns. This is the same as for a reference complex where tyrosine was replaced by alanine, which has no interaction with the excited state of the ruthenium complex. The synthesis of the reference complex was performed by the same procedure as that used for **22** but using *L*-alanine ethyl ester instead of *L*-tyrosine.[24]

Flash photolysis of the complex **22** in aqueous solution in the presence of a sacrificial electron acceptor, cobalt(III) pentamine chloride $[Co(NH_3)_5Cl]^{2+}$, led to the formation of $[Co(H_2O)_6]^{2+}$ and Ru^{III}. The latter was reduced by an intramolecular ET from the tyrosine unit, and a neutral tyrosine radical was generated, see Figure 6.[24] This was detected by direct observation of a transient absorption at 410 nm, where tyrosine radicals are known to absorb. This result was further confirmed by an EPR signal with a *g*-value of 2.0045, which is typical for a neutral tyrosine radical.[25] The electron transfer rate constant in **22** was found to be pH dependent. Increasing the pH value gave a higher electron transfer rate.[26]

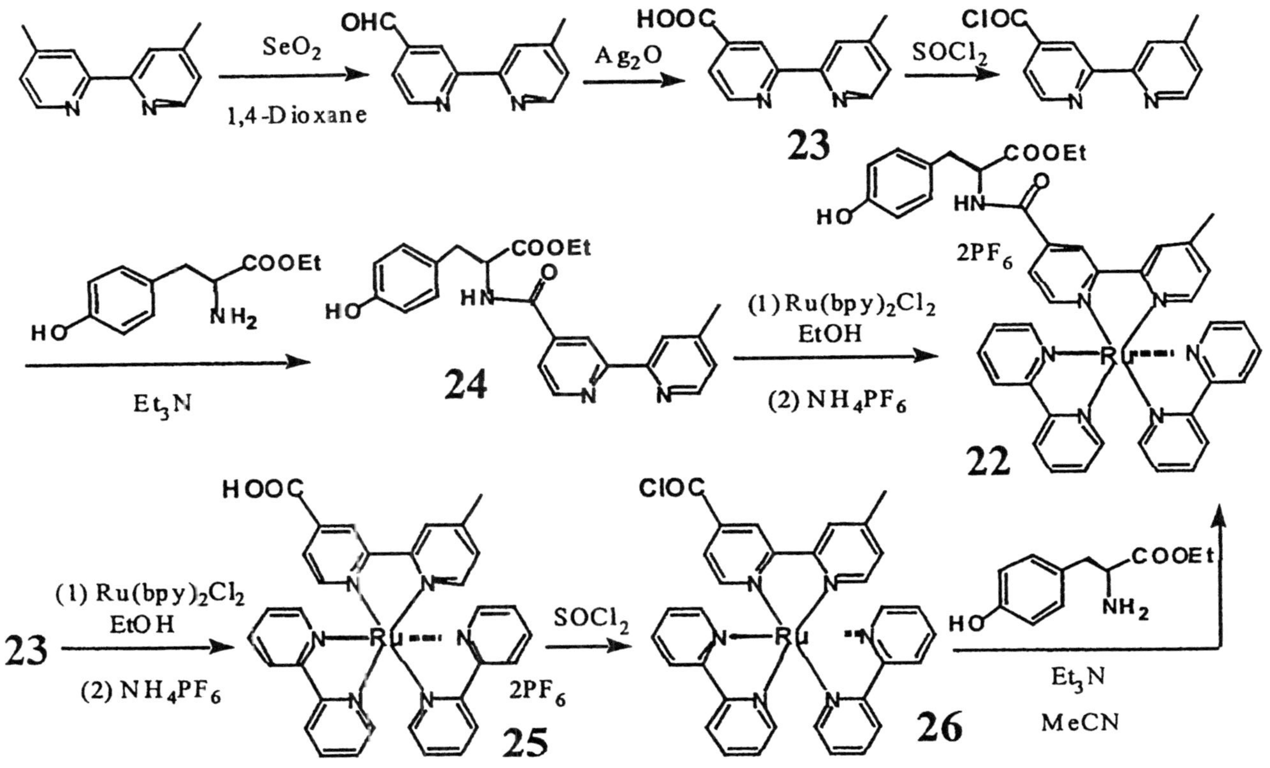

Scheme-7. Synthetic route of complex 22.

4. Synthesis of ruthenium *tris*-bipyridyl complex covalently linked to *L*-tyrosine with hydrogen-bonding substituents.

A detailed electron transfer study on complex **22** revealed that the rate of tyrosyl radical formation is pH dependent.[26] The electron transfer and the deprotonation of tyrosine are concerted reactions. The ET rate from tyrosine to the photogenerated Ru^{III} was low(k_{ET} *ca.* 10^5 s^{-1}) at neutral pH. Higher pH must be used to get a faster electron transfer. Natural Photosynthesis uses only neutral pH.[27,28] In Photosystem II, there is a histidine(His 190) located nearby the $Tyrosine_Z$. It was well accepted recently that His 190 forms a hydrogen bond with $Tyrosine_Z$. Some recent studies also suggested that the hydrogen bond might promote the electron transfer from $Tysrosine_Z$ to the photo-generated P_{680}^{+}.

Our results and comparison with PSII prompted us to construct a supramolecular complex in which the tyrosine is hydrogen bonded, in analogy with the proposed hydrogen bond between $Tyrosine_Z$ and His190 in PSII.[29] This could provide some indirect evidence that H-bonding would increase the electron transfer rate significantly. The supramolecular complex **27** [30] was designed for this purpose, as shown in Scheme-8.

The complex **27** is an extension of complex **22**. The H-bonds are introduced with two dpa-arms, which are connected to the *o*-positions of the phenol group of *L*-tyrosine. The synthesis of **27** was performed by a Mannich reaction. Complex **22** was refluxed with paraformaldehyde and dpa in EtOH/H_2O solution, giving complex **27**in 45% yield. The purification of complex **27** proved to be difficult, because the dpa arms had interacted strongly with the aluminium oxide used in the column chromatography. To overcome this problem, a mixture of hydrochloric acid and methanol was used as eluents. In this way, the dpa arms were protonated and therefore it was possible to purify the products by column chromatogrphy. Complex **27** was fully characterized by 2D-NMR and ESI-MS. The complex **27** could also be prepared in a reasonable yield by an alternative route. In this route, the dpa arms were first introduced by a Mannich reaction with the protected N-Boc-ethyl ester of *L*-tyrosine, giving compound **28** in a satisfactory yield. Deprotection of the Boc with acid gave compound **29** quantitatively. Compound **29** was then reacted with the "Ru-acid chloride" building block **26** in the presence of a base, yielding complex **27**, which was identical to the product as from the first route.

The dpa-arms in **27** form a strong hydrogen bond to the phenol group, see Figure 7, evidenced by ^{1}H NMR and NOESY spectroscopy.[30] There was no quenching effect for the excited state of ruthenium by the H-bonded tyrosine in **27**, actually complex **27** displayed a single exponential decay of emission in acetonitrile with a lifetime $\tau = 1350$ ns, slightly longer than that of $Ru^{II}(bpy)_3$.

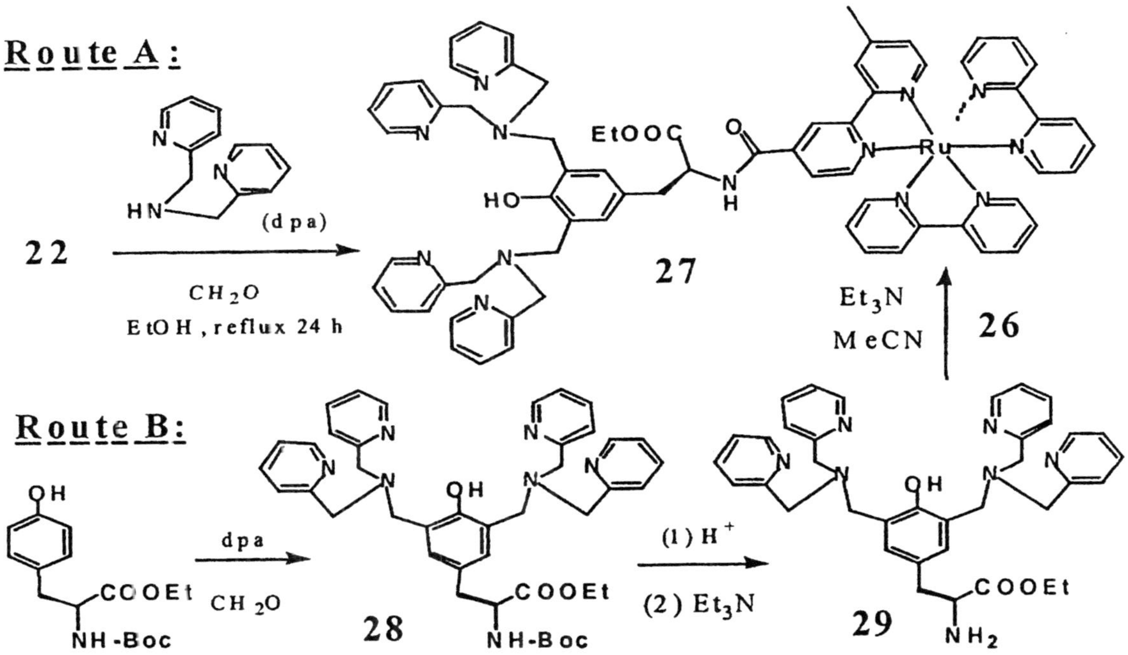

Scheme-8. Synthetic route of complex 27.

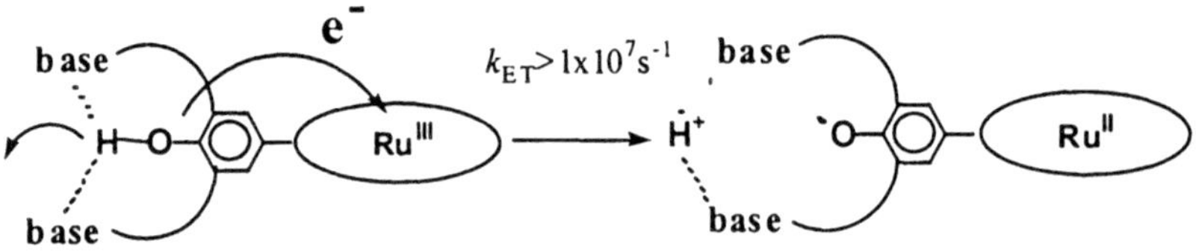

Figure 7. Illustration of internal H-bond promoted electron transfer and the formation of tyrosine radical in a modified artificial tyrosine-ruthenium complex. The electron transfer from the tyrosine to the photo-generated Ru^{III} has been much increased due to the hydrogen bonding of tyrosine.

The tyrosyl radical formation was evidenced by flash photolysis and EPR measurements on **27** in either acetonitrile or water in the presence of an electron acceptor (methylviologen, MV^{2+}, or cobalt pentaminechloride, $[Co(NH_3)_5Cl]^{2+}$).[30] An intramolecular electron transfer from the substituted tyrosine moiety to the photo-generated Ru^{III}, was observed with a rate constant $k_{ET} > 1 \times 10^7\ s^{-1}$, which is at least two orders of magnitude greater than the one from a related complex **22**, in which no dpa-arm is attached to the tyrosine unit. Therefore the hydrogen bonding between the substituted tyrosine and the dpa-arms in **27** was proposed to be responsible for the fast electron transfer as shown in Figure 7. This observation supports the proposed His190 and $Tyrosine_Z$ interaction at the donor side of PSII. The intramolecular electron transfer was so rapid that the overall rate was limited by the initial quenching of the Ru excited state by the acceptor. We are currently synthesizing a supramolecular system similar to **27** with an in-built electron acceptor, to determine the rate of intromolecular electron transfer from the tyrosine to the photo-generated Ru^{III}.

5. Synthesis of ruthenium *tris*-bipyridyl complex covalently linked to a Mn dimer via *L*-tyrosine.

When we were synthesizing complex **27**, two dpa arms were introduced. This was not only to have a strong H-bond, but also to form a site for coordinating two Mn ions. In order to oxidize water, it is necessary to store four oxidation equivalents in a Mn cluster. This might be hard to achieve with a mononuclear Mn complex. In PSII, the operating unit is a manganese cluster consisting of four manganese ions.[3] As model system, a Mn-dimer might be sufficient.[12] In accordance with our expectation, the reaction of complex **27** with manganese(II) acetate in methanol solution gave the trinuclear complex **30** in 94% yield. To our knowledge, this was the first complex, where a dinuclear manganese complex was covalently linked to a ruthenium *tris*-bipyridine type photosensitizer,[31] see the structure **30** in Figure 8. The characterization of complex **30** was made by ESI-MS and EPR. The valences of Mn ions were 2+, therefore the total charge of complex **30** was +6, with 3 PF_6^- counter ions. In the

ESI-MS spectrum of **30**, both mono-charged, double-charged and triple-charged molecular peaks were observed.

Despite the presence of the Mn-dimer, the excited state lifetime of the trinuclear complex **30** was found to be 110 ns in acetonitrile solution.[32,33] This is long enough for efficient bimolecular electron transfer to acceptors such as methylviologen(MV•$2PF_6$) and $[Co(NH_3)_5Cl]^{2+}$. In flash photolysis experiments with **30** in acetonitrile in the presence of MV^{2+}, the photo-generated Ru^{III} was immediately reduced back to Ru^{II} by a fast ($k_{ET} > 1\times10^7$ s^{-1}) intramolecular electron transfer from the manganese(II/II) moiety to give manganese(II/III). By repeated laser flashes in the presence of a sacrificial electron acceptor $[Co(NH_3)_5Cl]^{2+}$, the manganese cluster of **30** could be oxidized from the manganese(II/II) state to the manganese(III/IV) state by three step electron transfers, as shown by EPR experiments.[34] A possible ligand exchange of acetato bridges was found in the experimental condition. Analysis of the samples after laser irradiation by ESI-MS showed that one of the acetato bridges was replaced by two water molecules. Coordination of the water molecule(s) to the manganese ion(s) was the first step towards water oxidation. Therefore the above shown 3-steps electron transfer process closely mimics the 4-electron cycle in the PS II.[34]

6. Synthesis of ruthenium *tris*-bipyridyl complex covalently linked to a high valence Mn dimer with phenolate ligands.

In the repeated laser flash photolysis measurements with complex **30**, no higher oxidation state than $Mn_2^{III/IV}$ was detected. This could be due to the neutral ligands of dpa arms, which might be oxidized by high valent manganese. In our continued work, we are therefore trying to prepare some stable complexes of the type Ru-Mn_2, where the Mn is in a high oxidation state. In one set of experiments, modified version of the manganese part of complex **30** have been prepared. In these, one of the picolyl groups in the dpa arm has been replaced by a substituted phenol, see structure of **31**-**32**.[35]

The valences of the Mn dimer in complex **31** were III/III, proved by ESI-MS and EPR. Light driven multi-step electron transfer study of complex **31** are under investigation.

7. Synthesis of ruthenium *tris*-bipyridyl complex with anchors to nanostructured TiO_2.

For all Ru-Mn or Ru-TyrOH complexes which have been discussed above, an external electron acceptor (methylviologen, MV^{2+}, or cobalt pentaminechloride, $[Co(NH_3)_5Cl]^{2+}$) was used to the electron transfer

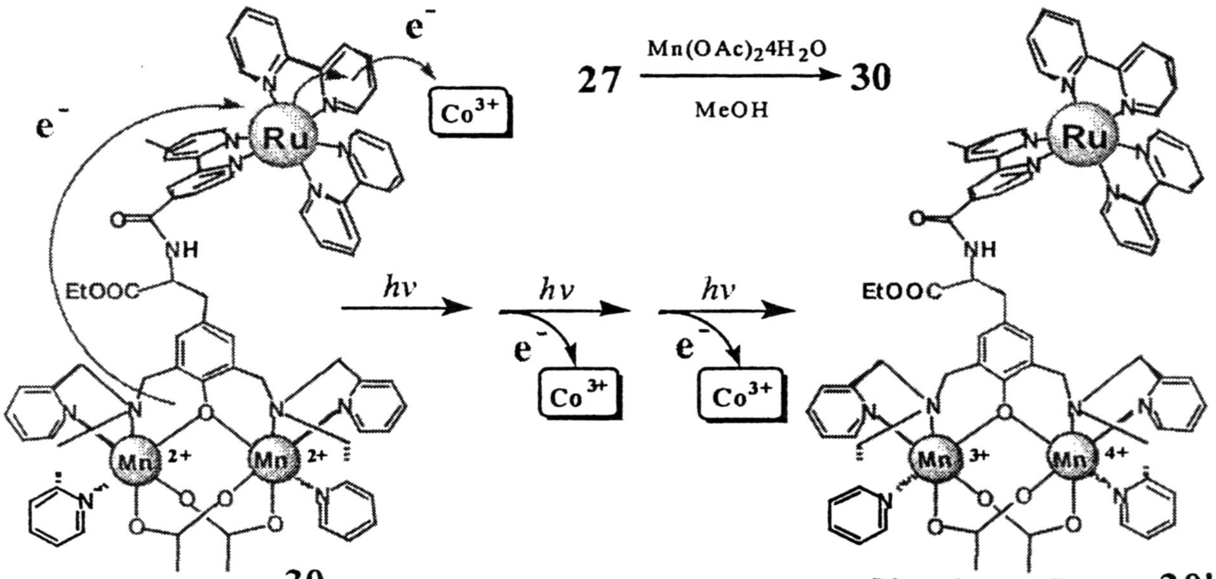

30

possible exchange of acetato ligands with water or OH^- **30'**

Figure 8. Photo-induced three steps intramolecular electron transfers from a Mn dimer to the photo-generated Ru^{III}(inside in the middle shows the synthesis of this trinuclear complex). Cobalt(III) pentaminechlorides are used as external electron acceptor.

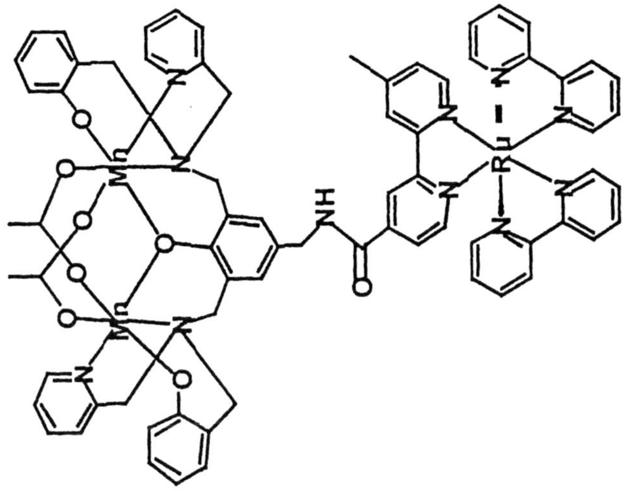

31

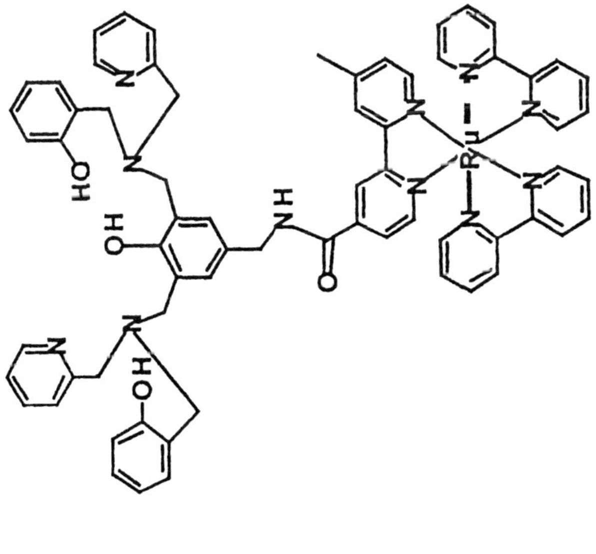

32

experiments. The external electron acceptors have a disadvantage in these systems, because the electron transfer rate from the excited state of ruthenium to the acceptor is limited by diffusion due to the bimolecular reaction. This requires that the lifetime of the excited state of ruthenium should not be too short. Also, in order to be able to resolve fast reactions from the donors, the initial electron transfer to the acceptor must not be the rate limiting step. One way to solve these problems is to link an electron acceptor covalently to the photosensitizer, but this will sometimes make the synthesis very difficult. Alternatively, we can use a nanostructured semiconductor such as TiO_2 as electron acceptor,[36] by binding the photosensitizer to the TiO_2 surface. Synthesis of such supermolecules are under the way in our lab.

It is likely that the use of TiO_2 as the electron acceptor will facilitate the construction of an artificial photosynthetic device. We hope that in the near future, solar energy could be converted into fuel(hydrogen or methanol) by such artificial photosynthetic devices.

Conclusions

In this review we have described the present level in a research project intended to develop artificial photosynthesis as a means to convert solar energy into a fuel. Our strategy is to design and synthesise multifunctional supramolecular complexes to achieve light-driven oxidation of water.[7,15] Our complexes are designed, using principles from the PSII reaction centre. In PSII, a triad composed of a multimer of chlorphyll molecules (P_{680}), a tyrosine residue (Tyr_Z) and a tetranuclear Mn complex are the active redox components in light driven water oxidation.

We have described how we approach our goal stepwise, by synthesis and characterisation of gradually more complex compounds. Our most recent supramolcular complexes contain analogues to all of the redox components on the donor side in PSII. Instead of P_{680} we use Ru tris-bipyridine complexes which are similar to P_{680} in terms of oxidative potential and which have a rich chemistry to use in synthetic attempts. Our complexes also contain an incorporated Mn complex and we have managed to synthesize molecules containing one Ru and two Mn ions. In addition we have introduced a tyrosine link between the Ru center and the Mn-dimer.

Our next goal, is to try to design a new generation of Ru-Mn supermolecules in which we have included our knowledge about the limitations of our present complexes. This is an exciting and difficult task for synthetic chemists, but it would be the key step in our attempts to construct a system for the production of hydrogen from water or the production of methanol from CO_2 in a true artificial photosynthesis system.

Achnowledgement

We thank our many earlier and present co-workers in the Swedish Consortium for Artificial Photosynthesis. This work was financially supported by grants from the Knut and Alice Wallenberg Foundation, Swedish National Energy Administration, the Swedish Natural Science Research Council, the Swedish Research Council for Engineering sciences, the European TMR Program (TMR Network CT96-0031) and the Nordic Energy Research Program. We also thank our colleagues and co-workers in the TMR program.

References

1. Licht, S., Wang, B., Mukerji, S., Soga, T., Umeno, M., Tributsch, M. *J. Phys. Chem. B* **2000**, *104*, 8920-8924.
2. Zouni, A., Witt, H.,-T.; Kern, J.; Fromme, P.; Krauss, N.; Saenger, W.; Orth, P. *Nature* **2001**, *409*, 739-743.
3. Renger, G. *Biochim. Biophys. Acta* **2001**, *1503*, 210-228.
4. Britt, R. D. S., K.; Yachandra, V. K. *Biochim. Biophys. Acta* **2001**, *1503(1-2)*, 2-6.
5. Gust, D., Moore, T. A., Moore, A. L. *Acc. Chem. Rec.* **2001**, *34*, 40-48.
6. Ruettinger, W., Dismukes, G. C. *Chem. Rev.* **1997**, *97*, 1.
7. Yagi, M., Kaneko, M. *Chem. Rev.* **2001**, *101*, 21-35.
8. Balzani, V. C., Sebastiano; Denti, Gianfranco; Juris, Alberto; Serroni, Scolastica; Venturi, Margherita. *Acc. Chem. Res.* **1998**, *31(1)*, 26-34.
9. Imahori, H., Sakata, Y. *Eur. J. Org. Chem.* **1999**, 2445-2457.
10. Sauvage, J.-P., Collin, J.-P., Chambron, J.-C., Guillerez, S., Coudret, C., Balzani, V., Barigelletti, F., De Cola, L., Flamigni, L. *Chem. Rev.* **1994**, *94*, 993.
11. Limburg, J., Vrettos, J. S.; Liable-Sands, L. M.; Rheingold, A. L.; Crabtree, R. H.; Brudvig, G. W. *Science* **1998**, *283*, 1524.
12. Limburg, J., Vrettos, J. S.; Chen, H.; de Paula, J. C.; Crabtree, R. H.; Brudvig, G. W. *J. Am. Chem. Soc.* **2001**, *123*, 423.
13. Binstead, R. A., Chronister, C. W.; Ni, Jinfeng; Hartshorn, C. M.; Meyer, T. J. *J. Am. Chem. Soc.* **2000**, *122*, 8464.
14. Britt, R. D. *Oxygenic Photosynthesis: The Light Reactions*; Kluwer Academic Publishers, Dortrecht, 1996.
15. Sun, L., Hammarström, L., Åkermark, B., Styring, S. *Chem. Soc. Rev.* **2001**, *30*, 36-49.
16. Burdinski, D., Wieghardt, K.; Steenken, S. *J. Am. Chem. Soc.* **1999**, *121*, 10781.
17. Burdinski, D., Bothe, E.; Wieghardt, K. *Inorg. Chem.* **2000**, *39*, 105.

18. Sun, L., Hammarström, L., Norrby, T., Berglund, H., Davidov, R., Andersson, M., Börje, A., Korall, P., Philouze, C., Almgren, M., Styring, S., and Åkermark, B. *Chem. Comm.* **1997**, 607-608.
19. Sun, L., . Berglund, H., Davidov, R., Norrby, T., Hammarström, L., Korall, P., Börje, A., Philouze, C., Berg, K., Tran, A., Andersson, M., Stenhagen, G., Mårtensson, J., Almgren, M., Styring, S., and Åkermark, B. *J. Am. Chem. Soc.* **1997**, *119*, 6996-7004.
20. Hammarström, L., Sun, L., Åkermark, B., Styring, S. *Biochim. Biophys. Acta* **1998**, *1365*, 193-199.
21. Berg, K. E., Tran, A., Raymond, M. K., Abrahamsson, M., Wolny, J., Redon, S., Andersson, M., Sun, L., Styring, S., Hammarström, L., Toftlund, H., Åkermark, B. *Eur. J. Inorg. Chem.* **2001**, 1019-1029.
22. Berglund-Baudin, H., Sun, L., Davidov, R., Sundahl, M., Styring, S., Åkermark, B., Almgren, M., Hammarström, L. *J. Phys. Chem. A* **1998**, *102*, 2512.
23. Hammarström, L., Sun, L., Magnuson, A., Frapart, Y., Berglund-Baudin, H., Åkermark, B., Styring, S. *Zeitschrift fuer Phys. Chem.* **1999**, *213*, 157-163.
24. Magnuson, A., Berglund, H., Korall, P., Hammarström, L., Åkermark, B., Styring, S., and Sun, L. *J. Am. Chem. Soc.* **1997**, *119*, 10720-10724.
25. Magnuson, A., Frapart, Y., Abrahamsson, M., Horner, O., Åkermark, B., Sun, L., Girerd, J.-J., Hammarström, L., Styring, S. *J. Am. Chem. Soc.* **1999**, *121*, 89-96.
26. Sjödin, M., Styring, S., Åkermark, B., Sun, L., Hammarström, L. *J. Am. Chem. Soc.* **2000**, *122*, 3932-3936.
27. Tommos, C., Babcock, G. T. *Acc. Chem. Rec.* **1998**, *31*, 18.
28. Tommos, C., Babcock, G. T. *Biochim. Biophys. Acta* **2000**, *1458*, 199-299.
29. Hauman, M., Junge, W. *Biochim. Biophys. Acta* **1999**, *1411*, 86-91.
30. Sun, L., Burkitt, M., Tamm, M., Raymond, M. K., Abrahamsson, M., LeGourriérec, D., Frapart, Y., Magnuson, A., Huang-Kenéz, P., Brandt, P., Tran, A., Hammarström, L., Styring, S. Åkermark, B. *J. Am. Chem. Soc.* **1999**, *121*, 6834-6842.
31. Sun, L., Raymond, M. K., Magnuson, A., LeGourriérec, D., Tamm, M., Abrahamsson, M., Mårtensson, J., Stenhagen, G., Hammarström, L., Styring, S. Åkermark, B. *J. Inorg. Biochem.* **2000**, *78*, 15-22.
32. Hammarström, L., Sun, L., Åkermark, B., Styring, S. *Catalysis Today* **2000**, *58*, 57-69.
33. Hammarström, L., Sun, L., Åkermark, B., Styring, S. *Spectrochimica Acta A* **2001**, *37*, 2145-2160.
34. Huang Kenéz, P., Magnuson, A., Abrahamsson, M., Lomoth, R., Tamm, M., van Rotterdam, B., Sun, L., Park, J., Hammarström, L., Åkermark, B., Styring, S. *J. Inorg. Biochem.* **2002**, *91*, 159-172.
35. Tran, A., Abrahamsson, M. L., Styring, A., van Rotterdam, B., Styring, S., Hammarström, L., Sun, L., Åkermark, B. *manuscript* **2002**.
36. He, J., Hagfeldt, A., Lindquist, S.-E., Grennberg, H., Korodi, F., Sun, L., Åkermark, B. *Langmuir* **2001**, *17*, 2743-2747.

Methane Valorization

Chapter 16

Catalytic Stability of Ni Catalyst for Partial Oxidation of Methane to Syngas

Zhenhua Li, Xiangyu Zhang, Fei He, and Genhui Xu

State Key Lab of C_1 Chemistry and Technology, School of Chemical Engineering and Technology, Tianjin University, Tianjin 300072, China

Ni catalysts supported on α-Al_2O_3, β-Al_2O_3 and γ-Al_2O_3 pellets for partial oxidation of methane were prepared by an impregnation method. Methane conversion and CO selectivity were measured for the partial oxidation of methane to syngas. On a 10% Ni/Al_2O_3 catalyst, methane conversion and CO selectivity were measured at reaction temperatures of 500-850 °C, at different high space velocities and at different CH_4/O_2 molar ratios in a continuous fixed bed reactor. The results confirmed that methane conversion and CO selectivity increase with the increase of reaction temperature and space velocity. The best CH_4/O_2 ratio for CO selectivity is 2. The catalytic stability of the supported Ni catalysts were also measured. XPS and TG analyses demonstrated that carbon deposition is one of the factors affecting catalytic stability.

In recent years, the catalytic partial oxidation of methane (POM) to syngas ($CO+H_2$) has been extensively investigated for natural gas utilization. Compared with steam reforming or carbon dioxide reforming for the production of syngas, POM has the following advantages: 1. mildly exothermic reaction; 2. syngas produced with a molar ratio of H_2/CO of 2/1, which can be directly used as feed for methanol synthesis or for Fischer-

Tropsch reaction; and 3. enhanced production of syngas operated at high space velocities.

Of the reported catalysts, a supported Ni catalyst is a promising catalyst for industrial production of POM since it has a high activity and a low price (*1-4*). However, the catalytic stability of Ni catalysts is not as good as noble metal catalysts such as Rh or Pt. In this work, Ni catalysts supported on 3-5 mm α-Al_2O_3, β-Al_2O_3 and γ-Al_2O_3 pellets were prepared using an impregnation method. The catalytic activity and stability were measured in a continuous fixed bed reactor. Methane conversion and CO selectivity were measured on a 10% Ni/Al_2O_3 catalyst under reaction temperatures of 500-850 °C, at different high space velocities and at different CH_4/O_2 molar ratios. Thermogravimetric (TG) and XPS analyses were conducted to analyze carbon deposits on the catalysts.

Experimental

Catalyst Preparation

The catalyst was prepared by an impregnation method. The required amount of $Ni(NO_3)_2 \cdot 6H_2O$ was slowly added to the support (diameter of 3-5 mm) to obtain the catalyst. After impregnation, the sample was dried at 120 °C for 2 h and calcined at 400 °C for 4 h, and then calcined at 800 °C for 8 h. Finally, the sample was reduced at 500 °C or 800 °C by hydrogen for 2 h.

Catalyst Evaluation

A quartz tube of dimension of 5 mm o.d.× 3 mm i.d. × 300 mm length was used as a reactor. The catalyst amounts were 60 mg. High purity Ar was used as the carrier gas for GC analysis of the inlet and outlet gas mixtures. The conversion and selectivity were calculated from the GC analysis results based on 100% C,H,O and N mass balance.

Specific Surface Area Measurement

The specific surface areas of the support and the catalyst before and after the reaction were measured using a pulse chemisorption instrument (CHEMBET-3000). The catalysts were degassed for 2 h at 200 °C before use. The adsorption at liquid nitrogen and desorption at room temperature were recorded when the partial pressures of N_2 in He was 0.1, 0.2 and 0.3. The BET equation was used to calculate the specific surface area.

Temperature Programmed Reduction (TPR) Measurement

TPR experiments were performed using an Automatic TPR/TPD 2910 Adsorption Instrument using 10% vol. H_2/Ar mixed gas with a flow rate of 20 mL/min. The amount of catalysts used was 100 mg and the temperature was increased at a rate of 10 °C/min to 900 °C.

Thermogravimetric(TG) Analysis

Thermogravimetric experiments were carried out in a TA-50 thermal analysis system. The experiment was conducted using N_2 gas (99.99%) and air (99.99%) at flow rates of 20 mL/min and 30 mL/min, respectively. Samples (about 10 g) were loaded into the thermobalance and heated to 800 °C at a controlled heating rate of 10 °C/min until a constant weight was achieved. The weight loss rate was recorded simultaneously.

XPS Characterization

XPS characterization was conducted to analyze the carbon deposits and the dispersion of Ni on the support. XPS analyses employed a multi-pass analyzer (PHI 1600) with pass energy of 188 eV for survey spectra and 29 eV for high-resolution spectra. X-rays were generated using a Mg K_α X-ray source.

Results and Discussion

The Catalytic Activity of 10% Ni/α-Al_2O_3, Ni/β-Al_2O_3 and Ni/γ-Al_2O_3 Catalysts after Reduction at 500 °C

The results showed that partial oxidation of methane does not occur on the 10% Ni/β-Al_2O_3 and Ni/γ-Al_2O_3 catalysts in the reaction temperature range of 500-850 °C and only the 10% Ni/α-Al_2O_3 catalyst shows a significant activity after these catalysts were reduced by H_2 at 500 °C for 2 hours. The TPR profile for these three catalysts is shown in Figure 1. The reduction did not occur on the 10% Ni/β-Al_2O_3 and Ni/γ-Al_2O_3 catalysts below 700 °C. For the 10% Ni/α-Al_2O_3 catalyst, three main reduction peaks occur and the reduction began at 350 °C. These results prove that the POM reaction does not occur on a NiO surface and that the reduced Ni is the active center for the POM reaction. This is consistent with the results obtained in other reports when powder alumina was used as the support (*5, 6*).

On the 10% Ni/α-Al_2O_3 catalyst, the methane conversion and CO selectivity were measured at reaction temperature of 500-850 °C, at different high space velocities and at different CH_4/O_2 molar ratios.

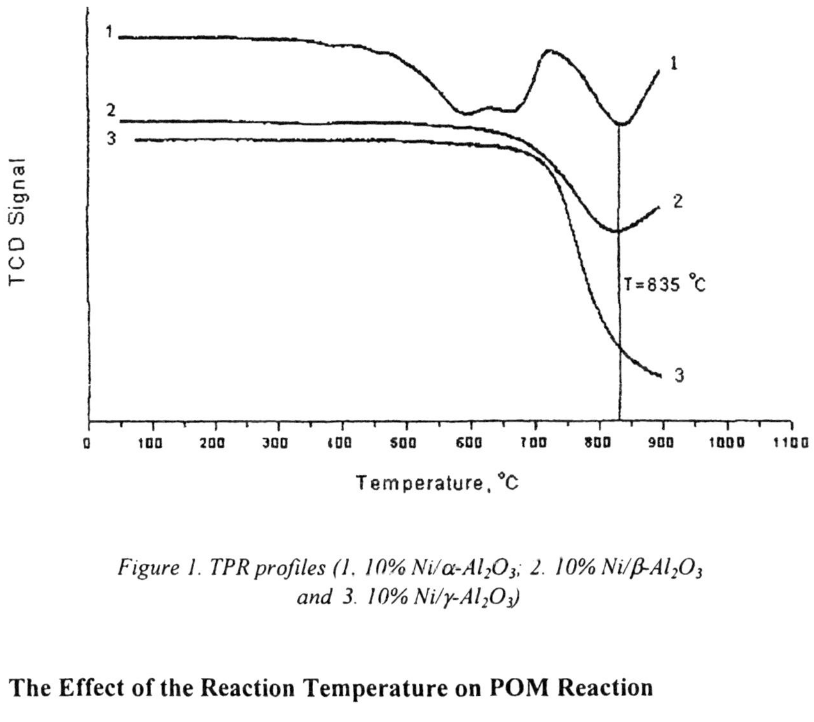

Figure 1. TPR profiles (1. 10% Ni/α-Al_2O_3; 2. 10% Ni/β-Al_2O_3 and 3. 10% Ni/γ-Al_2O_3)

The Effect of the Reaction Temperature on POM Reaction

It was found that on the 10% Ni/α-Al_2O_3 catalyst the POM reaction began when the reaction temperature was about 500 °C. The methane conversion and selectivity of CO and H_2 versus reaction temperature are shown in Figure 2. The methane conversion and selectivity of CO and H_2 increase with the increase of reaction temperature. After the reaction temperature is over 600 °C, the curve for selectivity versus temperature is steeper than the curve below 600 °C, which means that higher reaction temperature is more favorable for the POM reaction.

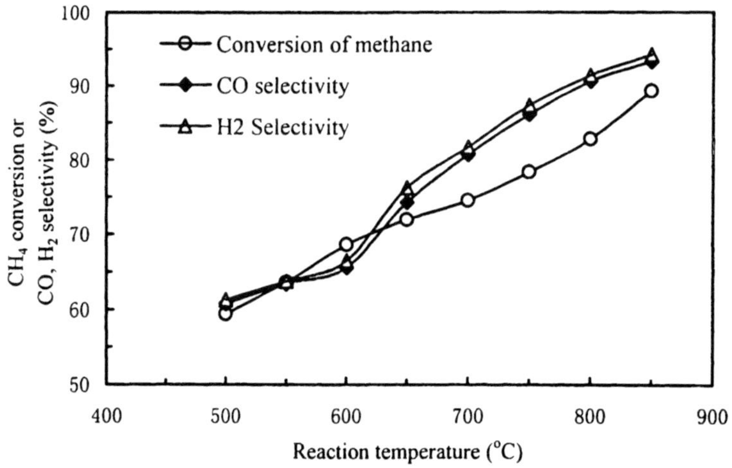

Figure 2. The catalytic activity of the 10% Ni/α-Al_2O_3 catalyst (prereduced at 500 °C by H_2 for 2 h) vs. reaction temperature,
Overall fow rate=260 mL/min, $n(O_2)$: $n(CH_4)$: $n(N_2)$=1: 2: 4

The Effect of $n(CH_4)/n(O_2)$ on POM Reaction

The methane conversion and selectivity of CO and H_2 versus $n(CH_4)/n(O_2)$ are shown in Figures 3 and 4, respectively, for $n(CH_4)/n(O_2)$ ratios of 1.0, 2.0, 4.0 and 8.0. The flow rates of CH_4, O_2 and N_2 were 38, 38, 140 mL/min; 76, 38, 140 mL/min; 76, 19, 140 mL/min and 76, 9.5, 140 mL/min, respectively. With the increase of $n(CH_4)/n(O_2)$, the methane conversion decreased. The selectivity of CO increased first and then decreased with the increase of $n(CH_4)/n(O_2)$. The optimum $n(CH_4)/n(O_2)$ for CO and H_2 selectivity is 2.

The Effect of Space Velocity on POM Reaction

Since higher space velocity is favorable for POM reaction (*7,8*), the effect of space velocity on methane conversion and CO selectivity for the 10% Ni/α-Al_2O_3 catalyst was investigated and the results are shown in Figures 5 and 6. With the increase of space velocity, both methane conversion and CO selectivity increased and reached the maximum value at a space velocity of 5.4×10^5 h^{-1}. This is consistent with the reported results (*7,*

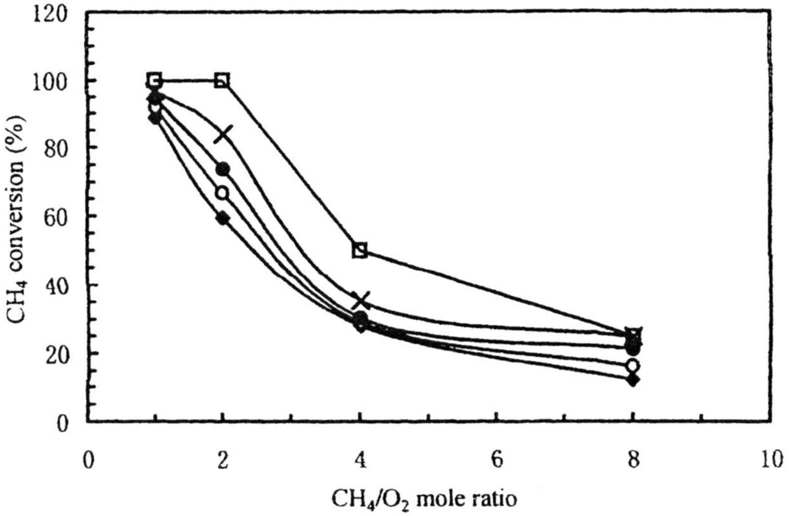

Figure 3. The catalytic activity of the 10% Ni/α-Al_2O_3 catalyst (prereduced at 500 °C by H_2 for 2 h) vs. mole ratio of CH_4 and O_2
Reaction temperature: ♦ 500°C, o 550°C, • 600°C, × 650°C;
□ the maximum methane conversion

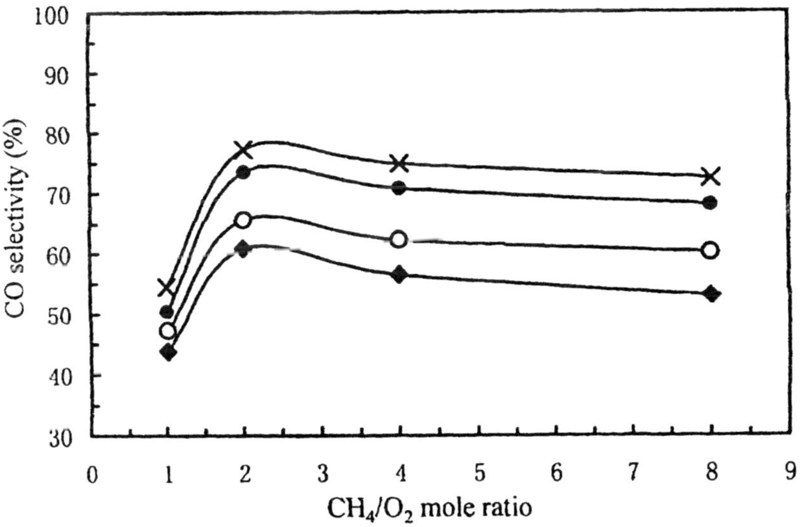

Figure 4. CO selectivity vs. mole ratio of CH_4 and O_2 on the 10% Ni/α-Al_2O_3 catalyst (prereduced at 500 °C by H_2 for 2 hr)
Reaction temperature: ♦ 500°C, o 550°C, • 600°C, × 650°C

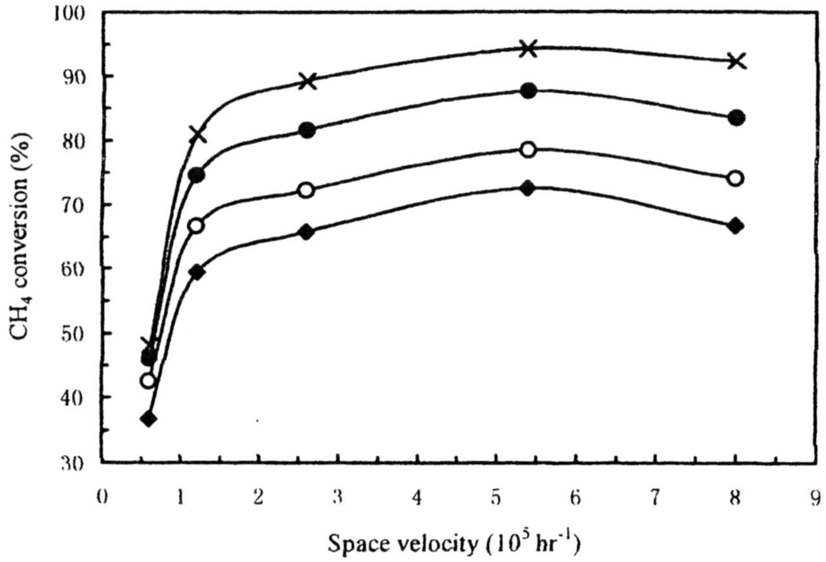

Figure 5. Effect of space velocity on methane conversion for POM
Reaction temperature: ◆ 500°C, o 550°C, • 600°C, × 650°C

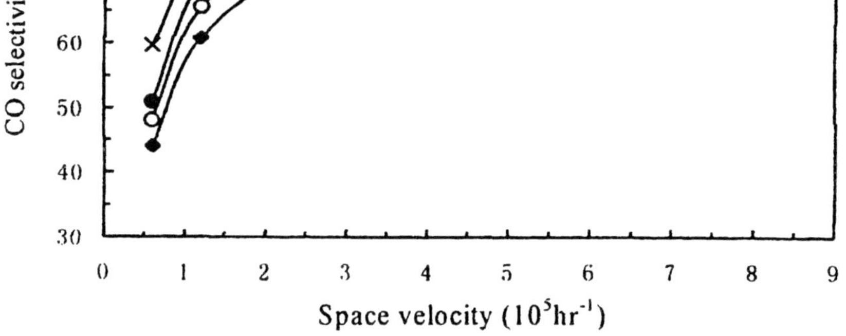

Figure 6. Effect of space velocity on CO selectivity for POM
Reaction temperature: ◆ 500°C, o 550°C, • 600°C, × 650°C

8). Bi et al (*7*) proposed that CH_4 and CO may be co-adsorbed on the same active site. A higher space velocity and temperature can stimulate CO desorption and prevent the complete oxidation of CO, which results in the increase of CH_4 conversion and CO selectivity.

The Effect of the Support on the Catalytic Activity and Stability

10%Ni/α-Al_2O_3, Ni/β-Al_2O_3 and Ni/γ-Al_2O_3 catalysts were tested for POM after reduction at 800 °C for 2 hours. As shown in Table 1, all catalysts exhibit good activity for POM and the 10% Ni/β-Al_2O_3 catalyst shows the best activity. It can be seen that the catalytic activity and the product selectivity increase in the order of the 10% Ni/α-Al_2O_3<Ni/γ-Al_2O_3<Ni/β-Al_2O_3. These were different from the results reported when the powder alumina was used. It has been reported (*7*) that the 8% Ni/α-Al_2O_3 is the best catalyst and the catalytic activity and the product selectivity increase in the order of 8%Ni/γ-Al_2O_3<Ni/β-Al_2O_3<Ni/θ-Al_2O_3 ≈Ni/α-Al_2O_3. This can be attributed to the differences of the support and the preparation method used.

Three catalysts were continuously tested for 100 hours at a reaction temperature at 700 °C and a space velocity of 1.5×10^5 hr^{-1} with the molar ratio of CH_4:N_2:O_2=2:4:1. The decay curves are shown in Figure 7. The results show that the catalytic stability increases in the order of the 10% Ni/γ-Al_2O_3< Ni/β-Al_2O_3< Ni/α-Al_2O_3.

The above results indicated that the property of alumina is a major factor in affecting the catalytic activity and stability. Therefore the specific surface area of the catalysts and the supports were measured and are shown in Table 2. It was found that for the 10% Ni/β-Al_2O_3 and Ni/γ-Al_2O_3 catalysts, the specific surface area decreased in the order of the support > (the catalyst before reduction at 800 °C) > (the catalyst after reaction at 650 °C for 4 hrs). As for the 10% Ni/α-Al_2O_3 catalyst, its specific surface area had a small increase compared to that of the support. It was deduced that two factors make the catalytic stability of the 10% Ni/β-Al_2O_3 and Ni/γ-Al_2O_3 less than that of the 10% Ni/α-Al_2O_3 catalyst. One is the thermal unstability of the supports β-Al_2O_3 and γ-Al_2O_3, whose pore structure may vary after higher temperature calcination or reaction at 650 -800 °C. The other is the amount of carbon deposits on the catalyst surface after the reaction. This was investigated by TG and XPS analyses.

Thermogravimetric Analysis

The weight loss data for different catalysts after the POM reaction are listed in Table 3. At the same level of Ni loading, the amount of carbon deposition increased in the order of 10%Ni/α-Al_2O_3<Ni/β-Al_2O_3 <Ni/γ-Al_2O_3.

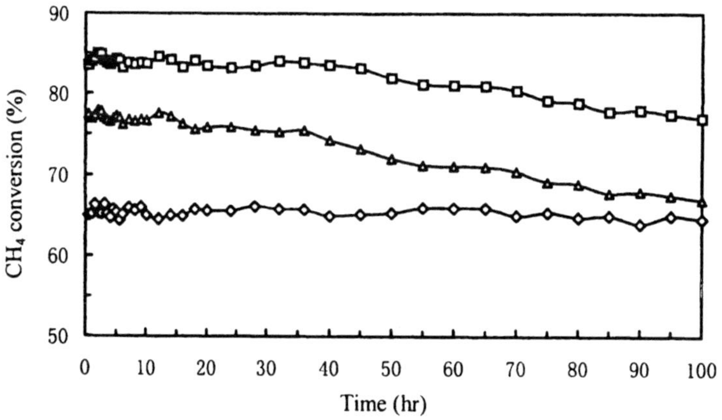

Figure 7. Decay curves of catalysts
◊: *10% Ni/α-Al_2O_3;* □: *10% Ni/β-Al_2O_3;* △: *10%Ni/γ-Al_2O_3*

Figure 8 shows the weight loss rate obtained under N_2 for the 10%Ni/α-Al_2O_3, Ni/β-Al_2O_3 and Ni/γ-Al_2O_3 catalysts after the POM reaction at a constant heating rate. The 10%Ni/α-Al_2O_3 and Ni/β-Al_2O_3 catalysts showed a narrow peak in the range between 500-650 °C with a maximum at 600 °C. The 10% Ni/γ-Al_2O_3 catalyst showed a broad temperature peak in the range of 500-750 °C with a maximum at 700 °C. These results indicate that the carbon material formed on the Ni/γ-Al_2O_3 catalyst is different from that formed on the Ni/α-Al_2O_3 and Ni/β-Al_2O_3 catalysts.

From Figure 8 and the results reported in the literature (*9*), it is clear that at higher temperatures and oxygen enriched condition, the oxidation and gasification of carbon would eliminate carbon deposits. Since carbon deposition is expected to occur easily at higher methane-to-oxygen ratios (*10*), the effect of the methane-to-oxygen ratio on the amount of carbon deposition for higher Ni loading catalysts needs to be investigated further.

Also, with the increase of Ni loading, the amount of carbon deposition increased but the weight loss data showed the same maximum peak. Zhu et al (*11*) has reported that the amount of carbon deposition on the 5% Ni-Ce(La)O_x, 10% Ni-Ce(La)O_x and 20% Ni-Ce(La)O_x catalysts were 0.064%, 2.90% and 18.00%, respectively. The 5% Ni-Ce(La)O_x catalyst showed a very stable performance and no deactivation was observed during a 100 h-on-stream test. Yu et al (*12*) reported that the saturated monolayer loading of Ni on α-Al_2O_3 is 5%. These results proved that the carbon deposition on the 5% Ni catalysts can be prevented and the catalytic stability is good.

Table 1. Catalytic Activities of Three Catalysts Reduced at 800 °C for POM

catalyst	O_2, N_2, CH_4 flow rate: 19, 70, 38 mL/min					O_2, N_2, CH_4 flow rate: 37, 140, 74 mL/min				
	X_{CH4}%	X_{O2} %	S_{CO} %	Y_{CO}%	H_2/CO	X_{CH4} %	X_{O2} %	S_{CO} %	Y_{CO}%	H_2/CO
10%Ni/α-Al_2O_3	50.64	98.80	75.72	38.30	2.03	62.42	98.59	82.75	51.67	2.02
10%Ni/β-Al_2O_3	69.33	98.59	82.85	57.45	2.08	73.57	98.59	87.21	64.18	2.05
10%Ni/γ-Al_2O_3	64.56	98.36	85.00	54.91	2.05	68.73	98.66	88.10	60.52	2.03

NOTE: reaction temperature: 650 °C, pressure: atmospheric pressure

Table 2. Specific Surface Area of the Support and the 10% Ni Catalysts

Catalyst	10% Ni/α-Al_2O_3			10% Ni/β-Al_2O_3			10%Ni/γ-Al_2O_3		
	1	2	3	1	2	3	1	2	3
Specific surface area, m^2/g	8.6	11.9	10.8	71.3	47.0	43.8	251.1	107.6	66.2

NOTE: 1.support, 2.catalyst before reduction, 3.catalyst after reaction at 650 °C for 4 hrs

XPS Characterization

XPS characterization was conducted for the 10%Ni/α-Al_2O_3, Ni/β-Al_2O_3 and Ni/γ-Al_2O_3 catalysts before reduction and after reaction at 700 °C for 100 hrs. The XPS spectra for the 10% Ni/α-Al_2O_3 catalyst, which are similar to the spectra for the other two catalysts, are shown in Figure 9.

The surface compositions for the three catalysts before reduction and after reaction at 700 °C for 100 hrs were determined by XPS and are shown in Table 4. It was noted that the amount of carbon on the 10%Ni/α-Al_2O_3 and the 10% Ni/γ-Al_2O_3 catalysts before reduction are almost the same, which is about double the amount of carbon on the 10% Ni/β-Al_2O_3 catalyst. After these catalysts were reduced at 800 °C by hydrogen for 2 hrs and reacted at 700 °C for 100 hrs, the carbon deposits on all the catalysts increased in the order of Ni/α-Al_2O_3<Ni/γ-Al_2O_3<10%Ni/β-Al_2O_3. This result that the Ni/α-Al_2O_3 catalyst is more resistant to carbon deposition than the Ni/β-Al_2O_3 and Ni/γ-Al_2O_3 catalysts is consistent with the TG results. Combined with the 100 hrs POM reaction result, it is clear that carbon deposition is a major factor resulting in the deactivation of Ni catalysts for POM.

Table 3. TG Analysis Results-Carbon Deposition on Catalysts after POM Reaction

Sample	*Reaction condition*	*Weight loss (wt%)*
5%Ni/γ-Al_2O_3	700 °C, 1.5×10^5 hr^{-1}, $CH_4:N_2:O_2$=2:4:1, 100 h	0.160
10%Ni/α-Al_2O_3	700 °C, 1.5×10^5 hr^{-1}, $CH_4:N_2:O_2$=2:4:1, 100 h	11.047
10%Ni/β-Al_2O_3	700 °C, 1.5×10^5 hr^{-1}, $CH_4:N_2:O_2$=2:4:1, 100 h	18.103
10%Ni/γ-Al_2O_3	700 °C, 1.5×10^5 hr^{-1}, $CH_4:N_2:O_2$=2:4:1, 100 h	25.752
20%Ni/α-Al_2O_3	700 °C, 1.5×10^5 hr^{-1}, $CH_4:N_2:O_2$=2:4:1, 10 h	17.750
20%Ni/β-Al_2O_3	700 °C, 1.5×10^5 hr^{-1}, $CH_4:N_2:O_2$=2:4:1, 10 h	23.957
20%Ni/γ-Al_2O_3	700 °C, 1.5×10^5 hr^{-1}, $CH_4:N_2:O_2$=2:4:1, 10 h	28.128

Table 4. Surface Compositions Obtained from XPS for Three Catalysts

	Atomic ratio, %					
	10%Ni/α-Al_2O_3		*10% Ni/β-Al_2O_3*		*10% Ni/γ-Al_2O_3*	
	Before reduction	*After reaction*	*Before reduction*	*After reaction*	*Before reduction*	*After reaction*
O	55.5	47.9	59.7	49.2	56.3	44.1
Al	25.5	23.5	30.8	26.7	28.2	22.5
Ni	5.1	2.7	2.1	0.9	1.1	0.8
C	13.9	26.0	7.4	23.2	14.4	32.6

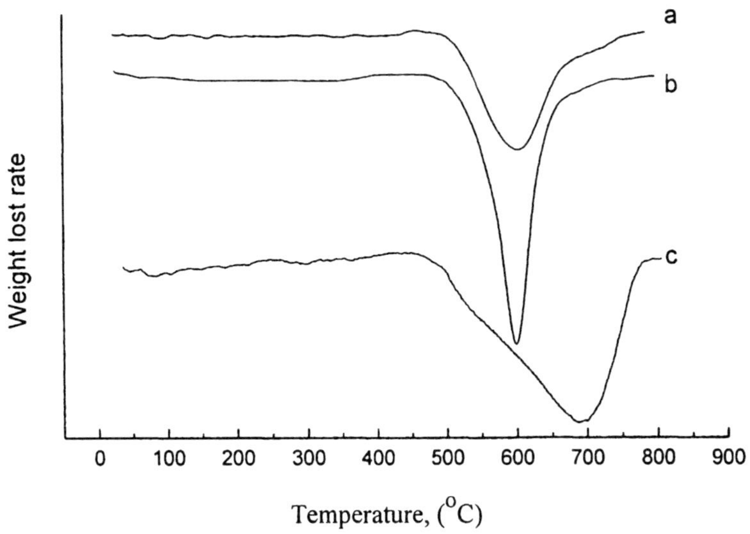

Figure 8. Weight loss rate profile in N_2 of the sample
a: 10% Ni/α-Al_2O_3; b: 10% Ni/β-Al_2O_3; c: 10%Ni/γ-Al_2O_3

Conclusions

From the results obtained, the catalytic activity of Ni/β-Al_2O_3 is better than that of Ni/α-Al_2O_3 and Ni/γ-Al_2O_3. The catalytic stability of the Ni/α-Al_2O_3 catalyst is better than that of the Ni/β-Al_2O_3 and Ni/γ-Al_2O_3 catalysts. The Ni/β-Al_2O_3 catalyst is the best catalyst considering both of the activity and stability. Carbon deposition is the major factor affecting the catalytic stability for POM.

Acknowledgment: We appreciate the financial support from Chinese National Natural Science Foundation (No: 20106013) and the 985 project of State Key Laboratory of C_1 Chemistry and Technology of Tianjin University.

References

1. Ashcroft, A.T.; Cheetham, A.K.; Foord, J.S.; Green, M.L.H.; Grey, C.P.; Murrell, A.J.; Vernon, P.D.F. *Nature* **1990,** *344,* 319-321.
2. Hickman, D.A.; Schmidt, L.D. *J. Catal.* **1992,** *138,* 267-282.

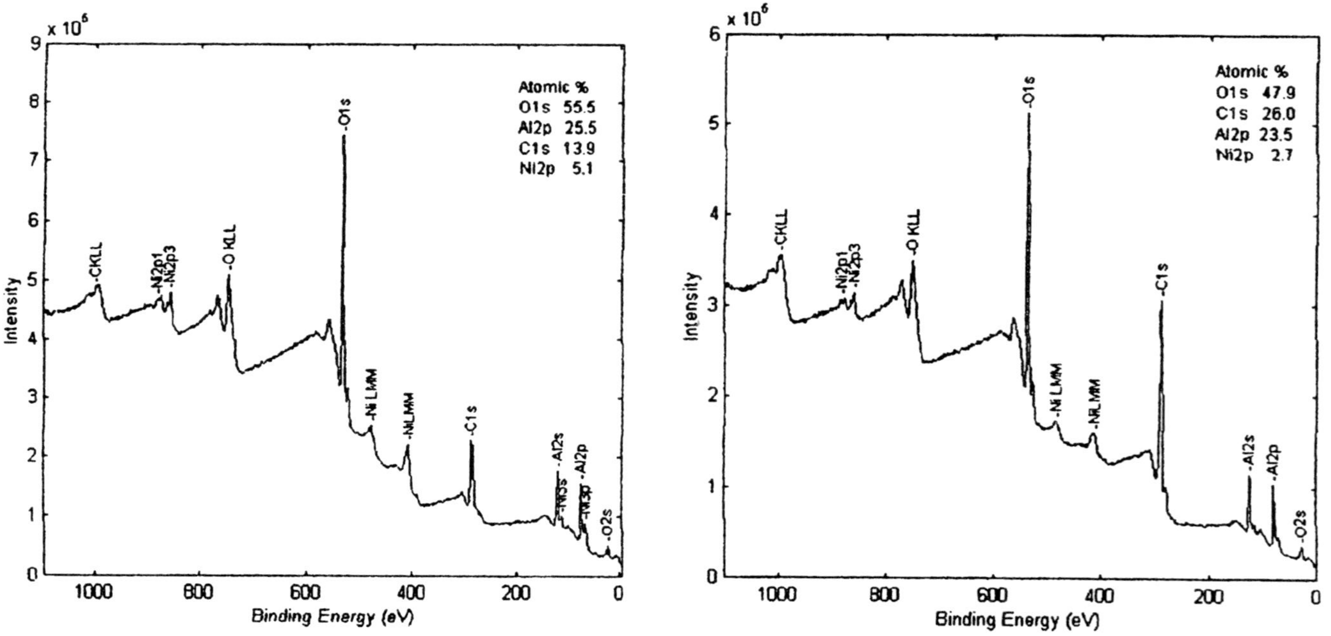

(a) (b)

Figure 9. XPS survey spectra for 10% Ni/α-Al_2O_3 catalyst
a: before reaction; b: after reaction

3. Tomiainen, P.M.; Chu, X.; Schmidt, L.D. *J. Catal.* **1994,** *146*,1-10.
4. Tsang, S.C.; Claridge, J.B.; Green, M.L.H. *Catal. Today* **1995,** *23*, 3-15.
5. Dissanayaka, D.; Rosynek, M.P.; Kharas, K.C.C.; Lunsford, J.H. *J. Catal.* **1991**,*132*,117-127.
6. Jin, R.C.; Chen, Y.X.; Li, W.Z.; Cui, W.; Ji, Y.Y.; Yu, C.Y.; Jiang, Y. *Appl. Catal. A: General* **2000**, *201*, 71-80.
7. Bi, X.J.; Hong, P.J.; Dai, S.S. *Mol. Catal. (Chinese)* **1998,** *12*, 342-348.
8. Cao, L.X.; Chen, Y.X.; Li, W.Z. *Mol. Catal. (Chinese)* **1994**, *8*, 375-382.
9. Nichio, N.N.; Casella, M.L.; Santori, G.F.; Ponzi, E.N.; Ferretti, O.A. *Catal. Today* **2000**, *62*, 231-240.
10. Bartholomew, C. H. *Catal. Rev.-Sci. Eng.* **1982,** *24*, 67-112.
11. Zhu, T.L.; Flytzani-Stephanopoulos, M. *Appl. Catal. A: General* **2001,** *208*, 403-417.
12. Yu, L.; Yuan, S.H.; Tian, J.Y.; Wang, S.; Chu W. *Chin. J. Catal.* **2001,** *22*, 383-386.

Chapter 17

Structure–Reactivity Studies of the Methane Partial Oxidation by Nitrous Oxide-Activated Iron Clusters in Various Zeolite Topologies

Peter-Paul H. J. M. Knops-Gerrits

Département de Chimie, Université Catholique de Louvain (UCL), CMAT, Batiment Lavoisier, Place L. Pasteur n°1, B–1348 Louvain-la-Neuve, Belgium (Telephone: +32–10–47 40 45; fax: +32–10–47 28 36; email: ppkg@chim.ucl.ac.be)

1. ABSTRACT

The introduction of iron in the lattice of zeolites with topologies such as CFI, MOR, FER, MFI and CHA, is performed via direct synthesis. Similar materials with high iron loading are made by post-synthetic sublimation of iron by mixing zeolites with $Fe^{III}(acac)_2$ and an oxygen calcination procedure. XRD, EXAFS, Mossbauer, DRS, magnetic measurements and EPR resolve and attribute presence of lattice iron and extra-framework clusters. XRD for as made and calcined materials reveals α-Fe_2O_3 particles are absent (no α-Fe_2O_3 reflection, d_{104} = 2.70 Å). From EXAFS structural parameters obtained in the k^3 weighted optimization give 1.7, 2.7 and 3.4Å distances for $Fe^{(III)}$-oxygen, $Fe^{(III)}$- $Fe^{(III)}$ and $Fe^{(III)}$-silicon bonds. In the EPR spectra the ratio of octahedral/tetrahedral iron can be assumed from the presence of lines at g= 2 and 4.3. In mononuclear iron complexes the Fe-O bond distances of 1.86 Å (2+, O_h) 6 coordinated and 1.90 Å (3+, T_d) 4 coordinated. The bond distances of 1.65 for Si-O and 1.90 Å for Fe-O are seen in the direction of the Fe-O-Si bridgeand the Fe-O-Si angles are close to 131.2°. Iron oxide clusters in appropriate zeolites realize oxidation of methane into methanol at room temperature after N_2O activation. In these experiments we work at 0.1 - 0.5 % level of the adsorption capacity of the zeolite. $Fe_{5.0}$-ZSM5 is optimally pretreated in a vacuum at various temperatures around 900°C, then the catalyst is allowed to cool and activation with N_2O occurs at 250°C. The reaction of α-oxygen with CH_4 occurs at 20°C followed by extraction with H_2O or D_2O.

2. INTRODUCTION

Theoretical modeling of enzyme active sites such as this of Methane Mono-Oxygenase studies how the dioxygen cleavage and methane activation on these diiron enzyme core models takes place [1]. Such active sites related to oxo- and hydroxo-bridged di- or poly-iron cores in biological systems have been recently subject to extensive theoretical modeling [1 and references therein]. It has been shown that model systems of MMO with only simple aqua and carboxylato ligands form intermediates with plausible energies and structures and a low-energy transition for the C-H abstraction. Topologic variation of zeolite structures remains a challenging task and allows us to tailor the porosity of zeolites to make them excellent hosts for metal oxide clusters with various sizes. Various new types of Structure Directing Agents (SDA's) can be made, but few lead to successful new structures. Panov *et al.* [2] recently studied the methane conversion into methanol with α-oxygen. α-Oxygen is an active surface oxygen species obtained by N_2O decomposition at 250°C on catalysts that have been pretreated at even higher temperatures under vacuum and/or under oxygen. Nitrous oxides are formed in the industrial synthesis of adipic acid. Per molecule of adipic acid a molecule of nitrous oxide is produced. The use of supported metal oxides tends to increase the activity of these metal oxides as result of a dispersion effect [3-4]. We investigate the use of small iron clusters stabilized inside the matrix of thermally activated Fe-ZSM-5, which can fix this new form of oxygen [5-10]. Whereas no shape-selective containment of methanol can be realized on bulk oxides, it can be performed on small metal clusters in lattices such as these of FeCIT-5, FeMOR, FeZSM-5 and FeCHA. A 14 MR material CIT-5 (CFI) has recently been prepared [1]. Benzene is also converted into phenol with α-Oxygen as demonstrated by Panov *et al.* [11-12, 17-20]. Methane is converted into methanol with α-Oxygen [13-17]. The use of N_2O in the benzene to phenol conversion with ZSM-5 zeolites has been accepted for commercial implementation and has passed pilot tests and is still to be commercialized by Solutia Inc. [18-20]. Substituting Al is to be avoided since the Fe(III) isomorphous substituted system at a SiO_2/M_xO_y of 90 gave an inferior selectivity of 90% versus 98% at a 16% conversion of benzene. With a Ga(III) isomorphous substituted system at a SiO_2/M_xO_y of 90 gives only 4% conversion of fluorobenzene and 75% selectivity for fluorophenol [25-27]. In nitrous oxide decomposition several routes exist depending on pressure and temperature. In situ Mossbauer studies imply the reduction of Fe^{III} to Fe^{II} in the stoichiometric oxidation reaction by nitrous oxide decomposition on these catalysts [21]. Fe-ZSM-5 type catalysts are more active than their Ga-ZSM-5 type counterparts [22]. The activation of benzene employs an acid mechanism and H-ZSM-5 catalysts [23-24], that of methane employs a redox mechanism and Fe-ZSM-5 catalysts [16-17].

3. EXPERIMENTAL

The zeolites are made from a basic silicate solution and the organic templates used. The low molecular weight silicate species such as Cab-O-Sil M5 (Cabot corporation) or LUDOX HS 30 (DuPont) can be used as a silicate sources, the $Fe(NO_3)_3$ salt (Aldrich) and $Fe^{III}(acac)_2$ complex (Aldrich) and $Al(NO_3)_3$ salt (JT-Baker) are used as iron and aluminum precursors, LiOH or NaOH are added. The basic silicate solution can be added to the acidified ferric salt solution, to give an overall basic gel. The acidified ferric salt solution contains oxygen coordination complexes such as nitrates, acetates or acetyl-acetonates of iron. These form stable complexes at low pH, but slowly dissociate in a basic medium liberating the metal ions. During the dissociation the silica can bind to form the ferri-silicate gel thus avoiding the precipitation of iron as (hydr)oxides. Once the ferri-silicate gel is formed, precipitation of iron is avoided even at higher pH. The organic structure directing agents are 1-adamanteinium-trimethyl-ammonium-hydroxide (ATMA, CHA), tetraethyl-ammonium-bromide (TEABr, MOR, Mordenite), tetra-propyl-ammonium-bromide (TPABr, MFI, ZSM-5), N(16)-methyl-sparteinium-hydroxide (NMS-OH, CFI, CIT-5).

Iron in the lattice of zeolites via direct synthesis.

The zeolites are synthesized under hydrothermal conditions in a temperature range of 423 K to 448 K. The syntheses were carried out in both sealed quartz tubes (75 mm x 15 mm I.D.) and in Teflon-lined Parr reaction vessels.

FeCHA. The gel contains the LUDOX HS-40 silica, Catapal Al, $Fe(NO_3)_3$ and LiOH and ATMAOH is added as the SDA. A mixture with following stoichiometry ATMAOH(4.5g) : 4 K_2O (11.22g) : Al_2O_3 (1.35g) : 0.1 $Fe(NO_3)_3$ (1.28g): 8 SiO_2 (30.05g): 200 H_2O (16.5g) is heated at 100°C at autogeneous pressure for 7 days to form FeCHA (Si/Fe = 80).

FeZSM-5 The gel contains $Fe(NO_3)_3$ acidified with sulfuric acid, LUDOX HS-30 silica and TPABr is added as the SDA. The mixture with following stoichiometry is heated at 150°C at autogeneous pressure for 6 days form ZSM-5 (Si/Fe = 68) : 2.4 TPABr (1.2g) : 13 NaOH (1g) : $Fe(NO_3)_3$ (0.75g): 68 SiO_2 (2.211g): 4 H_2SO_4 (0.75g) : 1000 H_2O (16.5g). A ZSM-5 support is prepared analogously, with following stoichiometry: 2 TPABr (0.98g): 2 NaOH (0.29g): 20 SiO_2 (2.21g): 500 H_2O (16.5g)

FeMOR The gel contains LUDOX HS-30 silica, $Fe(NO_3)_3$ acidified with sulfuric acid, and the SDA TEABr. The mixture with following stoichiometry is heated at 175°C at autogeneous pressure for 2 days to form Fe-MOR: 2 TEABr (1.0g) : 22 NaOH (0.29g) : 0.33 $Fe(NO_3)_3$ (0.460g): 20 SiO_2 (2.211g): 4 H_2SO_4 (0.75g) : 450 H_2O (15.0g).

FeCIT-5 The gel contains $Fe(acac)_2$ to which LiOH as the mineraliser and LUDOX HS-30 is added and the SDA. A reaction mixture with following composition is used to produce Fe-CIT-5 : 0.2 MeSPAOH (1.27 g): 0.1 LiOH

(0.18g): 0.02 $Fe(acac)_2$ (0.06g):1 SiO_2 (2.10g): 40 H_2O (3.47g) is heated at 175°C at autogeneous pressure for approximately 7 days. The recovered solid is washed with distilled, deionized water, and allowed to air dry. Either LiOH (Fisher) alone, or a mixture of LiOH together with either NaOH (Aldrich) or KOH (Aldrich) can be used, CIT-5 can be formed in the absence of Li^+ cations if the alkali cation (K^+ or Na^+) is present in very low concentration. ([ROH]/[SiO_2] less than 0.05) while the total [OH^-]/[SiO_2] ratio is held constant at 0.3. The gel requires SDA N(16)-methyl-sparteinium hydroxide (**I**) (MeSPAOH), and LiOH to form CIT-5, as in absence of LiOH, the product is SSZ-24 (Silica analog of $AlPO_4$-5). Fe concentrations should be kept low in order not to de-alkylate the SDA during the hydrothermal synthesis.

Iron Solid state Exchange on Zeolites

Commercial samples of ZSM-5 (PQ Company Si/Al ~ 81, VAW Company Si/Al ~ 11), MOR (TOSOH with Si/Al ~ 10), FER (PQ Company Si/Al ~ 10) were obtained and samples of CHA, CIT-5 were prepared according to a similar procedure as reported above. Prior to their use the samples are calcined at 550°C, and ammonium exchanged.

The solid state exchange was effected by mixing them with $Fe^{III}(acac)_2$ and a calcination at 550°C under O_2 by temperature programmed heating at 1°C/min to 175°C, 2 hours at 175°C, at 1°C/min to 550°C and 4 to 8 hours at 550°C. The amount of $Fe^{III}(acac)_2$ mixed with the zeolites was varied to obtain an iron loading ~ 5%, 2.5%, 2%, 1.25% and 0.5% respectively.

Catalytic oxidation of methane with N_2O is carried out in a specially designed quartz cell that is linked to a vacuum line with an oven used for heating of the catalyst at 900°C. A catalyst sample (0.5 to 1.0 g, particles with a mesh size of 35 to 60) is loaded in the specially designed quartz cell, with an isolated volume that has a negligible volume compared to the total reaction volume of the unit (210 cc), this to allow accurate adsorption of the gas phases. Before the experiment the samples are pretreated. GC-analysis programmed at 60°C for 10 min, at 4°C/min to 100°C on a HP-FFAP system with FID detection & parallel GCD analysis is used to quantify the products. The gas phase products include non-reacted methane and benzene, molecular oxygen from the decomposition of the α-O sites and an excess N_2 as carrier. In most cases no CO or CO_2 are detected in the 293K methane reaction experiments, furthermore no methanol is seen, this can only be obtained after extraction of the catalyst. In reactions with Fe-MOR and Fe-CFI the presence of ethane and some CO and CO_2 could be detected. The background reaction of oxidation of methane in the gas phase is circumvented by the short vacuum treatment that allows the non-reacted N_2O to be removed from the system and oxidation of methane occurs on the adsorbed catalyst. ^1H and ^{13}C NMR were used to characterize the products using Bruker NMR and MAS-NMR probes.

4. RESULTS

4.1. Synthesis

The introduction of iron in the lattice of zeolite topologies such as CFI, MOR, FER, MFI and CHA, is performed via direct synthesis and post-synthetic modification [31, 32, 35, 36]. When iron concentrations are kept low framework iron is seen. Extra-framework iron oxides are occluded by calcination of framework iron zeolites with high iron loading or physical mixtures of the aluminum containing zeolites with iron salts. When the latter procedure is applied on framework iron zeolites, the extra-framework iron oxide clusters can be anchored to the framework iron sites. The characterization of iron in the materials is performed with XRD, Mossbauer spectroscopy, SQUID, EPR, DRS and Fe-K edge X-ray absorption spectra. These techniques probe the nature of iron in FeCIT-5, FeMOR, FeZSM-5 and FeCHA.

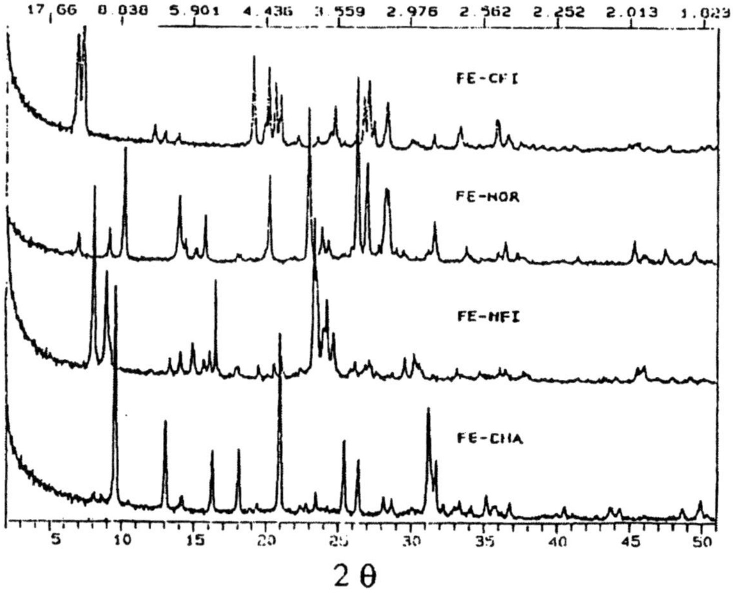

Figure 1. X-Ray Diffraction patterns of FeCIT-5, FeMOR, FeZSM-5 and FeCHA.

4.2. XRD characterization

(Reproduced with permission from J. Mol.Cat. A, ***2001,*** 166, *138. Copyright 2001 Elsevier Science)*

From XRD all materials, FeCIT-5, FeMOR, FeZSM-5 and FeCHA, are over 95% crystalline as can be seen in **Figure 1**. A 8 MR material FeCHA A 10 MR material FeZSM-5 A 12 MR material FeMOR. A new 14 MR material CIT-5 (CFI) has recently been prepared [22]. [Fe]-CIT-5 (Si/Fe ~ 25) with channel diameters of about 10.0x9.9 Å at intersections, has iron exclusively inside of the lattice after synthesis [32]. The framework structure has the 14 MR pores that are built around the SDA N(16)-methyl-sparteinium hydroxide [34]. Fe-CIT-5 (Si/Al ~ 10) with a 5 wt % loading is made by calcination of iron salts in Si- or Al-CIT-5. Here, the extra-framework iron present as small clusters of α-Fe_2O_3.

4.3. EXAFS characterization

To discriminate between mono-and polynuclear active sites [35], we prepared the [Fe]-ZEO with iron in the ZEOlite lattice via direct synthesis and Fe_X-ZEO, by dispersion of X wt% iron on the ZEOlite. Shape-selective formation of nano-clusters of iron oxides with various sizes is realized inside the pore-sizes varying from 10.0 over 8.0 and 6.3 to 4.3 Å of the CFI, MOR, MFI, and CHA zeolites. The Fe K edge X-ray absorption data were obtained for the Fe-CIT-5, Fe-ZSM-5, Fe-MOR and Fe-CHA zeolites containing iron clusters. EXAFS X-ray absorption spectra at the Fe-K edge were collected in transmission mode using a Si(220) crystal for the monochromator at the Stanford Synchrotron Radiation Laboratory. The samples are diluted in a 5:2 BN:catalyst ratio to ascertain correct optical thickness of the catalyst samples. Spectra are obtained at liquid nitrogen temperature under vacuum. The reference chamber contains an iron foil, which is scanned in transmission mode at room temperature.

The Fe K edge X-ray absorption data in **Figure 2.** were obtained for the [Fe]-CIT-5, [Fe]-ZSM-5, [Fe]-MOR and [Fe]-CHA zeolites containing iron clusters. In XAS the ferric nature of iron is clear from the edge shifts between +14.7 eV and +15.5 eV for these materials. The X-ray absorption data show the EXAFS and XANES region. The X-ray absorption data in the XANES region, the unweighted chi vs k plots (3.5 to 15.86 angstrom^{-1}) and k^3-weighted uncorrected Fourrier transforms of the data are obtained. The edge shift for the Fe occurs at 7111.2 eV and increases with +14.7 eV for FeZSM-5, with 15.3 eV for FeCIT-5 and FeCHA and with +15.5 eV for FeMOR. The standard deviation in the order of 0.5 eV. Since the energy resolution for the crystal, the slit size and the beam-line geometry is 3 eV these values are the same within the experimental error. In **Figure 3 and 4.** the EXAFS structure and the EXAFS structural parameters obtained as the k^3 weighted uncorrected FT intensities as function of the distance (r, Å) can be seen. These plots are used to inverse Fourier transform the data and subsequent curve fitting of the different peaks is performed to obtain corrected distances. The data is analyzed using XAMath, a package for reduction of XAS spectra. EXAFS data are analyzed using theoretical phase shifts and scattering amplitudes from FEFF 3.25. For iron in zeolites we use the phase and amplitude parameters of Fe_2O_3.

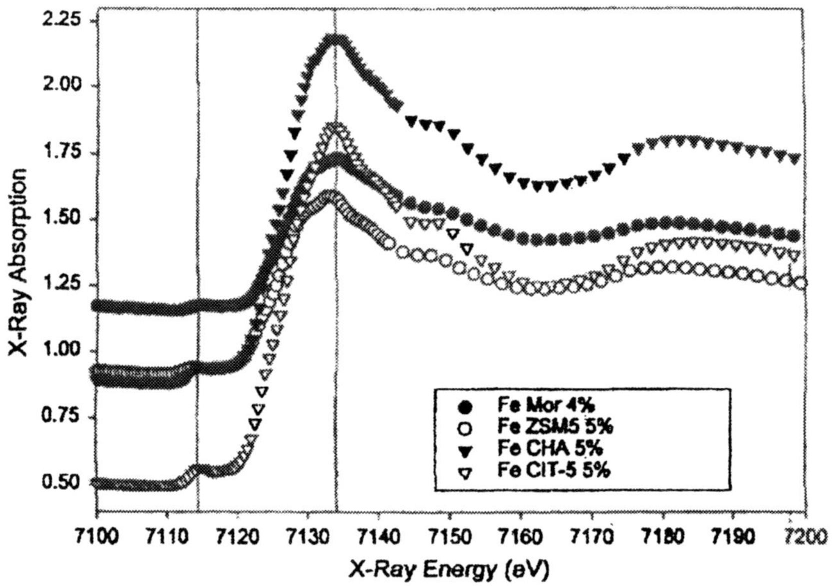

Figure 2. X ray absorption as function of the photon energy, showing pre-edge features and EXAFS region of Fe-CHA, Fe-ZSM5, Fe-MOR, Fe-CIT5 with 4 to 5 wt % Fe.

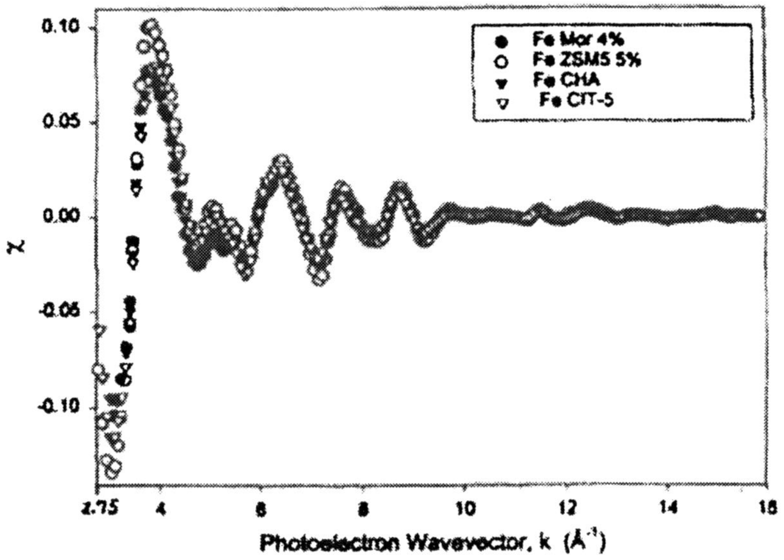

Figure 3. Extended X ray absorption Fine Structure (EXAFS) spectra as function of the photon energy of Fe-CHA, Fe-ZSM5, Fe-MOR, Fe-CIT5 with 5 wt % Fe.

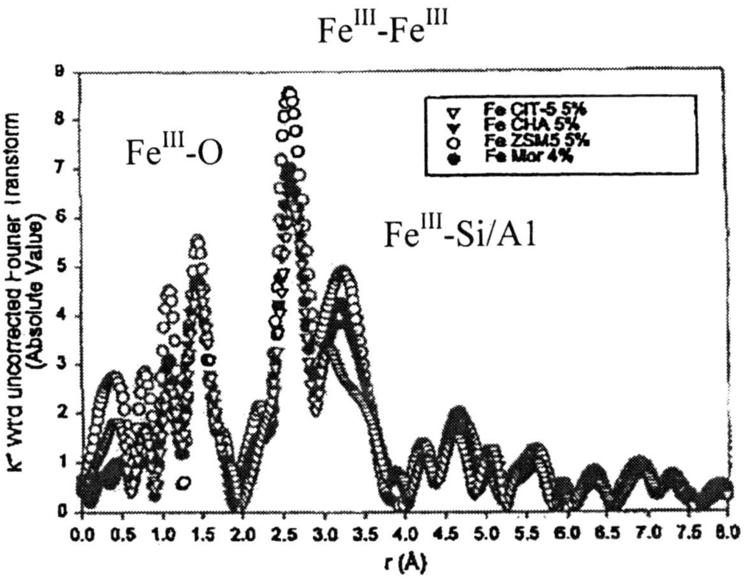

Figure 4. k^3-weighted fourrier transforms of the EXAFS data in Fig.3. Fe-CHA, Fe-ZSM5, Fe-MOR, Fe-CIT5. .(Reprinted from J.Mol.Cat.A., Vol.166., P.P.Knops-Gerrits & W.A.Goddard III, Methane partial oxidation on iron zeolites, pp140, Copyricht 2001, with permission from Elsevier Science)

4.4. *ab initio* calculations

The *ab initio* calculations used involve full geometry optimization of the clusters with density functional theory (dft) as implemented in Jaguar [15] (Jaguar 3.0, Schrodinger, *Inc*., Portland, Oregon, 1997) at the B3LYP method level (Becke3 hybridization functionals, Slater/Becke88 non-local exchange and Li, Yang, Parr local and nonlocal correlation corrections to the local potential energy functionals of Vosko, Wilk and Nusair) using the Los Alamos effective core potential and valence double Zeta for iron (LACVP** basis sets).

The molecular mechanics calculations involve a new molecular mechanics force field, the Universal force field (UFF) of Rappé *et al.* [13-14]. The force field parameters are estimated using general rules based only on the element, its hybridization and its connectivity. The force field functional forms, parameters, and generating formulas for the full periodic table have been published [13]. For charge equilibration used in molecular dynamics simulations the charges in the complexes were determined [16]. This can help to readjust the charges based on geometry and experimental atomic properties. We previously reported that upon interaction with CH_4 the geometry of a ferryl (O^{2-}) bound groups (intermediate Q) to the ferric (Fe^{III}-OH) and ferrous groups does not change substantially, but the Fe..O distances change. In a dinuclear intermediate the system becomes asymmetric as one OH group can show hydrogen bonding with the neighboring methoxy group. For a state with multiplicity 5 the bridging Fe-O distances change from 1.97 and 1.96 Å to 2.002 and 1.975 Å and the terminal Fe-O distances increase from 1.61 and 1.61 A to 1.748 ($Fe^{III}OH$) and 1.723 ($Fe^{III}OCH_3$) Å, the CH and OH distances are 1.097, 1.096, 1.094 and 0.982 Å, respectively. Overal the reaction on a model dinuclear iron complex with methane to the methoxy form is exothermic by 50.56 kcal/mol [1, 30, 33].

4.5. Mossbauer and magnetic characterization

In Mossbauer spectroscopy the absence and presence of a Hyperfine Magnetic Field (HMF) for [Fe]-CIT-5 and FeCIT-5 are seen. From SQUID measurements the magnetic character reveals paramagnetic as well as anti-ferromagnetic contributions. SQUID measurements in **Figure 5.** show a paramagnetic contribution that stems from isolated lattice iron ions. The anti-ferromagnetism is due to iron in small clusters. SQUID measurements in **Figure 6.** furthermore show at 260 K a low intensity Morin transition (weak ferromagnetism) is seen for the Fe-ZSM-5 (0.1 mm) crystals and at the elevated temperature of the Curie point for the weak ferromagnetic moments at (946 K) and the anti-ferromagnetic Néel point at (965 K) of this nano-material transitions are also observed.

4.6. DRS (Diffuse reflectance spectroscopy) and EPR characterization

From DRS the Al-containing samples show the strongest migration in and out of the lattice of iron upon thermal treatment. EPR shows the presence of tetrahedral Fe^{3+} coordination in the lattice for [Fe]-CIT-5, as seen in **Figure 7.**, and two different octahedral signals in post-synthetic modified materials are seen. In the EPR spectra the ratio of octahedral/tetrahedral iron can be assumed from the presence of lines at g= 2 and 4.3.

4.7. Formation of α-oxygen sites

The as synthesised [Fe]-ZSM-5 or CIT-5 samples are calcined under N_2 to 175°C and remain at 175°C for 2 hours, are heated to 550°C in 5 hours, remain at 550°C for 1 hour in a N_2 atmosphere and consecutively in air for 4 hours. The zeolites can be ammonia exchanged and re-calcined under air to create acid sites. The Fe-HAlZSM-5 or Fe-HAlCIT-5 are calcined under N_2 to 175°C, remain at 175°C for 2 hours, are then heated under N_2 to 550°C in 5 hours, remain at 550°C for 1 hour and then in air for 4 hours. The zeolites are evacuated for 2 to 12 hours at 550-800°C. They are then loaded with α-oxygen by adsorption of N_2O at 250°C for 1 to 2 hours at low pressure. Adsorption at higher temperatures e.g. 550°C results in efficiency decreases due to the formation of molecular oxygen. After loading of the α-oxygen the catalyst is shortly subjected to a vacuum of 10^{-3} atm for a variable time (10 minutes) to remove the nitrogen from the quartz cell. The sample is then cooled down to 20°C where the α-oxygen remains reactive.

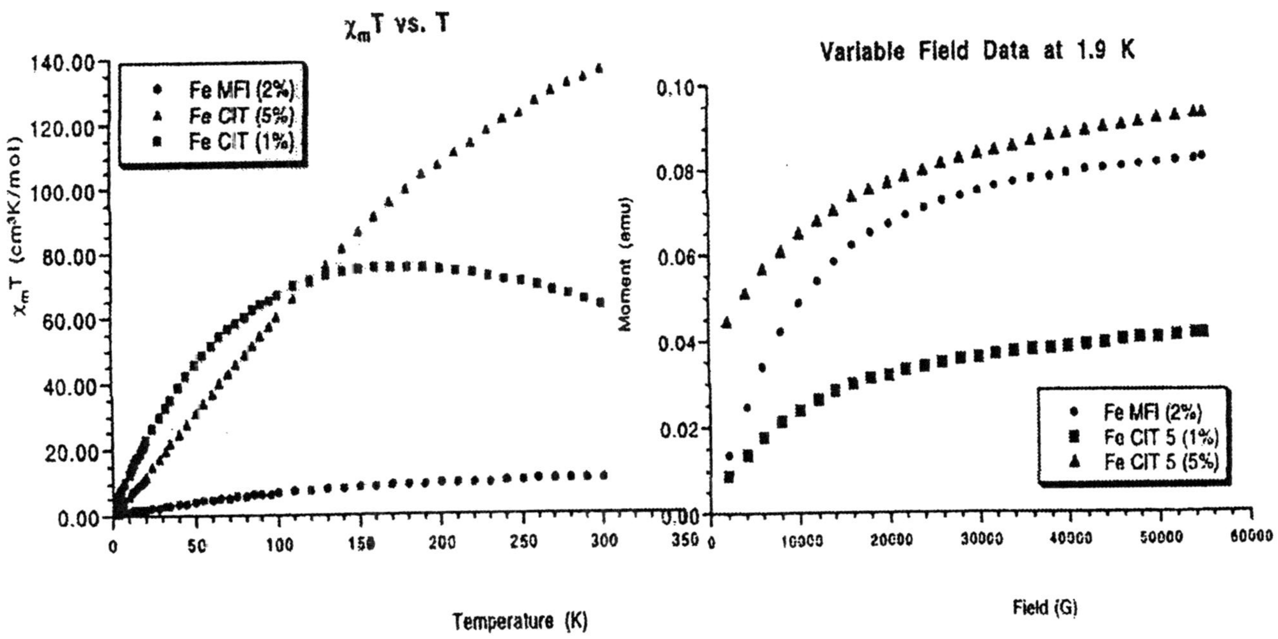

Figure 5. Magnetic Susceptibility data of FeCIT-5 (1% and 5%) and FeZSM-5 at various temperatures and at 1.9 K with various fields (in Gauss).
(Reproduced with permission from J. Mol.Cat. A, ***2001,*** 166, *142. Copyright 2001 Elsevier Science)*

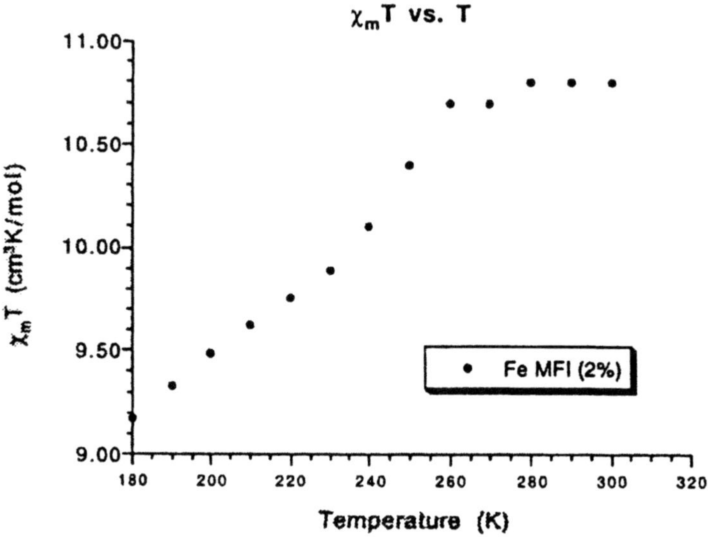

Figure 6. Magnetic Susceptibility data of FeZSM-5 around the Morin transition, which is shifted for the nano-magnetite phase from the bulk value of 260K.

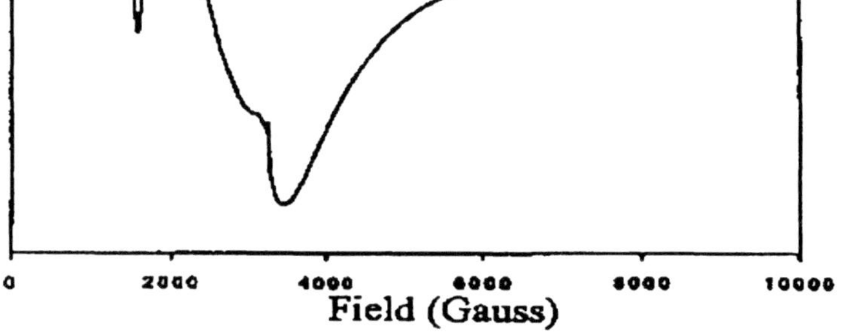

Figure 7. EPR signal at g=4.3 of tetrahedral Fe^{3+} coordination in the lattice of the [Fe]-CIT-5.

(Reproduced with permission from J. Mol.Cat. A, ***2001,*** 166, *142. Copyright 2001 Elsevier Science)*

4.8. Catalytic conversion of methane

After loading of the α-oxygen the catalyst is reacted with methane at 20°C for a variable time (10 minutes to 4 hours). Subsequently it is extracted with various solutions, e.g. acetonitrile-water or pure water H_2O or D_2O/acetone. GC-MS analysis shows that the mixture contains methanol as indicated by the mass fragments. The solvent acetonitrile and acetone are used for the extraction of methanol. The parameters, that affect the catalytic results seen in **Table 1**., are the amount of iron in the sample, the size of the iron clusters,

Table 1: Partial oxidation of methane with α-Fe sites in ZSM-5 and other zeolites.

Zeolite	%Fe	Oxidation	µmol Fe	µmol. MeOH
ZSM-5 Si/Al 11	0.5%	5 Torr N_2O/ 250°C	75	-
ZSM-5 Si/Al 11	2.0%	5 Torr N_2O/ 250°C	360	97
ZSM-5 Si/Al 11 recycle	*2.0%	5 Torr N_2O/ 250°C	360	76
ZSM-5 Si/Al 11	5.0%	5 Torr N_2O/ 250°C	800	163
ZSM-5 Si/Al 11	5.0%	**10 Torr N_2O/ 250°C	800	208
ZSM-5 Si/Al 11	5.0%	**10 Torr N_2O/ 250°C/CD_2H_2/ 20°C	800	186
ZSM-5 Si/Al 81	5.0%	5 Torr N_2O/ 250°C	800	49
ZSM-5 Si/Al 81	5.0%	5 Torr O_2/550°C	800	< 1
ZSM-5 Si/Al 81	5.0%	5 Torr O_2/800°C	800	< 1
ZSM-11	5.0%	5 Torr N_2O/ 250°C	800	123

Fe-Zeolite, 5 Torr N_2O is equilibrated on the catalyst for 2 hours at 250°C & 5 Torr CH_4 is equilibrated on the catalyst for 1 to 2 hours at 20°C; *catalyst recycled by calcination at 550 °C & evacuation at 900 °C; **equilibration pressures increases to 10 Torr N_2O, 10 Torr CH_4;***methanol & various (oxygenated) coupling products were formed

the crystallite size, the Si/Al ratio. Furthermore the activity is function of the pretreatment conditions and the use of N_2O. In order to study the pretreatment effects, the various Fe containing sample is pretreated with O_2 at 550°C, a vacuum at 550°C, 750°C, 800°C or 900°C. After this thermal pretreatment, the activation of the iron with N_2O is performed at 250°C. The reaction with benzene and CH_4 is performed at 20°C. Extraction with acetone and water, respectively, of the zeolites and consequent GC-analysis is used to quantify phenol and methanol. The observed results are absence of activity at 550°C, a moderate activity at 750°C, a similar activity at 800°C and 900°C. In further experiments we chose the evacuation at 800°C. In a second series of experiments the activation effects are probed with a pretreatment under O_2 at 550°C and a sample evacuation at 800°C. Activation in absence of oxidant is used as a blank control, when we take the pretreatment temperature up to 800°C under O_2 and then allow to cool to 20°C, always under O_2, before purging and reacting with CH_4 at 20°C and extraction with H_2O and quantification by GC shows that methanol is not formed and reacting with benzene at 20°C and extraction with acetone and quantification by GC shows that phenol is not formed. In absence of activation (blank control) no products are observed. The reduction of Fe^{3+} to Fe^{2+} is claimed [21] as a requisite to the formation of methanol and phenol. In a third series of experiments different zeolite topologies loaded with Fe are tested and the metal loading effects are probed, with N_2O at 250°C and reaction with CH_4 at 20°C is performed. Extraction and quantification are similar as for series 1, for the highest yield case, methanol can be observed and quantified by ^1H NMR of solution. Catalysis with the Fe-ZSM-5 (Si/Al ratio 11) at different iron loading gives different results, the $Fe_{0.5}$-ZSM5 shows no activity, whereas $Fe_{2.0}$ - ZSM5 shows 97 μmol CH_3OH / g catalyst and the $Fe_{5.0}$ - ZSM5 shows 163 μmol CH_3OH / g catalyst. Analysis of 1g contains 20 mg or 360 μmol Fe/g catalyst. Reactions with benzene at 20°C and extraction with acetone and quantification by GC shows that phenol is formed and with $Fe_{2.0}$ - ZSM5 shows 168 or 173 μmol phenol / g catalyst. Catalysis with the FeCIT-5, FeMOR, FeZSM-5 and FeCHA gives various methane oxidation products, respectively. The first two catalysts also give various (oxygenated) coupling products (such as ethanol, acetone, 1-propanol,...) containing however only a few μmol of methanol. It has previously been emphasized that the reductive versus oxidative conditions during the high-temperature treatment affect the active site. The various other non-selective products were formed due to formation of surface methyl radicals, by reaction of reduced O- in stead of (O)α.

Regeneration experiments have currently been investigated and show the multiple use of the catalyst in several cycles of this process with a limited decay in activity of about 20%. On the study of effects of the Si/Al ratio on the iron distribution pattern occurring and its impact on oxygen activation, a formal positive correlation between the Al content and the methanol formation activity, occurs. Quantification of the methanol formed by GC shows sub-stoichiometric activity, based on a diiron sites.

5. DISCUSSION

For the synthesis of structures with iron in the lattice, in which $Fe(NO_3)_3$ is acidified with sulfuric acid, before addition to the gel containing the template. The post-synthetic modification of zeolites allows us to obtain materials with a

Table 2. Pretreatment is performed by evacuation at 800°C, activation Table 2: Shape-selective effects on the partial oxidation of methane with α-Fe sites in different topologies. (methane has a size of 4.2 Å)

	CAT (μm) Crystal size	channel diameter (Å)	Conversio of Methane μmol	MeOH μmol	EtOH μmol	PrOH μmol	Aceton μmol
5.0%	CFI* 3-7 μr	10.0x9.9 Å	39	4	12	13	10
5.0%	MOR* 1-3 μm	6.5x7.0 Å & 4.8x5.7 Å	33	10	8	9	6
5.0%	MFI 0.1 μm	5.1x5.5 Å & 5.3x5.6 Å	163	163	< 1	< 1	< 1
2.5%	FER 0.3 μm	4.2x5.4 Å & 4.8x3.5 Å	54	54	< 1	< 1	< 1
5.0%	CHA 0.1 μm	3.8x3.8 Å	10	10	< 1	< 1	< 1

In all experiments 1 g catalyst Fe-zeolite catalyst was used and about +-5 Torr N_2O is equilibrated on the catalyst for 2 h at 250°C and, +-5-100 Torr CH_4 is equilibrated on the catalyst for 1-4 h at 20°C and *4 h at 100°C. ** solely 1-propanol and no 2-propanol is observed

higher iron loading. These are made by mixing zeolites with various amounts of $Fe(acac)_2$ and an oxygen calcination procedure. Fe-MOR and FeZSM-5 are prepared with the structure directing agents (SDA) TEABr and TPABr, respectively. Fe-CIT-5 is prepared with a gel that requires the SDA N(16)-methyl-sparteinium and LiOH. FeCHA is prepared with a gel with SDA 1-adamanteinium-trimethylammoniumhydroxide [31-323].
Characterization with XRD, EXAFS, Mossbauer, DRS, magnetic measurements and EPR can resolve and attribute presence of lattice iron and extra-framework clusters. XRD for as made and calcined materials with iron in the lattice reveals the absence of α-Fe_2O_3 particles that are > 10 Å due to the absence of an α-Fe_2O_3 reflection (d_{104} = 2.70 Å) as seen in **Figure 1**. In XAS the ferric nature of iron is clear from the edge shifts between +14.7 eV and +15.5 eV for these materials as seen in **Figure 2.** From EXAFS as seen in **Figures 3 and 4.** and **Table 3** structural parameters obtained in the k^3 weighted optimization give 1.7, 2.7 and 3.4Å distances for $Fe^{(III)}$-oxygen, $Fe^{(III)}$- $Fe^{(III)}$ and $Fe^{(III)}$-silicon bonds. In the XANES region a notable pre-edge feature of Fe^{3+} is seen. This feature is quite analogous to the Fe_2O_3 signature observed. The coordination geometry, i.e. for a tetrahedral cluster the t_2 orbital (3d derived) is mostly p-like and the white line at the edge in the region of the 1s->εd dipole forbidden transition is seen. In the

Table 3: EXAFS evaluation of the Fe-Fe and Fe-Si backscattering for the zeolite samples.

	Back-scatterer	Ref. file	Inter atomic distance	Corrected coordination number	Debey-Waller factor (Å^2)
Fe Chabazite	FeFe	α-Fe_2O_3-Fe1	2.989	4.07	0.007
	FeSi	$Ca_3Fe_2Si_3O_{12}$-FeSi	3.569	1.91	0.002
	FeFe	α-Fe_2O_3-Fe5	4.682	1.33	0.002
Fe ZSM-5	FeFe	α-Fe_2O_3-Fe1	3.002	5.88	0.008
	FeSi	$Ca_3Fe_2Si_3O_{12}$-FeSi	3.587	2.61	0.001
	FeFe3	FeO-Fe4	0.840	0.10	0.007
Fe Mordenite	FeFe	β-Fe_2O_3-Fe1	3.043	5.46	0.010
	FeSi	$Ca_3Fe_2Si_3O_{12}$-FeSi	3.581	2.70	0.001
	FeFe3	FeO-Fe4	4.982	4.04	0.006
Fe CIT-5	FeFe	α-Fe_2O_3-Fe1	3.017	4.42	0.008
	FeSi	$Ca_3Fe_2Si_3O_{12}$-FeSi	3.578	0.55	0.004
	FeFe4	α-Fe_2O_3-Fe5	5.496	3.40	0.006
	FeFe3	FeO-Fe4	4.967	3.64	0.005

octahedral cluster this white line is practically not observed. A second shoulder can be assigned to the treshold of the 1s->εp dipole allowed transitions. In the EXAFS region curve-fitting of the inverse transformed data gives an Fe-O distance of 1.7 Å in the r-window of 1.5-2.0 Å and an Fe-Fe distance of 2.7 Å in the r-window of 2.0-3.0 Å. The inverse transformed data in the r-window of 3.0 - 4.0 Å gives Fe-Si distances from 3.1 to 3.5 Å for CIT-5 and around 3.3 Å for CHA, ZSM-5 and MOR with a high Debye-Waller factor (0.0060 A^2). For Fe-CIT-5 the shell of Si and Al shows the biggest deviation in distances due to the side pockets in the 10 Å linear channels, so iron oxide clusters exist in the 10 Å - 12 Å cages.

The relatively high FeFe coordination numbers between 4.07 and 5.88 for the polynuclear Fe complexes hide the information about possibly formed dinuclear sites. Furthermore with respect to the actual catalysts under operando conditions these nano-α-Fe_2O_3 samples may very well undergo decomposition during the high temperature pretreatment. It is therefore unlikely that after we pass the elevated temperature of the Curie point for the weak ferromagnetic moments at (946 K) and the anti-ferromagnetic Néel point at (965 K) of this nano-material the nature is still the same as this in the reported EXAFS structural characterization. Dilatometric experiments have shown that a dilatation of 1.3 to 1.7 is observed in the different axis of the nano-α-Fe_2O_3 samples, and in the context of the ZSM-5 supported nano-clusters this temperature effect around 950 K gives a favorable spreading of the iron phase over the interior walls of the channel system at the elevated temperature of the Curie point for the weak ferromagnetic moments and the anti-ferromagnetic Néel point of these nano-materials. In larger channel topologies this cannot be effectuated. It should also be noted that the pretreatment conditions higher than 1000 K are more severe than the synthesis conditions around 823 K of these clusters. Therefor we prefer not to make correlations between the observed EXAFS data of the precursor materials and the active pretreated catalysts [33, 35, 38].

SQUID measurements in **Figure 5.** show a paramagnetic contribution stems from isolated lattice iron ions and anti-ferromagnetism is due to iron in small clusters. SQUID measurements in **Figure 6.** furthermore show at 260 K a low intensity Morin transition (weak ferromagnetism) for the Fe-ZSM-5 (0.1 mm) crystals. In the EPR spectra such as this in **Figure 7**. the ratio of octahedral/tetrahedral iron can be assumed from the presence of lines at g= 2/4.3, and depends on sample hydration. Mössbauer spectroscopy, is sensitive to the HMF of anti-ferromagnetic coupled dimers, valence and spin-state of paramagnetic $Fe^{(III)}$ ions and Morin-transitions are not seen in the magnetic sub-spectrum e.g. changes in the quadrupole splitting, since the fraction of the clusters that is > 20 nm is extremely small.

In the catalytic experiments as seen in **Table 1 & 2.** we work at 0.1 - 0.5 % level of the adsorption capacity of the zeolite, since 20 μmol is the maximum amount of methanol that can be obtained from a 0.1g zeolite extraction, versus the 40-200 μmol methanol observed in experimental 1g zeolite extractions. α-Oxygen is formed on iron sites on FeZSM-5 [2-8] by N_2O decomposition at elevated temperature [9-10]. We speculate that the medium-sized pores are the most active sites for methanol formation. On the basis of the sorption properties and molecular dimensions for CHA the effective size of the pore openings is determined to be between 4.89 and 5.58 A, the cage has a length of 11 A and a diameter of 6.5 A. The low selectivity result of the CHA sample is clearly influenced by the calcinations procedures, which if to severe as in our case, could produce some collapse of the structure [37]. The larger zeolites MOR, FER and CIT-5 may not offer the same hydrogen bonding environnement for the small molecules of the type of methane an methanol and oxidation as well as coupling reactions are observed [29]. On the study of effects of the Si/Al ratio on the iron distribution pattern occurring and its impact on oxygen activation, a formal negative correlation between the Al content and the methanol formation activity, is seen. Since the proton density is lower there are fewer clusters with a formal charge of +1 can be formed and replace a proton and a lower number of clusters means fewer active sites. [17, 35, 36].

6. CONCLUSIONS

Mono-nuclear iron sites in [Fe]-ZSM-5, dinuclear iron clusters stabilized inside the matrix of thermally activated Fe-ZSM-5 and iron sites associated with nano-hematite phases in ZSM-5 and various other zeolites are compared using theoretical tools as well as experiments. For [Fe]-CIT-5, [Fe]-MOR, [Fe]-ZSM-5 and [Fe]-CHA tetrahedral Fe^{3+} coordination in the lattice is observed by EPR and different (distorted) octahedral signals in the impregnated FeCIT-5, FeMOR, FeZSM-5 and FeCHA materials are seen. EXAFS data on highly loaded samples with 5 wt% iron gives an Fe-O distance of 1.7 Å and an Fe-Fe distance of 2.7 Å in and Fe-Si distances from 3.1 to 3.5 Å for CIT-5 and around 3.3 Å for CHA, ZSM-5 and MOR. For tetrahedral iron and silicon atoms bond distances of 1.65 for Si-O and 1.90 Å for Fe-O as obtained from QM, in the direction of the Fe-O-Si bridge, the Fe-O-Si angles are close to 131.2°. For the dinuclear iron complexes bond distances of 1.95 and 1.99 Å obtained from QM, in the direction of the Fe-O groups, accompanied by Fe-O-Fe angles close to 90-93° and an O-Fe-O bite angle of 74.5°. Where isolated species coexist with nanoclusters, the averga cluster size will be larger and this can explain the higher Fe-Fe coordination numbers seen for the 823 K calcined, but not 950 K pretreated catalysts for which we propose the EXAFS data in this publication. We think both mononuclear ions and oxo nanoclusters are the two active types of iron species. With respect to the actual catalysts under operando conditions these rhombohedrical nano-α-Fe_2O_3 samples, with an atomic arrangement resembling that of the cubic iron oxides seen in our EXAFS data, may very well undergo decomposition during the high temperature pretreatment. It is therefore unlikely that after we pass the elevated temperature of the Curie point for the weak ferromagnetic moments at (946 K) and the anti-ferromagnetic Néel point at (965 K) of this nano-material the nature is still the same as this in the reported EXAFS structural characterization. A dilatation of 1.3 to 1.7 is observed in the different axis of the nano-α-Fe_2O_3 clusters supported on ZSM-5 may give a favorable spreading of the iron phase over the interior walls of the channel system at the elevated temperature of the Curie point for the weak ferromagnetic moments and the anti-ferromagnetic Néel point of these nano-materials.

Methane can be converted into methanol with α-Oxygen at room temperature using N_2O as a selective oxidant inside the pores of various zeolites [31]. Extraction of the catalysts with water is preferred over acetonitrile-water mixtures [32]. The iron oxide clusters realize the sub-stoichiometric oxidation of methane, favoring oxygen insertion chemistry over coupling chemistry. Whereas the [Fe]- ZSM-5 and the [Fe]- CIT-5 are inactive, $Fe_{2.0}$ – ZSM-5 and $Fe_{5.0}$ – ZSM-5 are the best zeolites, indicating the necessity of an iron loading > 0.5% wt Fe to observe methanol formation. In these catalysts the iron oxide dimers and small clusters realize the sub-stoichiometric (based on a diiron sites) oxidation of methane and benzene after N_2O activation, the dinuclear Fe active sites may favor oxygen insertion chemistry over coupling chemistry. $Fe_{5.0}$ – CIT-5 and other zeolites with rings larger than 10MR give various products, the selectivity for methanol is lower, other products are also seen, due to absence of product shape selective containment. We speculate that the medium-sized pores are the most active sites for methanol formation. The low selectivity result of the CHA sample is influenced by the severe calcinations procedures, which produces the collapse of the structure during the pretratment. Larger clusters may have a either a formal charge of +1 and and some might be expected easily replace a proton or a formal charge of 0 and can be associated with acidic protons via hydrogen bonding. The larger zeolites MOR, FER and CIT-5 may not offer the same hydrogen bonding environnement for the small molecules of the type of methane an methanol and the reactivity is not restricted to oxidation reactions alone since also coupling type reactions are observed. In situ Raman studies of the operando catalysts after pretreatment, oxidation and reaction are now under way to give more evidence for the study of the interaction of the iron species with the zeolite framework and with the acidic protons, which is complex and not fully understood.

7. ACKNOWLEDGMENTS

PPKG thanks the UCL FSR for financial support of a project on materials synthesis and the FNRS FRFC for financial support of a project on the oxidation of alkanes. I wish to thank Dr.Jesse Goellner and Prof.Bruce Gates, UC Davis, Dept. of Chemical Engr. & Materials Science for help with the XAS.

8. REFERENCES

[1.]P.P. Knops-Gerrits, P.A. Jacobs, A. Fukuoka, M. Ichikawa, F. Faglioni, W.A. Goddard III, *J.Mol.Cat., A.*, 166, 3-15, 2001, Special Issue, P.P. Knops-Gerrits, M. Witko, R. Millini & W.A. Goddard III, Eds.
[2.] G.I. Panov, V.I. Sobolev, A.S. Kharitonov, J.Mol.Catal. 61 (1990) 85.
[3.] P.P. Knops-Gerrits, W.J. Smith, Stud.Surf.Sci.Catal., 130, Proceedings of the 12th Int. Congress on Catalysis, 2000, 3531-3539.
[4.] J.S.J. Hargreaves, G.J. Hutchings, R.W. Joyner, Nature, 348 (1990) 428.
[5.] J.R. Anderson, P. Tsai, J.Chem.Soc., Chem. Commun., 1435 (1987).
[6.] L.E. Iton, R.B. Beal, D.T, Hodul, J.Mol.Catal., 21 (1983) 151.
[7.]J. Cejka, A. Vondrova, B. Wichterlova, G. Vorbeck, R. Fricke, Zeolites, 14 (1994) 147.
[8.]Y.F. Chang, J.G. McCarty, Y.L. Zhang, Catal. Letters, 34 (1995) 163.
[9.]X.B. Feng, W.K. Hall, J.Catal. 166 (1997) 368.
[10.]V.I. Sobolev, G.I. Panov, A.S. Kharitonov, A.M. Volodin, V.N. Romanikov., K.G. Ione, J. Catal., 139 (1993) 435
[11.]A.S. Kharitonov, G.I. Panov, K.G. Ione, V.N. Romanikov., G.A. Sheveleva, L.A. Vostrovikova, V.I. Sobolev, US Patent 5,110,995 (1992).
[12.]G.I. Panov, A.S. Kharitonov, G.A., Sheveleva, WO 95/27691 (1994).
[13.]V.I. Sobolev, K.A. Dubkov, O.V. Panna, G.I. Panov, Catal.Today, 24 (1995) 251
[14.]K.A. Dubkov, V.I. Sobolev, G.I. Panov, Kinet. and Catal., 39 (1998) 72 & 299
[15.]K.A. Dubkov, V.I. Sobolev, E.P., Talsi, M.A. Rodkin, N.H. Watkins, J.Mol.Cat.A 123 (1997) 155.
[16.]V.I. Sobolev, K.A. Dubkov, E.A. Paukshtis, L.V. Pirutko, M.A. Rodkin, G.I. Panov, Appl. Catal. A, 141 (1996) 185.
[17.]G.I.Panov, V.I.Sobolev, K.A.Dubkov, A.S.Kharitonov, Stud.Surf.Sci.Catal. 101 (1996) 493.
[18.]A.K. Uriarte, Rodkin,M.A., Gross,M.J., A.S. Kharitonov, G.I. Panov, Stud. Surf. Sci. Catal., 110 (1997) 857.
[19.]G.I.Panov, A.K. Uriarte, M.A. Rodkin, V.I. Sobolev, Catal. Today, 41 (1998) 365.
[20.]N.S. Ovanyesan, A.A. Shteinman, K.A. Dubkov, V.I. Sobolev, G.I. Panov, Kinet.Catal., 39 (1998) 792
[21.]G.I. Panov, V.I. Sobolev, K.A. Dubkov, A.E. Parmon, N.S. Ovanyesan, A.E. Shilov, A.A. Shteinman, React. Kinet. Catal.Lett., 61, 251 (1997).
[22.]P. Wagner, M. Yoshikawa, M. Lovallo, K. Tsuji, M. Tsapatsis, M.E. Davis, J.Chem.Soc., Chem.Commun., 22 (1997) 2179,.
[23.]M. Hafele, A. Reitzmann, E. Klemm, G., Emig, Stud. Surf. Sci. Catal.,110 (1997) 847.

[24.]E. Suzuki, K. Nakashiro, Y. Ono, Chem. Lett., (1988) 953.
[25.]Y. Ono, K. Tohmori, S. Suzuki, K. Nakashiro, E. Suzuki, Stud.Surf.Sci.Catal., 41 (1988) 75.
[26.]M.H. Gubelmann, P.J. Tirel, EP 341165 (1989).
[27.]M.H. Gubelmann, J.M. Popa, P.J. Tirel, EP 406050 (1990).
[28.]P.J. Tirel, M.H. Gubelmann, J.M. Popa, Rec.prog.rep., 9th.Int.Zeol.Conf. Montreal, (1993).
[29.]T. Ito, J.H. Lunsford, Nature, 314 (1985) 721.
[30.]P.P.Knops-Gerrits, Xiao-yuan Li, W.A.Goddard III, Raman Spectroscopy, Wiley, 782 (2000) S-L Zhang & B-F Zhu, Eds.
[31.]P.P. Knops-Gerrits & W.A. Goddard III,*J.Mol.Cat.,* A. 166 (2001) 127, W.A. Goddard III, P.P. Knops-Gerrits, R. Millini, M. Witko, Eds.
[32.] P.P. Knops-Gerrits, W.A. Goddard III, Stud. Surf. Sci. Catal., 135 (2001) 265. A.Galarneau, F.Di Renzo, F.Fajula, J.Védrine, Eds.
[33.] P.P. Knops-Gerrits, A. Fukuoka, W.A. Goddard III, Journal of Inorganic Biochemistry, 86 (2001) 297.
[34.] P.P. Knops-Gerrits, M.Cuypers, Stud. Surf.Sci.Catal., 142 (2002) 263.
[35.] R. Joyner, M. Stockenhuber, J.Phys.Chem.B., 103 (1999) 5963.
[36.] E.M. El-Malki, R.A. van Santen, W.M.H. Sachtler, J.Phys.Chem.B., 103 (1999) 4611.
[37.]R. Shah, M.C.Payne, M.H. Lee, J.D.Gale, Science, 271 (1996) 1395.
[38.] B.C. Gates, Chem.Rev, 95 (1995) 511.

Chapter 18

Direct Partial Oxidation of Methane to Methanol in a Specially Designed Reactor

Qijian Zhang[1,2], Dehua He[1,2,*], Xin Zhang[1,2], Qiming Zhu[1,2], and Shuiliang Yao[3]

[1]State Ley Laboratory of C_1 Chemistry and Technology, Department of Chemistry, Tsinghua University, Beijing 100084, China
[2]Tsinghua University and ABB Joint Chemistry Laboratory on Greenhouse Gas Control Technology, Beijing 100084, China
[3]Research Institute of Innovative Technology for the Earth, Kyoto 619–0292, Japan

Studies on direct gas-phase partial oxidation of methane to methanol were carried out in a specially designed reactor. A comparatively high yield of methanol was obtained over a wide temperature range. When the O_2 in the feed gas was depleted, the reaction would quickly terminate and the secondary reaction of the produced methanol would not take place significantly. The total feed gas flow rate had little effect on methanol yield while the pressure was an important influencing factor.

Introduction

The direct partial oxidation of methane to methanol was first found possible in the early 20th century (*1-3*). Since then, it has been sporadically investigated for about 80 years because of the great industrial potential for the use of natural gas. During the 1980's, the investigation was stimulated by the two "energy crises" that occurred in the 1970s and early 1980s. In 1984, Liu et al. (*2*) and

Gesser et al. (*3*) reported high selectivity of methanol (more than 80%) at methane conversion of higher than 8% from the catalytic and non-catalytic oxidation of methane. These promising results inspired more researchers to carry out further studies into this reaction in the following decades. Some quite good and important results have been achieved and several reviews printed (*4-9*).

Over temperature ranges of 400-500℃ and pressure ranges of 3.0-5.0 MPa, homogeneous partial oxidation can result in quite good methane conversion and methanol selectivity. Till now, the homogeneous reaction has produced the most promising yields of methanol plus formaldehyde. The yields are at least as good as those reported for catalytic processes (*5*). Therefore studies on homogeneous reactions have attracted much attention. The effect of many factors, such as temperature, pressure, feed gas ratio, residence time, reactor type and reactor wall material, have been studied, but different researchers reported different results. There is no general agreement on how to control the oxidation of methane to favor the production of methanol, and how the reaction conditions affect the product distribution. Some results and explanations have even been contradictory (*6-9*).

In our previous work, a comparatively high yield of methanol from the partial oxidation of methane was obtained (*7*). The high yield was attributed to the special structure of the reactor, in which the feed gas was isolated from contact with the metal wall of the stainless steel. In our recent works, the reaction has been further investigated. In this paper, the results are reported and a "one pass" concept is proposed for producing methanol from pipeline natural gas.

Experimental

The gas phase partial oxidation of methane was carried out in a specially designed quartz lined stainless steel tubular reactor at 300-500℃ and 1.0-5.0MPa. The inner quartz line (460 mm long, 6.7 mm i.d.) was tightly fixed in the stainless steel line (Figure 1). A Viton O-ring pressed by a locking nut was used in the top of the quartz line for preventing the gas from leaking into the ringed gap between the quartz line and the stainless steel line. The reactor temperature (outer-wall temperature of the stainless steel) was controlled by a temperature controller with a thermocouple. The reaction temperature (center temperature of the inner quartz line) was measured by a thermocouple inside a ϕ3 stainless steel tube which was covered by a quartz tube (5.5 mm o.d.) with an upper dead end (called shelth tube). The assembly was designed to prevent the contact of the reactants with the metal walls of the stainless steel so as to eliminate the influence of the metal wall on the reaction and thus avoid the deep oxidation caused by the metal surface.

The flow rate of each reactant was controlled by a mass controller and the pressure was controlled by a back-pressure regulator. The reaction effluent was kept at about 150 °C to prevent condensation of liquid products. The products were analyzed by an on-line gas chromatography (SHIMADZU GC-8A, TCD, H_2 as carrier gas).

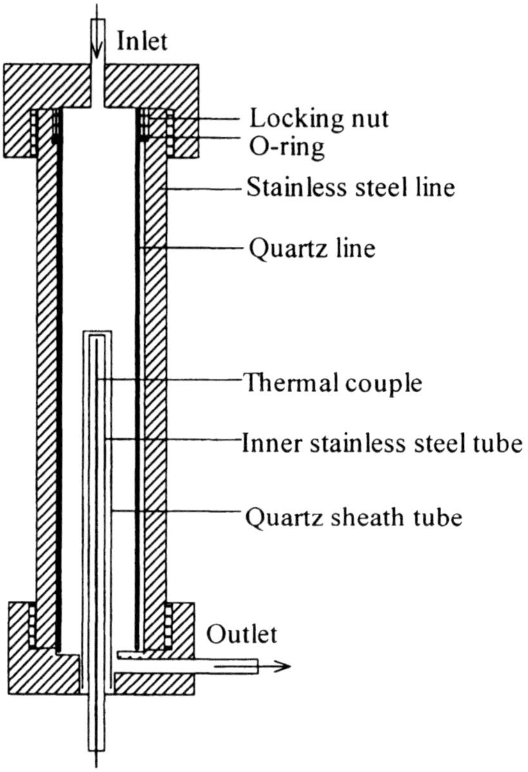

Figure 1. Illustration of the structure of tubular reactor

Results and Discussion

In the specially designed reactor, comparatively high yields of methanol (ca 7-8%) were obtained at 430-470℃ (Table 1). At the temperature of 422℃, only 1% CH_4 and 20% O_2 was converted. When the temperature was increased to 429

Table 1 Effect of rezction temperatures on the conversion of methane and oxygen and the selectivities of products .

$T/℃^*$	Conversion /%		Selectivity /%				CH_3OH Yield /%
	O_2	CH_4	CO	CO_2	CH_3OH	C_2H_6	
422	20	1.0	80.1	0	19.9	0	0.2
429	100	13.0	34.4	5.0	60.7	0.0	7.9
435	100	13.0	33.8	4.4	61.8	0.1	8.0
445	100	12.7	33.7	4.5	61.7	0.2	7.8
455	100	13.1	33.7	4.8	62.3	0.2	8.2
466	100	12.8	34.4	6.4	59.0	0.3	7.6
475	100	11.7	38.3	6.3	57.0	0.4	6.6
485	100	9.6	47.8	4.6	47.0	0.6	4.5
495	100	7.3	65.7	7.1	26.3	0.9	1.9

P=5.0 MPa, $CH_4/O_2/N_2$=100/10/10 (mL/min), id=6.5 mm

*The thermocouple was placed in the center of the inner quartz line at the point 14 cm long from the furnace bottom.

℃ (just 7℃ higher), the conversion of oxygen increased quickly to 100% and conversion of methane increased to 13%. Such results showed typical character of a free radical reaction. The very rapid rise in conversions has also been observed by many other researchers (*8-11*).

It has also been observed by other researchers that a further increase in reactor temperature, generally results in a slight drop in the conversion and methanol selectivity (*8-11*). In our experiments, oxygen was depleted at temperatures above 430℃, but methane conversion was kept constant at ca. 13% and methanol selectivity was kept at ca. 60% before the decreasing at temperatures above 470 ℃. It is supposed that the reaction mainly took place in a very short zone, and the oxygen in the feed was used up in this zone. In the downstream of the zone, there was no oxygen that could be used. There might be no oxidizing free radical to react with produced methanol which leads to the deep oxidation.

We measured the axial temperature profiles (inside temperatures of the quartz line) during the reactions by moving the thermocouple inside the long quartz sheath tube along the axis of the reactor (Figure 2A). The temperature profiles obtained at different reactor temperatures (outer-wall temperatures of the stainless steel) are shown in Figure 2B. Considerable reaction began to occur in

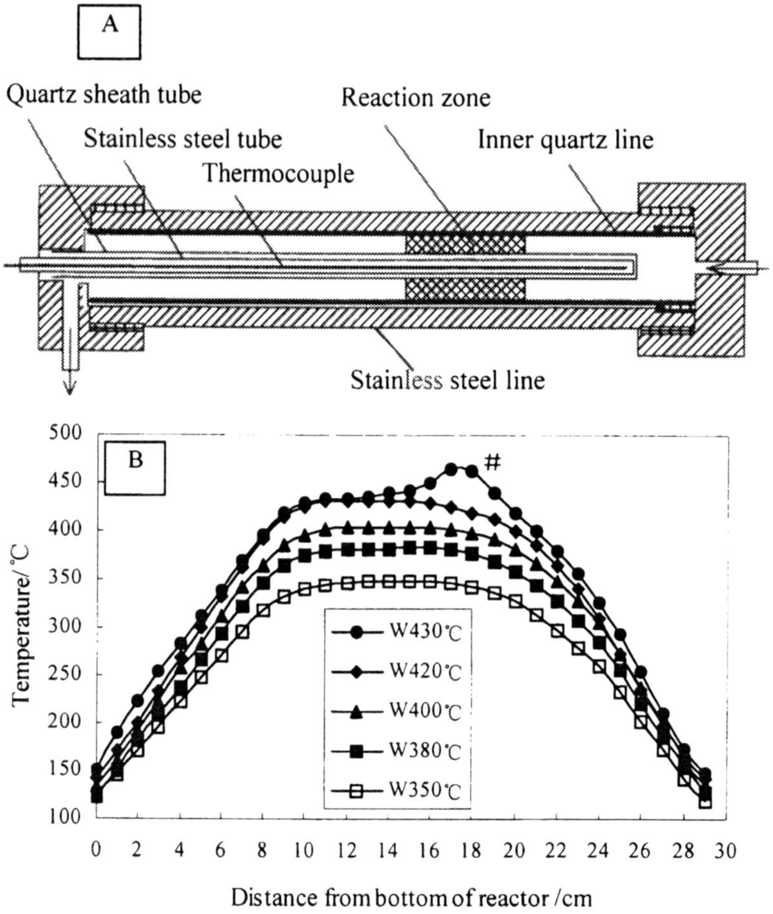

Figure 2. Temperature profiles of reactor at different reaction temperatures
reaction condition: $CH_4/O_2/N_2$=100:10:10 (mL/min), P=5.0MPa
'W' stands for outer-wall temperature of the stainless steel line

the inner quartz line when the reactor temperature (outer-wall of the stainless steel) increased to 420 ℃ (while the measured temperature in the center of the quartz line at the point 14 cm long from the furnace bottom was 432 ℃). At this temperature, oxygen was only partially consumed, and the temperature profile had little difference with those measured at lower reactor temperatures. When the reactor temperature (outer-wall temperature of the stainless steel line) increased to 430℃, the oxygen was depleted, and there appeared an abrupt temperature increase in the inner quartz line (as shown in Figure 2B Curve #). Because the oxidation of methane is a exothermic reaction, such an increase in temperature may indicate that the reaction proceeded primarily in a short zone along the reactor (reaction zone, Figure 2A) and a great deal of heat was released in this zone causing the temperature to increase. Because the CH_4/O_2 ratio of the feed gas was high, the oxygen was the limiting species. It was used up in the fierce reaction zone. In the downstream of the zone, there was no oxygen existing so little secondary reaction could occur.

Table 2 Partial oxidation of methane in stainless steel line

*T/℃**	*Conversion /%*		*Selectivity /%*				*CH_3OH Yield /%*
	O_2	*CH_4*	*CO*	*CO_2*	*CH_3OH*	*C_2H_6*	
386	5.3	0.5	31.0	61.1	7.9	0.0	0.040
417	100.0	6.9	54.8	29.8	15.2	0.3	1.047
438	100.0	7.2	55.5	29.5	14.4	0.6	1.044
459	99.9	7.7	54.2	30.2	14.7	0.9	1.126
480	100.0	8.1	52.3	28.4	18.2	1.2	1.477

P=5.0MPa, $CH_4/O_2/N_2$=100/10/10 (ml/min), id=6.5mm
*The thermocouple was placed at the point 14 cm long from furnace bottom.

It also can be seen that CO was the initial product and CO_2 was produced only when oxygen was completed converted (Table 1). In the experiment with just the stainless steel line reactor (no inner quartz line), CO_2 was found to be produced at the temperature when no significant reaction occured, and its selectivity was higher than CO selectivity (Table 2). In our earlier research (*9*), CO_2 was the initial product at low temperatures, and this was explained by the catalytic deep oxidation by the metal surface in the ringed gap between the inner lining quartz tube and the stainless steel tube. This indicated that the feed gas could leak into the ringed gap and result in the contact of the feed gas with the metal wall. The selectivity of methanol was about 39% when oxygen was completely consumed (*9*), which was lower than that obtained in the specially deigned reactor. Compared to the reaction conditions, the ultimate controlling factor was the reactor structure. The leak of feed gas into the ringed gap might be an important factor that influences the partial oxidation of methane. In one of the our papers, it has been stated that the isulation of the ringed gap between the inner quartz line and the SS tube is of great importance (*7*).

The effect of feed gas total flow rates on the partial oxidation of methane was also investigated. The results are shown in Table 3. When the total flow rate was increased from 120 mL/min to 180 mL/min, the conversion of CH_4 and yield of CH_3OH decreased slightly. Further increase in the total flow rate did not affect the conversion of CH_4 and yield of CH_3OH markedly. Therefore, the total gas flow rate had very little effect on the oxidation of methane when the conversion of oxygen reached 100%. It is worth noting that the loss of methanol resulted in the increase of CO while the selectivity of CO_2 was not affected. Such results indicate that CO and CH_3OH were produced via the same intermediate(s).

Table 3. Effect of total flow rate on the gas phase partial oxidation of methane to methanol.

Total flow rate /(mL/min)	*Conversion/%*		*selectivity/%*				*CH_3OH yield/%*
	CH_4	*O_2*	*CH_3OH*	*CO*	*CO_2*	*C_2H_6*	
120	13.1	100	62.3	33.7	4.8	0.2	8.2
180	11.4	100	57.9	37.1	4.4	0.5	6.6
240	11.5	100	57.4	37.4	4.5	0.7	6.6
360	11.9	100	55.7	35.1	8.2	1.0	6.7

$CH_4/O_2/N_2$=100:10:10 (mL/min), P=5.0 MPa, T=450 ℃

We also measured the axial temperature profiles of the reactor during the reaction under different flow rates (Figure 3). At the reactor temperature of 430 ℃, the oxygen was completely consumed for the flow rates of 120 and 240 ml/min, and the temperature increases in the reaction zone were observed during the reaction. It can be seen from the temperature profiles that the length of the reaction zones, in which the temperature changes were observed, were almost the same (about 4 cm long) for both 120 and 240 mL/min (Figure 3). But the temperature increase in the zone was more fiercely in flow rate of 240 mL/min than in that of 120 mL/min and the hotspot temperature in the reaction zone reached 475 ℃ for the 240 mL/min flow rate. These results indicate that increasing total flow rate only resulted in the increase of the amount of feed gas

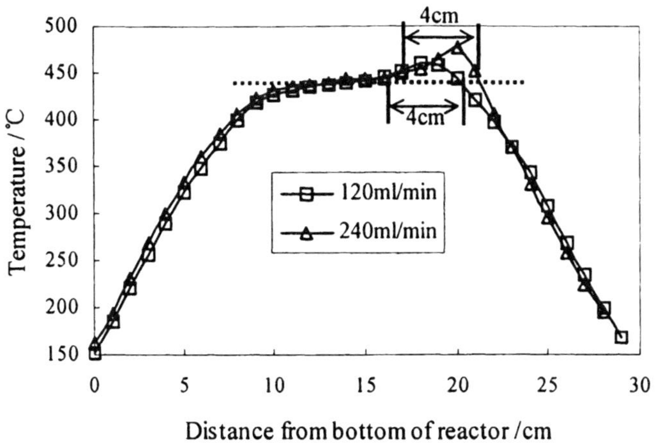

Figure 3. Temperature profiles of reactor under different flow rates. Reaction conditions: P=5.0MPa, $CH_4/O_2/N_2$=10/1/1, Wall Temperature=430 ℃

reacted in the reaction zone which caused the temperature to increase to a higher value. The phenomenon of the temperature changes in the reaction zone under different flow rates also suggests that the oxidation of methane occurred primarily and proceeded fiercely and very quickly in the zone.

The reaction pressure is an important factor for partial oxidation of methane to methanol (Figure 4). At pressure of 1.0MPa, the oxygen was not consumed significantly (that is the reaction did not take place significantly) until 470℃, while at pressure of 2.0MPa, the oxygen was consumed completely at less than 450℃. And increasing the pressure resulted in the complete oxygen conversion at even low temperature (420-430℃). These results were in agreement with the general conclusion drawn by Foulds et al. (*8*) and Burch et al. (*10*).

The effect of pressure on the products distribution are shown in Table 4. The selectivity and yield of methanol increased as the pressure increased and began to level off when the pressure was above 4.0 MPa. A similar trend has been observed (*8*) though Burch et al. (*10*) reported that no smooth trend about the effect of pressures on methanol selectivity was found. At pressure of 1.0 MPa, the complete oxygen conversion occurred at temperature as high as 500 ℃. The dominant product was CO, and there was significant C_2H_6 observed in the products. Bearing in mind that the oxidative coupling of methane to ethane and ethene was carried out at low pressure and high temperature, high pressure should be favorable to the production of methanol from partial oxidation of methane.

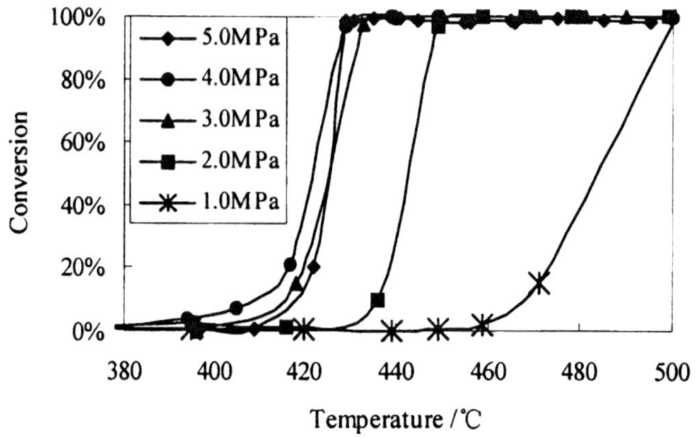

Figure 4. Effect of pressure on O_2 conversion on the partial oxidation of methane at $CH_4/O_2/N_2$=100/10/10(mL/min).

Table 4. Effect of pressure on the products distribution for the partial oxidation of methane to methanol

Pressure /MPa	*Conversion/%*		*Selectivity/%*				*CH_3OH yield /%*
	CH_4	*O_2*	*CH_3OH*	*CO*	*CO_2*	*C_2H_6*	
5.0	13.1	100	62.3	33.7	4.8	0.2	8.2
4.0	12.3	100	53.9	35.1	10.6	0.4	6.6
3.0	9.3	100	32.4	52.2	14.6	0.9	3.0
2.0	8.0	100	17.1	66.6	15.1	1.2	1.4
1.0[a]	8.6	100	6.3	77.3	9.6	5.8	0.5

$CH_4/O_2/N_2$=100:10:10 (mL/min), T=450 ℃, [a]T=500 ℃

At present, natural gas is mainly used as a fuel for the public or for manufacturers of electricity. Based on the above-mentioned results, we proposed a new concept called "one pass" for producing methanol from the pipeline natural gas (as shown in Figure 5). The natural gas taken from the pipeline is mixed with oxygen, heated and passed through the reactor to produce methanol. The effluent mixture is then passed through a condenser to separate the liquid products. Then the outlet gas could be pumped back to the pipeline.

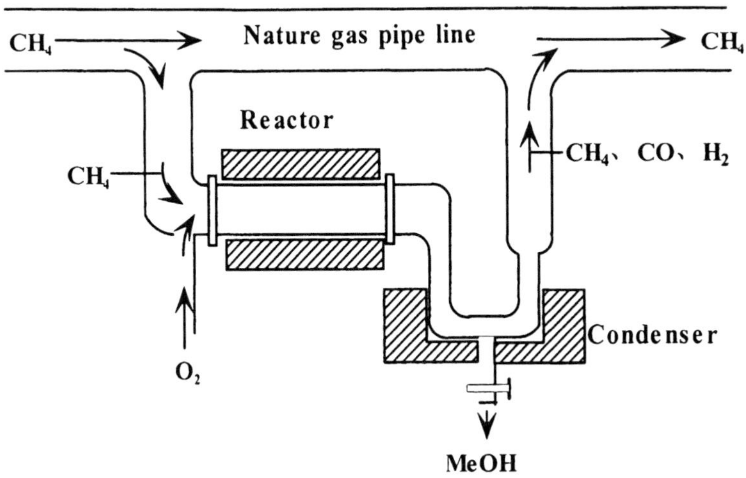

Figure 5. "One pass" concept for producing methanol from pipe-line natural gas.

Conclusion

Comparatively high yields of methanol were obtained in a specially designed reactor. The reaction was completed in a very short zone, and when oxygen in the feed was depleted, the production would be stable. Therefore, high yields can be obtained over wide temperature ranges. The total feed gas flow rate had little effect on the partial oxidation while pressure was an important influencing factor. A "one pass" concept was provided for producing methanol from pipeline natural gas.

Acknowledgement

We acknowledge the financial support from the National Key Fundamental Research Project Foundation of China (G199902240-06), the Research Foundation of Tsinghua University and the Analytical Foundation of Tsinghua University.

References

1. Bone, W.A. and Vheeler, R.V. *J. Chem. Soc.* London **1902**, *81*, 535.

2. Liu, H.F.; Liu, R.S.; Lunsford, J.H. *J. Am. Chem. Soc.* **1984**, *106*, 4117-4121.
3. Gesser, H.D. *International Chemical Congress of Pacific Basin Societies*, 1984, Honololu, Paper 03002.
4. Foster, R. *Appl. Catal.* **1985**,*19*,1.
5. Feng, W.Y.; Knopf, F.C.; Dooley, K.M. *Energy Fuels* **1994**, *8*, 815-822.
6. Omata, K.; Fukuoka, N.; Fujimoto, K. *Ind. Eng. Chem. Res.* **1994**, 33,784-789.
7. Zhang, Q.; He, D.; Li, J., et al. *Appl. Catal. A: General* **2002**, *224*, 201-207.
8. Foulds, G.A.; Gray, B.F.; Miller, S.A.; Walker, G.S. *Ind. Eng. Chem. Res.* **1993**, *32*, 780-787.
9. Han, Z. *Ph.D. thesis*, Department of Chemistry, Tsinghua University, China, 1999.
10. Burch, R.; Squier, G.D.; Tsang, S.C. *J. Chem. Soc. Faraday Trans.* **1989**, *1*, 3561-3568.

Conversion of Methane and Other Greenhouse Gases via Plasmas or Microwave Heating

Chapter 19

Plasma Catalytic Hybrid Reforming of Methane

Thomas Hammer, Thomas Kappes, and Wolfgang Schiene

Siemens AG, Corporate Technology Department CT EN3, Paul-Gossen-Strasse 100, 91052 Erlangen, Germany

Non-thermal plasma induced steam reforming of methane has been investigated applying a dielectric barrier discharge (DBD) reactor for pure plasma processes and a dielectric packed bed (DPB) reactor for plasma catalytic hybrid processes. Neither the H_2O conversion nor the yield and energy efficiency of H_2-formation of the DBD reactor showed to be sufficient for practical application. This could be explained by numerical simulation showing low H_2O dissociation rates and high losses due to vibrational excitation of H_2O and radical recombination. In contrast by plasma catalytic hybrid reforming using nickel as a catalyst good H_2O conversion and high selectivity towards H_2 were obtained. Compared to pure DBD treatment at a temperature of 400 °C the energy requirements for H_2-generation were reduced for a factor 10 down to about 700 kJ/mol. The other reaction products at that temperature were CO_2 and small amounts of CO. Between 400 °C and 600 °C the CO_2-yield increased less than the H_2-yield and increasing amounts of CO were formed, whereas at lower temperatures substantial amounts of C_2H_6 were detected.

Introduction

Hydrogen enriched fuel gases offer the potential for efficient low NOx combustion processes. The reduction of NOx-emission is caused by a reduction of the adiabatic flame temperature due to stable, lean combustion and in some cases due to steam addition or exhaust gas recirculation. In the past reduced NOx-emissions both from gas turbines (*1*,*2*) and internal combustion engines (*3*) were reported. At the same time the energy efficiency of internal combustion engines (*3*) was improved for 15-50 % compared to pure gasoline operation. In stationary applications there is an additional potential for the reduction of specific CO_2-emissions: If the hydrogen is generated by integrated reforming of fossil fuels, carbonaceous products of the reforming reaction like CO_2 (*4*), solid carbon (*5*) or in the case of methane higher hydrocarbons (*6*) may be separated prior to combustion and utilized as raw material e.g. in the chemical industry. For mobile application of PEM fuel cells requiring hydrogen as a fuel on-board generation of hydrogen is desirable because of the high energy storage density of gasoline or diesel fuel.

Large scale production of hydrogen is performed e.g. by catalytic steam reforming of methane. However, this process requires a temperature of about 900 °C. At lower temperatures catalyst coking and poisoning can get a problem. For the generation of hydrogen enriched fuel gases a low temperature reforming process which is not sensitive to coking or poisoning would be desirable.

Recently plasma reforming has been proposed for the efficient generation of hydrogen and higher hydrocarbons in a compact, light weight reactor. For large scale application arc based plasma torches heating the gas very rapidly to temperatures of several thousand degrees Celsius for complete fuel conversion are utilized. However, for small scale application and for incomplete fuel conversion non-thermal plasmas which avoid excessive gas heating are a much better choice. Non-thermal plasma (NTP) reforming induced by dielectric barrier discharges (DBD) has been shown to have the potential for the generation of hydrogen and higher hydrocarbons.

In facilities where waste heat can be utilized to support NTP-reforming an endothermic reforming process like methane steam reforming

$$CH_4 + H_2O \rightarrow CO + 3\ H_2,\ \Delta H^0 = 206\ \text{kJ/mole} \qquad (1)$$

rather than an exothermic one will be applied: However, neither the selectivity nor the energy efficiency of DBD induced methane steam reforming showed to be sufficient for practical application. Instead the results indicated that non-oxidative coupling occurred:

$$2\ CH_4 \rightarrow C_2H_6 + H_2,\ \Delta H^0 = 66\ \text{kJ/mole} \qquad (2)$$

This can be explained by numerical model calculations of the plasma induced chemical kinetics predicting insufficient water conversion and high energy losses due to vibrational excitation of water. For this reason plasma catalytic hybrid steam reforming of methane was investigated in the temperature range 200-600 °C.

Experiments on Plasma Induced Steam Reforming of Methane

The conversion of methane and water as well as the yields of hydrogen, carbon monoxide, carbon dioxide, methanole, and higher hydrocarbons up to C_4H_n were investigated as a function of the gas temperature, plasma input power, gas flow, and feed ratio of methane to water.

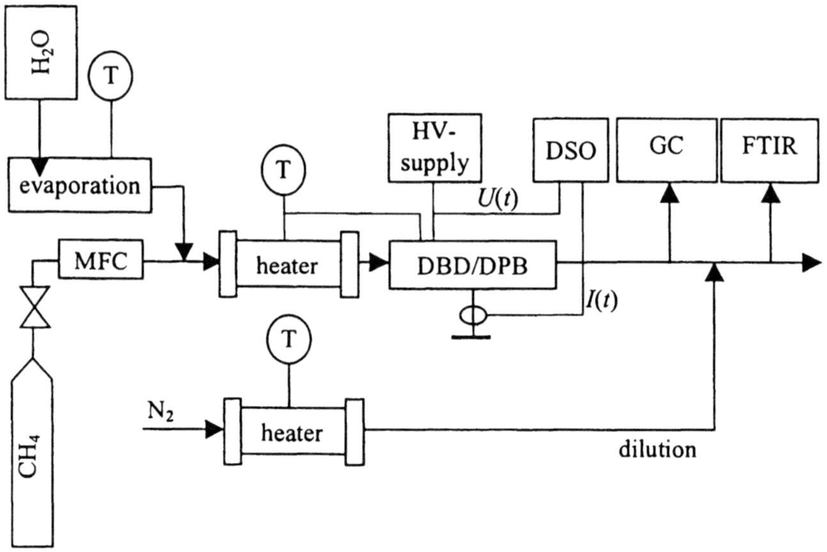

Figure 1. Experimental set-up for the investigation of methane steam reforming. MFC – mass flow controller; DSO – Digital Storage Oscilloscope; DBD/DPB – Dielectric Barrier Discharge/Dielectric Packed Bed Reactor; U(t) – High Voltage Probe; I(t) – Pulse Current Transformer; GC – Gas Chromatograph; FTIR – FTIR Absorption Spectrometer

The experimental set-up (Figure 1) consisted of a gas mixing manifold, electrical heaters for gas conditioning, a thermostatically heated plasma reactor electrically excited by a sinusoidal high voltage power supply, electrical diagnostics for determination of the average plasma input power, and gas analysis.

Methane was fed from a pressurized gas cylinder using a mass flow controller, water was spray injected and evaporated at 250 °C. The gas mixture was heated to the desired temperature using a tubular heat exchanger.

For gas analysis a gas chromatograph (GC: Shimadzu GC 14B equipped with a combination of a HayeSep R/Q column and a mole sieve 13X column) and an FTIR-absorption spectrometer (FTIR: Perkin Elmer System 2000 with a 1 m White cell heated to 190 °C) were applied. Prior to FTIR-analysis the gas was diluted for a factor 10 with pure nitrogen.

From the measured input and output concentrations $c_{species,in}$ and $c_{species,out}$ and the total input and output molar gas flows Q_{in} and Q_{out} of the plasma reactor conversions $\eta_{species}$ and yields $\xi_{species}$ were calculated according to

$$\eta_{species} = \frac{c_{species,in} - c_{species,out}}{c_{species,in}} \tag{3}$$

$$\xi_{species} = \frac{c_{species,out} \cdot Q_{out}}{\sum_{species,in} \alpha_{species,in} \cdot c_{species,in} \cdot Q_{in}} \tag{4}$$

where $\alpha_{species,in}$ gives the stochiometric ratio of output species to input species. Thus the yield depends on the reaction equation assumed for the formation of the product species.

Dielectric Barrier Discharge Reactor

Reference measurements on NTP-induced steam reforming of methane were performed using a coaxial DBD reactor (Figure 2) which was thermally isolated and thermostatically heated to the desired temperature. An alumina tube with an outer diameter of 25 mm and an inner diameter of 20 mm was applied as a dielectric barrier between the ground electrode and the high voltage electrode. The ground electrode was manufactured by flame coating the outer tube surface. The structured high voltage electrode consisted of sharp edged circular discs with a diameter of 16 mm fed on a stainless steel rod. The number and spacing

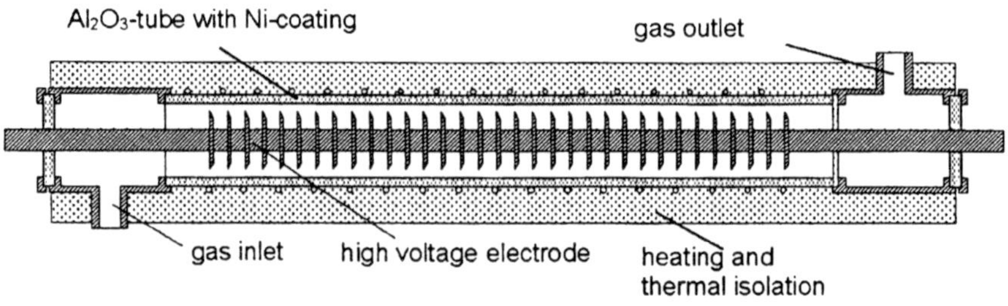

Figure 2. DBD-reactor set-up for methane steam reforming.

of the discs was chosen such that the active region of the reactor had a length of 200 mm.

Dielectric Packed Bed Reactor

Plasma catalytic experiments were carried out in a thermostatically heated dielectric packed bed (DPB) reactor, which combines the features of a catalytic fixed bed tube reactor and of a DBD reactor (Figure 3). Its ground electrode and barrier were identical to those of the DBD-reactor. A stainless steel tube with an outer diameter of 10 mm was used as inner electrode. Thus a discharge gap 10 mm was obtained, which was filled with the catalyst packing consisting of nickel dispersed on ceramic pellets having diameters between 2 and 4 mm. For electrical discharge excitation a sinusoidal high voltage power supply with a maximum output power of 600 W was applied.

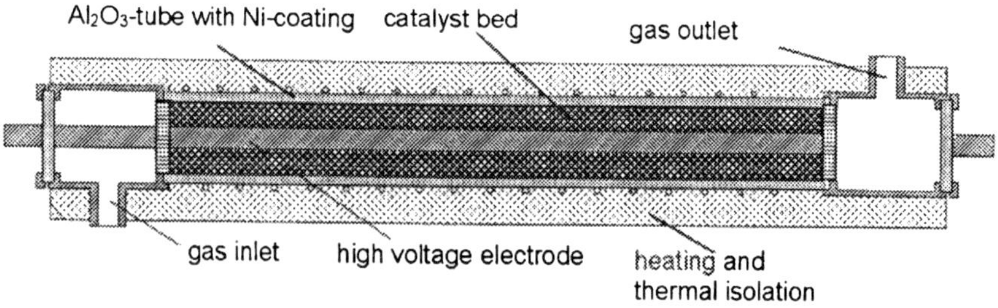

Figure 3. DPB-reactor set-up for methane steam reforming.

Experimental Results

DBD-Induced Reforming

We found that DBD-induced CH_4-conversion and product yields depended on all parameters under investigation. As a reference the dependencies on temperature and plasma input power are shown in Table I and Table II, respectively.

Table I. Dependence of Conversion and Product Yields on the Temperature

T in	CH₄-Conversion and Product Yields in %									
°C	CH_4	H_2	CH_3OH	CO	CO_2	C_2H_6	C_2H_4	C_3H_8	C_3H_6	C_4H_{10}
200	2.3	1.3	0.23	0.43	0.02	1.4	0.03	0.39	0.00	0.45
400	4.9	1.7	0.11	0.31	0.05	2.6	0.11	0.68	0.01	0.37
600	7.9	2.7	0.00	0.21	0.19	4.1	0.71	0.73	0.12	0.26

Experimental conditions: Plasma input power 50 W, flow rate 1 Nliter/min, feed ratio CH_4:H_2O = 1:2

Table II. Dependence of Conversion and Product Yields on the Plasma Input Power

P in	CH₄-Conversion and Product Yields in %									
W	CH_4	H_2	CH_3OH	CO	CO_2	C_2H_6	C_2H_4	C_3H_8	C_3H_6	C_4H_{10}
50	2.3	1.3	0.23	0.43	0.02	1.4	0.03	0.39	0.00	0.45
100	6.9	2.4	0.43	0.79	0.06	2.2	0.06	0.67	0.00	1.03
150	12.1	4.1	0.46	0.96	0.12	3.8	0.12	1.20	0.04	1.78
200	17.3	6.5	0.57	1.12	0.22	5.1	0.28	1.64	0.12	1.35

Experimental conditions: Temperature 200 °C, flow rate 1 Nliter/min, feed ratio CH_4.H_2O = 1:2

Because the net water conversion was negligibly small, the hydrogen yield was calculated under the assumption that only CH_4 but not H_2O contributes to H_2-formation (non-oxidative coupling, eq 2). It depends much stronger on power than on temperature. The conversion of CH_4 is roughly proportional both to temperature and plasma input power, however, the yields of H_2 and higher hydrocarbons (mainly C_2H_6) are not.

Increasing the CH_4:H_2O feed ratio from 1:3 to 1:1 reduced the yields, however, the molar flows of the reaction products remained nearly unchanged. When the flow rate was increased at constant specific plasma energy input (ratio of power to gas flow rate) both the CH_4-conversion rate and the product yields increased.

The energy requirements for H_2-generation did not depend substantially on the specific plasma energy input and decreased with increasing temperature. They ranged from 2 to 8 MJ/mole H_2.

Plasma-Catalytic Hybrid Reforming

In contrast, when plasma-catalytic hybrid treatment was applied both methane and water were converted (Figure 4). Carbon was nearly completely oxidized to CO_2. This observation fits much better to steam reforming followed by a water shift reaction resulting in the gross reaction

$$CH_4 + 2H_2O \rightarrow CO_2 + 4\ H_2,\ \Delta H^0 = 165\ \text{kJ/mole} \qquad (5)$$

Therefore now the H_2-yield was calculated taking into account both the CH_4 and the H_2O-feed according to eq 5.

At temperatures below 400 °C no catalytic conversion was observed when the plasma was switched off. In this temperature range mainly H_2, CO_2, and small amounts of C_2H_6 were formed by plasma-catalytic hybrid reforming. At

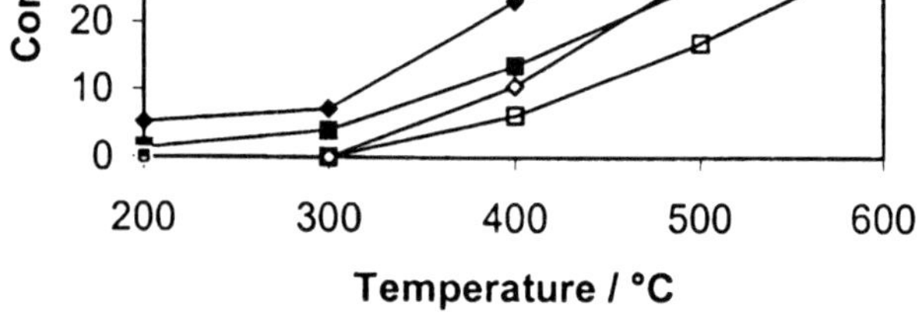

Figure 4. Plasma-catalytic hybrid reforming of methane – conversion as a function of the temperature (flow rate 1 Nliter/min, plasma input power 150 W, feed ratio CH_4:H_2O = 1:2)

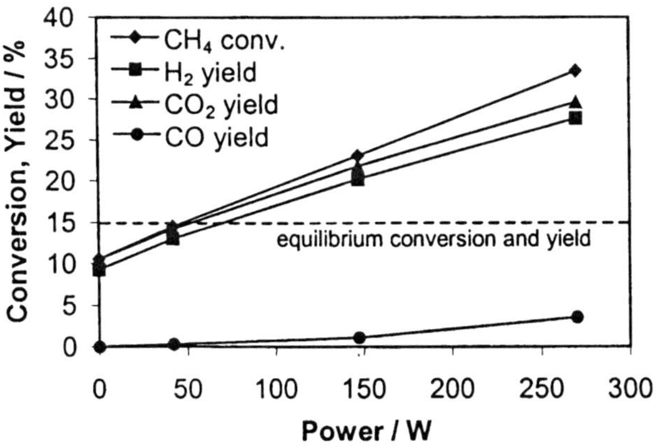

Figure 5. Plasma-catalytic hybrid reforming of methane – conversion as a function of the plasma input power (flow rate 1 Nliter/min, temperature 400 °C, feed ratio CH_4:H_2O = 1:2)

temperatures above 400 °C substantial CO-yields were achieved, and the CO_2-yield dropped relative to the H_2-yield. In contrast to DBD-reforming the C_2H_6-yield got negligible at higher temperatures. Neither carbons deposits nor formation of wax was observed.

Conversion and yields increased linearly with the plasma input power. At temperatures around 400 °C and input powers above 60 W the yields of H_2 and CO_2 exceeded the values calculated from thermodynamic equilibrium relations. With increasing power an increasing selectivity towards CO-formation was observed whereas the selectivity for CO_2-formation decreased.

The dependency of CO- and CO_2-concentrations on specific energy and temperature may be explained by the assumption that CO is formed by plasma induced gas phase processes and by catalytic processes, whereas CO_2 is mainly formed due to a low temperature catalytic water shift reaction converting CO to CO_2.

For all temperatures above 200 °C by the combination of plasma and catalyst the energy requirements were reduced compared to DBD-treatment. Values as low as 315 kJ/mole H_2 were achieved at 600 °C. A question which cannot be answered until now is whether the increased conversions and yields were purely caused by non-thermal plasma effects like ionization, radical formation, and generation of ultraviolet radiation influencing catalytic reactions, or if they at least partly were caused by catalyst heating.

Numerical Model Calculations

Numerical model calculations were performed (*7*) based on a program code named RAWI which was supplied from the Faculty for Electrical Engineering, Karlsruhe University, Germany (*8*). This code allows to simulate homogeneous gas discharges of the DBD-type coupled to a simple electrical circuit model and to the gas discharge induced chemical kinetics. The kinetic model to be used for the model calculations is given by a set of reaction equations and temperature dependent rate coefficients (Arrhenius parameters). Electron collision reaction rate coefficients are calculated from a numerical solution of the electron energy distribution function collision cross sections as a function of the collision energy. Further input data of the model are relative permittivity and thickness of the dielectric barrier, thickness of the discharge gap, external reactor voltage, input gas temperature, pressure, and composition.

Simulations of DBD-induced steam reforming of gas mixtures containing H_2O and CH_4 with feed ratios between 1 and 3 showed, that at 400 °C 35 % to 50 % of the input energy are lost due to vibrational excitation of H_2O and less than 8 % are spent in H_2O dissociation, while another 30 % to 40 % of the input energy are spent in dissociation of CH_4. Therefore the most important reaction under these conditions is

$$CH_4 + e \rightarrow CH_3 + H \qquad (6)$$

followed by the abstraction reaction

$$CH_4 + H \rightarrow CH_3 + H_2 \qquad (7)$$

and by the recombination of methyl radicals forming C_2H_6

$$CH_3 + CH_3 \rightarrow C_2H_6 \qquad (8)$$

However, according to the numerical simulation about 80 % of the radicals are lost by the radical recombination

$$CH_3 + H \rightarrow CH_4 \qquad (9)$$

Therefore not only the product spectrum of DBD-induced steam reforming but the poor energy efficiency can be understood, too.

Conclusions

NTP-reforming of methane in CH_4-H_2O-mixtures was investigated. Dielectric barrier discharges mainly induce formation of H_2 and C_2H_6 by decomposition of CH_4 at negligible H_2O-conversion. Numerical simulations of the DBD-induced chemical kinetics showed that less than 50 % of the energy dissipated in discharge filaments are spend for CH_4- and H_2O-dissociation – 38 % and 7 % respectively. High losses are caused by electron collision vibrational excitation of H_2O. Another source of losses is the radical recombination of CH_3 and H to CH_4.

By the combination of the DPB-plasma and a Ni-catalyst high H_2O-conversion rates and selectivities towards H_2- and CO_2-formation were achieved. Small amounts of CO were formed at high temperatures > 400 °C, whereas substantial amounts of C_2H_6 could be detected at low temperatures. For temperatures above 200 °C the energy requirements for H_2-formation dropped for an order of magnitude down to values as low as 315 kJ/mole H_2 at 600 °C. Further improvements can be expected if thermal losses e.g. due to barrier heating could be avoided, which may be well above 60 % of the plasma input power (*9*).

References

1. Phillips, J.N.; Roby, R.J. *Power Engineering* **2000**, *May*, 36-40.
2. Maughan, J.R.; Bowen, J.H.; Cooke, D.H.; Tuzson, J.J. *Proc. ASME Cogen-Turbo Conference*, **1994**, 381-390.
3. Jamal, Y.; Wyszynski, M.L. *Int. J. Hydrogen Energy*, **1994**, *19(7)*, 557-572.
4. Nihous, G.C.; Mori, Y.; Masutani, S.M.; Vega, L.A.; Kinoshita, C.M. *Int. J. Hydrogen Energy*, **1994**, *19(4)*, 387-94.
5. Fulcheri, L.; Schwob, Y. *Int. J. Hydrogen Energy*, **1995**, *20(3)*, 197-202.
6. Kogelschatz, U.; Zhou, L. M.; Xue, B.; Eliasson, B. In *Greenhouse Gas Control Technologies, (Proc. 4th Int. Conf., Meeting Date 1998)*; Eliasson, Baldur; Riemer, Pierce; Wokaun, Alexander, Eds.; Elsevier, Oxford, UK, 1999, pp 385-390.
7. Schiene, W. Ph.D. thesis, Faculty of Electrical Engineering, Ruhr-University Bochum, Germany, 2002.
8. Müller, H. Ph.D. thesis, Faculty of Electrical Engineering, University of Karlsruhe, Germany, 1991.
9. Nozaki, T; Unno, Y; Miyazaki, Y; Okazaki, K *J. Phys. D - Applied Physics*, **2001**, *34(16)*, 2504-2511.

Chapter 20

Methane Coupling and Reforming Using Non-Equilibrium Pulsed Discharge at Room Temperature: Catalyst–Pulsed Discharge Combined System

S. Kado[1], K. Urasaki[2], H. Nakagawa[3], K. Miura[3], and Y. Sekine[2]

[1]Department of Applied Chemistry, School of Engineering, The University of Tokyo, Hongo, Bunkyo-ku, Tokyo 113–8656, Japan
[2]Department of Applied Chemistry, School of Science and Engineering, Waseda University, Okubo, Shinjyuku, Tokyo 169–8555, Japan
[3]Department of Chemical Engineering, Graduate School of Engineering, Kyoto University, Kyoto 606–8501, Japan

The comparison between DC corona and pulsed discharge was made for the formation of C2 hydrocarbons. Using the corona discharge, the consecutive dehydrogenation from ethane was observed. On the other hand, in the pulsed discharge reaction, acetylene was produced with 95% selectivity independent of methane conversion. In addition, the combination of catalyst and pulsed discharge had a great effect on the selectivity. The use of Ni-catalyst with the pulsed discharge promoted carbon dioxide reforming selectively at room temperature. And the yield of ethane reached 49% by using 1 wt% Pt/SiO_2.

Introduction

Natural gas has attracted considerable attention as an alternative energy source to petroleum due to its abundant reserves and cleanness. Chemical utilization of natural gas, however, is limited because methane, which is a major component of natural gas and one of the greenhouse gases, is one of the most stable molecules. The current process to convert methane is steam reforming (*1*) and/or partial oxidation (*2*) to produce hydrogen and carbon monoxide though these reactions still have some problems.

On the other hand, the conversion of methane into the valuable products directly has been intensively investigated in recent years. The oxidative coupling of methane to form ethane and ethylene is considered to be a promising route, and it is well known that basic metal oxides such as MgO, Sm_2O_3, La_2O_3, etc. are effective catalyst for this reaction (*3-6*). Although many researches to improve the selectivity to C2 hydrocarbons have been done (*7-9*), there are few reports to obtain C2 yield higher than 30%. It is difficult to control the selectivity under the condition of high methane conversion in the presence of oxygen since the consecutive oxidation of products proceeds to form carbon monoxide and carbon dioxide. To improve the overall yield, a number of recycle/separation arrangements (*10*) and separative chemical reactors (*11-13*) have been applied to this problem, and they succeeded in obtaining much higher C2 yield than 30%.

We have directed our attention to the non-equilibrium pulsed plasmas, which is able to keep gas phase temperature as low as room temperature, to activate methane and quench the activated species rapidly formed by direct dehydrogenation. There are many studies on plasma methane conversion (*14-18*). We have succeeded in producing acetylene with the selectivity of 95% under the condition of high methane conversion (*19*) and carbon dioxide reforming (*20*) at room temperature using this method. In this paper, the difference between this pulsed discharge and corona discharge was examined. And catalyst – pulsed discharge combined system was investigated to form ethane selectively and to promote reforming reaction.

Experimental Section

A reaction apparatus of non-equilibrium pulsed discharge used in this study was almost the same as that reported previously (*19*). A flow type reaction apparatus composed of a Pyrex glass tube with 4.0 mm internal diameter was used as the reactor. All experiments were conducted under the conditions of room temperature and atmospheric pressure. Stainless steel electrodes of 1.0 mm diameter were inserted from each end of the reactor and fixed with a certain gap.

As shown in Figure 1, a quartz tube with 6.0 mm internal diameter was used in the catalytic reaction, and the gap distance was fixed at 10 mm. The catalyst bed was put between stainless meshes, and the height of catalyst bed was varied.

The pulsed discharge was initiated by supplying a negative high voltage with a DC power generator (MATSUSADA Precision Inc. HAR-30R10). The waveforms of discharge current and voltage were measured by digital oscilloscope (LeCroy Japan Corp. 9314C, 400 MHz bandwidth) using a voltage probe (PMK-20kV, 100 MHz, LeCroy) and current transformer (AP015, 50 MHz, LeCroy). The current width of one pulse was far shorter than 1 μs, and the peak value of the pulse current reached about 40 A. Owing to the short width of current, the pulsed spark discharge belongs to the non-equilibrium plasma.

Almost the same apparatus was used when corona discharge was applied to the reaction. A point of difference was insertion of a resistor of 1 MΩ between the power supply and the cathode.

A catalyst of 1 wt% Pt/SiO_2 was prepared by the incipient wetness method using $[Pt(C_5H_7O_2)_2]$ and CARiACT Q-15 SiO_2 (10-20 mesh, Fuji Silysia Chemical Ltd.), and $Ni_{0.1}Mg_{0.9}O$ solid solution catalyst was prepared by the same method as Tomishige et al. (*21*) using $Ni(NO_3)_2{\bullet}4H_2O$ and $Mg(NO_3)_2{\bullet}6H_2O$. And a catalyst of 48 wt% Ni loaded on the surface of porous carbon in highly dispersed state ($Ni_{0.48}/C_{0.52}$) was prepared by Nakagawa in Kyoto University (*22*) and applied to carbon dioxide reforming of methane with the pulsed discharge at room temperature in this study.

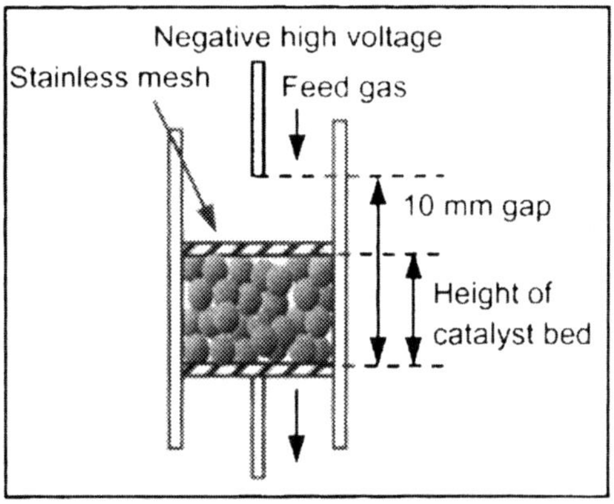

Figure 1. Catalyst – pulsed discharge combined reactor.

All products were analyzed with two gas chromatographs of FID-GC-14A and TCD-GC-8A. The conversions of the reactants and the selectivities were defined as follows.

Reactant conversion (%) = (moles of reactant consumed) / (moles of reactant introduced)∗100

Product selectivity (%) = (carbon based moles of product derived from methane) / (moles of CH_4 consumed)∗100

Results and Discussion

Comparison between Corona Discharge and Pulsed Discharge

Formation of C2 Hydrocarbons from Methane

Table I shows the experimental results of the effect of the current in DC corona discharge and the effect of the pulse frequency in pulsed discharge when $CH_4/O_2/Ar$=5/1/4 sccm was fed under the condition of 1.5 mm gap distance.

Table I. Comparison of Experimental Results between Corona and Pulsed Discharge

DC Corona Discharge						
Current	*CH_4 Conv.*	*Selectivity (%)*				
/ mA	*(%)*	*C_2H_6*	*C_2H_4*	*C_2H_2*	*CO*	*CO_2*
0.5	3.4	44.2	12.4	6.4	32.2	3.6
4.3	26.7	14.0	28.5	22.6	32.1	1.3
8.0	36.3	5.2	17.4	41.3	34.3	1.4
Pulsed Discharge						
Frequency	*CH_4 Conv.*	*Selectivity (%)*				
/ Hz	*(%)*	*C_2H_6*	*C_2H_4*	*C_2H_2*	*CO*	*CO_2*
36	27.0	1.7	3.9	60.0	33.7	0.7
91	43.2	1.2	3.4	60.9	33.9	0.7
200	58.6	0.7	3.0	61.3	34.3	0.6

Conditions: $CH_4/O_2/Ar$=5/1/4 sccm, 1.5 mm gap distance, ambient temp., 0.1 MPa.
Applied voltage in corona discharge: -0.7 kV to –0.3 kV.
Spark voltage in pulsed discharge: -5.2 kV to -3.5 kV.

When the corona discharge was applied to the reaction, methane conversion increased with the increase in the current while the conversion depended on the pulse frequency in the pulsed discharge reaction. In both methods, the selectivity to carbon monoxide and carbon dioxide kept constant at about 34% and 1% respectively independent of the conversion. In this study, the use of the non-equilibrium plasmas was able to prevent the consecutive oxidation of C2 hydrocarbon products.

On the selectivity to each C2 hydrocarbon (ethane, ethylene and acetylene), there was the difference between the corona and the pulsed discharge. In the corona discharge, the consecutive dehydrogenation of C2 hydrocarbons was observed, and the selectivity to each C2 hydrocarbon depended on methane conversion strongly. Under the condition of the low conversion, ethane was the main product in C2 hydrocarbons. With the increase in the conversion, the selectivity to ethylene increased, and acetylene became the main product at 36% methane conversion. These results mean that the reaction mechanism to form C2 hydrocarbons in the corona discharge is the same as that in the conventional thermal method. The dissociation of C-H bond to produce $CH_3\bullet$ was the initiation step, and then C_2H_6 was produced by the coupling of CH_3 radicals. Ethylene and acetylene were formed from ethane by the consecutive dehydrogenation. This also indicates that the formation of the specific product selectively, ethylene only for example, is difficult in the corona discharge.

$$
\begin{array}{lllll}
2CH_4 \rightarrow & 2CH_3\bullet \rightarrow & C_2H_6 \rightarrow & C_2H_4 \rightarrow & C_2H_2 \\
 & + 2H\bullet & & + H_2 & + H_2
\end{array}
$$

Table II. Effect of Gap Distance in Pulsed Discharge

Gap Distance / mm	*CH_4 Conv. (%)*	*Selectivity (%)*		
		C_2H_6	*C_2H_4*	*C_2H_2*
1.5	24.0	1.7	4.4	93.7
10.0	82.8	0.1	2.6	97.4

Conditions: CH_4 10 cm^3min^{-1}, ambient temp., 0.1 MPa, 100 Hz, spark voltage: -4.3 kV(1.5 mm gap), -7.4 kV(10 mm gap).

On the other hand, when the pulsed discharge was used, the main product remained acetylene independent of methane conversion, and its composition in C2 hydrocarbons was about 93%. The pulsed discharge was useful for the selective formation of acetylene. We have reported that acetylene was produced with about 95% selectivity when pure methane was fed into the discharge region (*19*), and in this study, the effect of the gap distance was investigated under the conditions of 10 cm^3min^{-1} methane flow rate, 100 Hz pulse frequency, room temperature and 0.1 MPa. The gap distance had a great effect on the conversion and yield of acetylene as shown in Table II.

Methane conversion dramatically increased up to 83%, and the selectivity to acetylene slightly increased. The one path yield of acetylene reached 81% when the gap distance was 10 mm.

Formation of Oxygen-Containing Liquid Compounds with Corona Discharge

As mentioned above, the selectivity to the product was independent of the conversion with the pulsed discharge. The same phenomenon was observed when methane conversion was controlled by varying the total flow rate. In addition, no oxygen-containing liquid product such as methanol or formaldehyde was detected under any conditions.

The use of the corona discharge, however, was able to produce formaldehyde and methanol directly from methane, under the condition of the short residence time, that is low methane conversion. Table III shows the experimental results including the selectivity to methanol and formaldehyde under the conditions of 1.0 mA current and 2.0 mm gap distance.

Table III. Partial Oxidation of Methane with Corona Discharge

CH_4/O_2 / sccm	*CH_4 Conv. (%)*	*Selectivity (%)* CO	CO₂	C2	CH_3OH	HCHO
50/10	1.5	28.1	2.6	44.0	6.2	19.1
50/10	1.6	28.3	2.3	42.3	7.6	19.5
100/20	0.6	29.8	2.5	55.4	1.4	10.9
40/20	3.7	34.4	2.4	46.5	3.3	13.4

Conditions: ambient temp., 0.1 MPa, 1.0 mA, 2.0 mm gap distance.
Applied voltage: -1.6 kV.

Under the condition of CH_4/O_2=50/10 sccm, methane conversion was 1.5%, and the selectivity to methanol and formaldehyde was 25%. The high reproducibility was confirmed under the same condition, 1.6% methane conversion and 27% selectivity to the oxygenates. The partial oxidation of methane into methanol and formaldehyde proceeded at room temperature and atmospheric pressure.

When the residence time was reduced at the same CH_4/O_2 ratio of 5, the selectivity to the oxygenates decreased to 12%, and the selectivity to C2 hydrocarbons increased. In order to increase the concentration of $CH_3\bullet$, the feed gas condition was changed into CH_4/O_2=40/20 sccm. The selectivity to the oxygenates decreased to 17% and the selectivity to carbon monoxide increased with the increase in methane conversion. The decrease in the selectivity to the oxygenates was caused by the consecutive oxidation of oxygen-containing liquid products.

Emission Spectrum of Corona Discharge and Pulsed Discharge

Figures 2-(a) and (b) are the emission spectrums of the corona discharge and the pulsed discharge measured by MS257™ (ORIEL INSTRUMENTS, 220.00 mm input focal length, 300 l/mm grating, 20 s exposure time) under the condition of pure methane feed.

Although obtained spectra are noisy due to the problem in concentrating light, there are distinct differences between the corona discharge and the pulsed discharge. First, the black-body radiation was observed in the corona discharge, which was not observed in the pulsed discharge. This indicates that there is no carbon deposition in the discharge channel of the pulsed discharge, and that the growth process of carbon is different. Actually, in the corona discharge, the concentrated carbon deposition on the cathode and the growth along the discharge channel were observed. On the other hand, in the pulsed discharge, there was no carbon deposition in the discharge channel, but a small amount of carbon was deposited on the wall of the reactor and the electrodes besides the discharge channel.

Secondly, the emission belonging to rotational band of CH (431.5 nm) was observed in the corona discharge, which was not clearly observed in the pulsed discharge. Judging from the difference in the experimental results and the emission spectra between the corona discharge and the pulsed discharge, it is natural that the reaction mechanism to produce acetylene with the pulsed discharge is different from that with the corona discharge.

From the experiments using some kinds of the isotope, the reaction mechanism in the pulsed discharge was mostly clarified (*23*). It is possible that methane decomposes into C by the electrons impact, and the coupling of C forms C_2 which is considered to be the precursor of acetylene. The produced acetylene also decomposes into C by the collision of electrons, and then acetylene is formed by the same path again. This repetition mechanism of the decomposition and the recombination among C, C_2 and C_2H_2 leads to the selective formation of acetylene.

Possible main reactions in the pulsed discharge

$$CH_4 \rightarrow C + 4H$$
$$C + C \rightarrow C_2$$
$$C_2 + H \rightarrow C_2H$$
$$C_2H + H \rightarrow C_2H_2$$

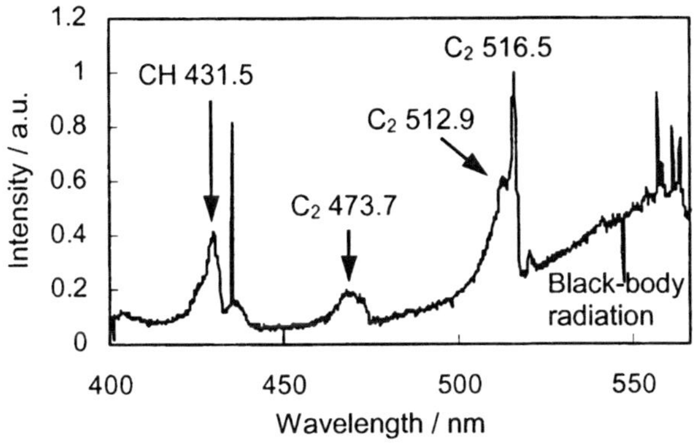

(a) Corona discharge

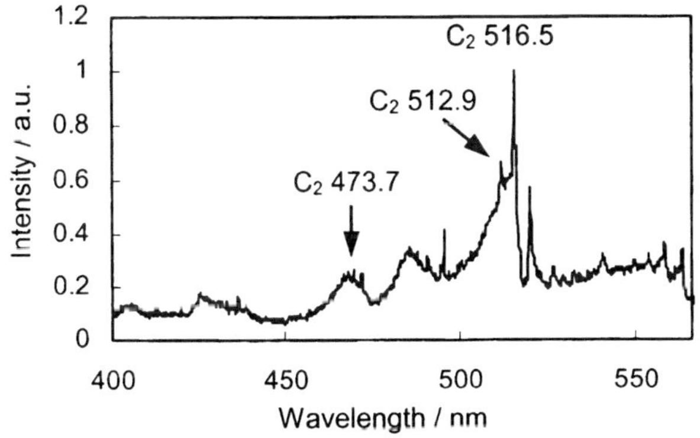

(b) Pulsed discharge

Figure 2. Emission spectrum of corona discharge (a) and pulsed discharge (b) under the conditions of 50 cm^{3}min^{-1} pure methane feed, ambient temp., 0.1 MPa, and 2.0 mm gap distance.

Catalyst – Pulsed Discharge Combined System

Selective Formation of Ethane with Pt/SiO_2 – Pulsed Discharge

To investigate the catalytic pulsed discharge reaction, 1 wt% Pt/SiO_2 was inserted between the electrodes. The height of catalyst bed from the downstream electrode was varied from 0 to 10 mm under the conditions of 10 mm gap distance and 100 Hz pulse frequency. The results are shown in Figure 3. The height of 0 mm means non-catalytic reaction, and the discharge region was filled with the catalyst at 10 mm height (0.1 g-Cat.). Pt/SiO_2 was used without pretreatment like reduction by hydrogen. To prevent the carbon deposition, methane was diluted with 80% hydrogen.

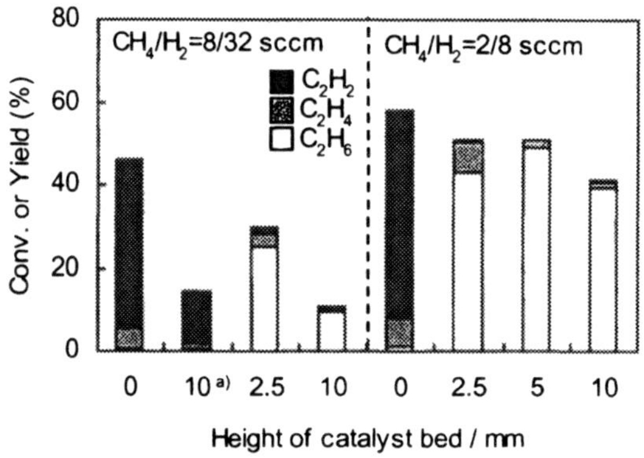

Figure 3. Effect of Pt/SiO_2 and height of catalyst bed ([a] SiO_2 without Pt) under the conditions of CH_4/H_2=1/4, ambient temp., 0.1 MPa, 10 mm gap distance, and 100 Hz pulse frequency.

As compared with results in Table II, to dilute methane with hydrogen did not affect the selectivity so much. The selectivity to acetylene in Figure 3 was 87% in non-catalytic reaction. On the other hand, the use of Pt/SiO_2 had a great influence on the product composition, and the selectivity to ethane increased drastically, while the conversion decreased. Under the condition of CH_4/H_2=8/32 $cm^3 min^{-1}$, methane conversion was 46% in non-catalytic reaction, which decreased to 14.5% when the discharge region was filled with SiO_2 (void fraction: 65%). When the discharge region was filled with Pt/SiO_2, ethane was produced with the selectivity of 94% and the conversion of 10.5%. Methane conversion increased to 33% by decreasing the amount of the catalyst (2.5 mm height), and 25% yield of ethane was obtained.

Under the condition of CH_4/H_2=2/8 cm^3min^{-1}, the decrease in the conversion by the insertion of catalyst was not so large compared with that under the condition of higher flow rate. Although methane conversion at 5 mm and 2.5 mm height were almost the same at 51%, the selectivity to ethylene slightly increased to 15% at 2.5 mm. The highest ethane yield of 49% was obtained at 5 mm height of catalyst, much higher than 30%, which is the upper limit in the conventional thermal chemistry.

Promotion of Carbon Dioxide Reforming of Methane

Even in non-catalytic reaction, carbon dioxide reforming of methane proceeded by the use of the pulsed discharge at room temperature, but carbon-based selectivity to carbon monoxide, which was slightly affected by the pulse frequency and the gap distance, was as low as 66% at 100 Hz and 10 mm gap distance. To increase carbon monoxide selectivity, it was necessary to dilute methane with carbon dioxide, and about 90% selectivity was obtained at CH_4/CO_2=1/4.

Table IV. Effect of NiMgO and $Ni_{0.48}/C_{0.52}$ on CO_2 Reforming

Catalyst	*Height / mm*	*Conv.(%)*		*Sel.(%)*		*H_2/CO ratio*
		CH_4	*CO_2*	*C2*	*CO*	
non	0	59.9	53.1	33.0	65.9	1.0
NiMgO[a)]	10	29.6	23.9	37.0	63.0	1.1
NiMgO[b)]	10	29.5	26.2	15.5	84.5	1.0
NiMgO[a)]	2.5	52.3	44.7	34.4	65.6	1.1
$Ni_{0.48}/C_{0.52}$[a)]	2.5	48.5	46.8	13.9	86.1	1.1

a) without pretreatment, b) pretreated by H_2 reduction at 1123 K for 30 min.
Conditions: CH_4/CO_2=10/10 sccm, ambient temp., 0.1 MPa, 10 mm gap distance, 100 Hz pulse frequency.

Table IV shows the effect of $Ni_{0.1}Mg_{0.9}O$ and $Ni_{0.48}/C_{0.52}$ on carbon dioxide reforming at CH_4/CO_2=10/10 sccm. Although no remarkable effect was observed by using without pretreated NiMgO, the use of pretreated one had a considerable effect on the selectivity. Carbon monoxide selectivity increased to 85% and the content of ethylene in C2 hydrocarbons became 78%. Carbon dioxide reforming of methane was promoted by the use of pretreated NiMgO solid solution catalyst.

Even if the height of the catalyst bed was reduced to 2.5 mm (0.08 g-cat.), the pretreatment of hydrogen reduction at high temperature was necessary to promote the reforming reaction in spite of the low reaction temperature owing to the pulsed discharge. The use of $Ni_{0.48}/C_{0.52}$ catalyst had a possibility to resolve this problem. The catalyst of $Ni_{0.48}/C_{0.52}$ was used without the pretreatment, but the selectivity to carbon monoxide was as high as that obtained with the pretreated NiMgO catalyst. The selectivity to carbon monoxide reached 86%, and methane conversion was 49% at 2.5 mm height of the catalyst bed (0.07 g-cat.).

Produced H_2/CO ratio was almost constant at 1.0 to 1.1 independent of the experimental conditions as shown in Table IV. In non-catalytic reaction and in the presence of NiMgO catalyst without pretreatment, acetylene was produced as well as carbon monoxide to cause higher H_2/CO ratio than that of stoichiometric carbon dioxide reforming of 1.0. Under those conditions, the contribution of reverse water gas shift reaction was so high that H_2/CO ratio was not so higher than 1.0.

Conclusion

The comparison between the corona discharge and the pulsed discharge have been made in this study. The difference was in the reaction mechanism to produce C2 hydrocarbons. Methane decomposes into C by electrons impact, and it is possible C_2 radical is the precursor of acetylene in the pulsed discharge, while in the corona discharge, acetylene was formed by the consecutive dehydrogenation from ethane.

The catalyst – pulsed discharge combined system had a considerable effect on the products composition. Ethane was produced from methane selectively by the use of Pt/SiO_2. The highest ethane yield in this study was 49% with the selectivity of 96%, much higher than 30%, which is the upper limit in the conventional thermal chemistry.

The use of Ni-catalyst promoted carbon dioxide reforming of methane at room temperature, and the selectivity to carbon monoxide reached about 85%. Especially, $Ni_{0.48}/C_{0.52}$ catalyst had a great effect without the pretreatment, though no remarkable effect was observed when NiMgO without the pretreatment was used.

References

1. Barbieri, G.; Violante, V.; Di Maio, F. P.; Criscuoli, A.; Drioli, E. *Ind. Eng. Chem. Res.* **1997,** *36,* 3369-3374.
2. Morioka, H.; Shimizu, Y.; Sukenobu, M.; Ito, K.; Tanabe, E.; Shishido, T.; Takehira, K. *Appl. Catal. A: Gen.* **2001,** *215*, 11-19.

3. Otsuka, K.; Jinno, K.; Morikawa, A. *Chem. Lett.* **1985,** 499-500.
4. Lin, C. H.; Campbell, K. D.; Wang, J. X.; Lunsford, J. H. *J. Phys. Chem.* **1986,** *90,* 534-537.
5. Burch, R.; Dost, A. A.; Rice, D. A.; Webb, G. *Catal. Lett.* **1989,** *3,* 331-338.
6. Takenaka, S.; Kaburagi, T.; Yamanaka, I.; Otsuka, K. *Catal. Today* **2001,** *71,* 31-36.
7. Al-Saeed, S. M.; Lobban, L. L. *Ind. Eng. Chem. Res.* **1995,** *34,* 1060-1073.
8. Zeng, Y.; Lin, Y. S. *Ind. Eng. Chem. Res.* **1997,** *36,* 277-283.
9. Mleczko, L.; Pannek, U.; Niemi, V. M.; Hiltunen, J. *Ind . Eng. Chem. Res.* **1996,** *35,* 54-61.
10. Jiang, Y.; Yentekakis, I. V.; Vayenas, C. G. *Science* **1994,** *264,* 1563-1566.
11. Tonkovich, A. L.; Carr, R. W.; Aris, R. *Science* **1993,** *262,* 221-223.
12. Tonkovich, A. L.; Carr, R. W. *Chem. Eng. Sci.* **1994,** *49,* 4647-4656.
13. Bjorklund, M. C.; Kruglov, A. V.; Carr, R. W. *Ind. Eng. Chem. Res.* **2001,** *40,* 2236-2242.
14. Eliasson, B.; Liu, C. J.; Kogelschatz, U. *Ind. Eng. Chem. Res.* **2000,** *39,* 1221-.1227
15. Liu, C. J.; Mallinson, R. G.; Lobban, L. L. *Appl. Catal. A.* **1997,** *164,* 21-33.
16. Liu, C. J.; Mallinson, R. G.; Lobban, L. L. *J. Catal.* **1998,** *179,* 326-334.
17. Suib, S. L.; Zerger, P. P. *J. Catal.* **1993,** *139,* 383-391.
18. Bromberg, L.; Cohn, D. R.; Rabinovich, A. *Energy Fuels* **1998,** *12,* 11-18.
19. Kado, S.; Sekine, Y.; Fujimoto, K. *Chem. Commun.,* **1999,** *24,* 2485-2486.
20. Kado, S.; Urasaki, K.; Sekine, Y.; Fujimoto, K. *Chem. Commun.,* **2001,** *5,* 415-416.
21. Tomishige, K.; Himeno, Y.; Matsuo, Y.; Yoshinaga, Y.; Fujimoto, K. *Ind. Eng. Chem. Res.,* **2000,** *39,* 1891-1897.
22. Miura, K.; Nakagawa, H.; Kitaura, R.; Satoh, T. *Chem. Eng. Sci.* **2001,** *56,* 1623-1629.
23. Kado, S.[1]; Urasaki, K.[1]; Sekine, Y.[2]; Fujimoto, K.[3] [1]The University of Tokyo; [2]Waseda University; [3]The University of Kitakyushu *unpublished.*

Chapter 21

Evaluation of a High-Energy Transform Efficiency Pulse Power Supply for Methane Plasma Conversion

Mamoru Okumoto, Shuiliang Yao, Akira Nakayama, and Eiji Suzuki

Catalysis Science Laboratory, Research Institute of Innovative Technology for the Earth, 9–1 Kizugawadai, Kizu-cho, Soraku-gun, Kyoto 619–0292, Japan

A new high frequency pulse power supply of a μs order rise time is evaluated. The discharge efficiency (ratio of power from the pulse power supply to power to the discharge gas) has achieved 80 %. Methane conversion is carried out using a point-to-point type of reactor at a pulse frequency up to 6 kHz and at atmospheric pressure. Acetylene is the main product. Methane conversion efficiency (defined as moles of methane converted per power injected into the gap) is 0.51 mmol/kJ at maximum. The discharge and methane conversion efficiencies can possibly be further improved after comparison with the results using a pulse power supply of a rise time of 10 to 20 ns.

I. Introduction

In recent years, researchers that use plasma discharges for methane conversion to fuels and chemicals have been focused *(1-10)*. Those discharges are typically classified to 5 types according to the voltage waveform and discharge properties, 1) corona or arc (AC or DC) *(11-12)*, 2) surface discharge (AC+ insulator) *(13-14)*, 3) partial discharge (AC+ ferroelectrics) *(15-17)*, 4) microwave *(18-19)*, and 5) pulsed plasma (pulsed corona, streamer or spark) *(20-26)*. Since a pulsed plasma of high frequency can give a pulse electric discharge with a high energy injection rate over a short pulse discharge duration (100 ns order), this kind of pulsed plasma is useful for methane conversion to acetylene and hydrogen, which requires a high reaction temperature *(26)*. Experimental results already showed that the energy efficiency and acetylene have achieved 3 mmol/kJ and 90 %, respectively, these figures are competitive with those of the conventional acetylene production processes. For the industrialization of the pulsed plasma conversion of methane, the efficiency of the pulse power supply becomes important *(27)*.

In this study, a new inverter pulse power supply has been used to evaluate the transform efficiency of the power supply and the energy efficiency of methane conversion. This power supply generates a high-repetition rate of an intermittently positive oscillation pulse voltage by controlling an inverter circuit and pulse transformer. Comparison with a conventional pulse power supply of short rise time (10 - 20 ns) and the new inverter pulse power supply is given about methane conversion.

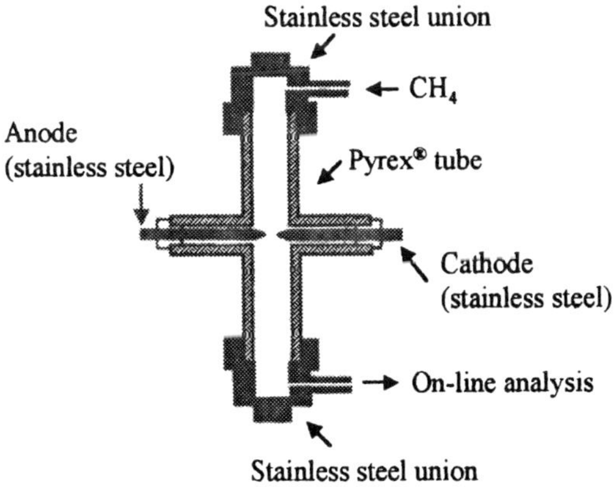

Figure1 Schematic diagram of plasma reactor

II. Experimental Setup

II. A. Reactor and pulse power supply

Figure 1 shows a point-to-point type of plasma reactor for methane conversion. The plasma reactor consisted mainly of a Pyrex® tube {15 mm (outer diameter: o.d.), 12 mm (inner diameter: i.d.), 0.8 m (length)}, a quartz tube {12 mm (o.d.), 8 mm (i.d.), 0.4 m (length)}, and two stainless steel electrodes {1.5 mm (o.d.), 0.18 m (length)} with sharp-pointed discharge terminals. The gap distance between the two electrodes is fixed at 5.0 mm.

The schematic layout of the pulse power supply (PEEC. Japan) is shown in Figure 2. This inverter pulse power supply can generate an intermittently positive polarity pulse high voltage having a rise time of 1.0 μsec and a pulse width of about 3.5 μsec (in the case of resistance load) by controlling an inverter circuit.

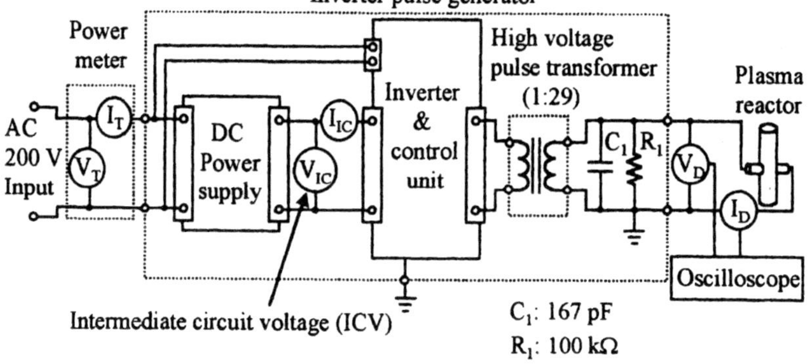

Figure 2 Schematic diagram of inverter pulse power supply and plasma reactor

II. B. Measurement system

Discharge voltage (V_D) and currents (I_D) are measured with a voltage divider (EP-50K, PEEC, Japan) and current transformer (model 2-1.0, Stangenes Industries Inc.), respectively. Signals from voltage and current transformers are recorded and averaged with a digital oscilloscope (Tektronix TDS 754D, Sony Tektronix) of a 500 MHz bandwidth.

Methane (99.9995 %, Liquid Gas co.) is introduced into the upper part of the reactor. Hydrocarbon products from the lower part of the reactor are analyzed with an online chromatograph (GC 103, FID, Okura Riken) equipped with a 2-m Porapak N column and a methanizer prior to FID to convert hydrocarbons to alkanes. All experiments are performed at atmospheric pressure.

II. C. Powers

The transform efficiency of the power supply is evaluated by comparison of the electric powers P_{in} (input energy to the power supply), P_{out} (output energy from the power supply to plasma forming capacitors), and P_D (discharge energy to background gases). P_{in} is measured with a power meter (HIOKI 3193 Power Hitester, Hioki), which calculates the power using Eq. (1). P_{out} and P_D are calculated by Eqs. (2)-(3).

$$P_{in} = |V_T| \cdot |I_T| \cos\theta \text{ [W]} \tag{1}$$

Where V_T, I_T, and θ are effective value of voltage (V), effective value of current (A), and phase angle between V_T and I_T (rad), respectively.

$$P_{out} = f \times \frac{1}{2} \cdot C_1 \cdot (V_D)^2 \text{ [W]} \tag{2}$$

Where C_1 and f are the plasma forming capacitance of the pulse power supply (F) and pulse repetition frequency (Hz), respectively. V_D is the peak value of the pulse voltage (V).

$$P_D = f \times \sum_i \frac{V_{Di} + V_{Di+1}}{2} \frac{I_{Di} + I_{Di+1}}{2} (t_{i+1} - t_i) \text{ [W]} \tag{3}$$

Where, V_{Di} and I_{Di} are discharge voltage (V) and cathode current (A) at discharge time t_i (s), respectively.

II.D Transform efficiency

The transform efficiencies of the pulse power supply are defined as charge efficiency η_c (Eq. (4)) and discharge efficiency η_d (Eq. (5)).

$$\eta_c = \frac{P_{out}}{P_{in}} \times 100 \ [\%] \tag{4}$$

$$\eta_d = \frac{P_D}{P_{out}} \times 100 \ [\%] \tag{5}$$

II.E Methane conversion

Equations (6) to (7) are used to calculate product selectivity and methane conversion.

$$\text{Selectivity} = \frac{2 \times \text{moles of } C_2}{\text{moles of } CH_4 \text{ converted}} \times 100 \ [\%] \tag{6}$$

$$\text{Methane conversion} = \frac{\text{moles of } CH_4 \text{ converted}}{\text{moles of } CH_4 \text{ before reaction}} \times 100 \ [\%] \tag{7}$$

The electric energy efficiency of the pulsed plasma is represented by a methane conversion efficiency R in mol /J (Eq. (8))

$$R = \frac{\text{Moles of methane converted per second}}{P_D} \ [\text{mol / J}] \tag{8}$$

III. Results and Discussion

III. A. Characteristic of the pulse power supply

The discharge characteristics are first evaluated. Figure 3 shows the typical waveforms of discharge voltage and current. The rise time of V_D is about 1 μs and peak value is 5.8 kV. The voltage falls to a ground level due to the breakdown of the gap. The cathode discharge current has a peak value of 42 A. The whole pulse discharge finishes within 300 ns.

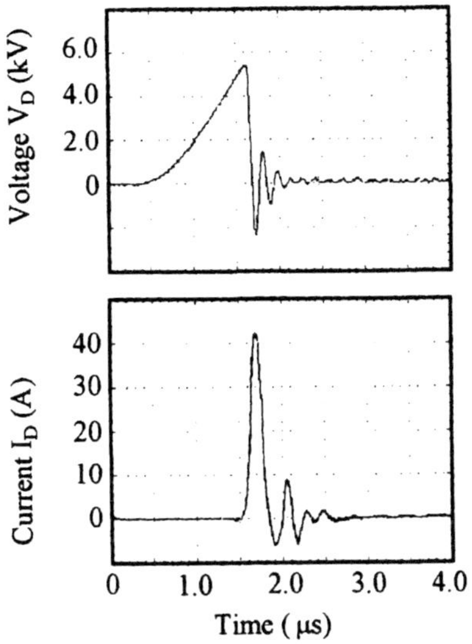

Figure 3 Waveforms of discharge voltage and current (CH_4 flow rate: 100 ml/min, Intermediate circuit voltage: 150V)

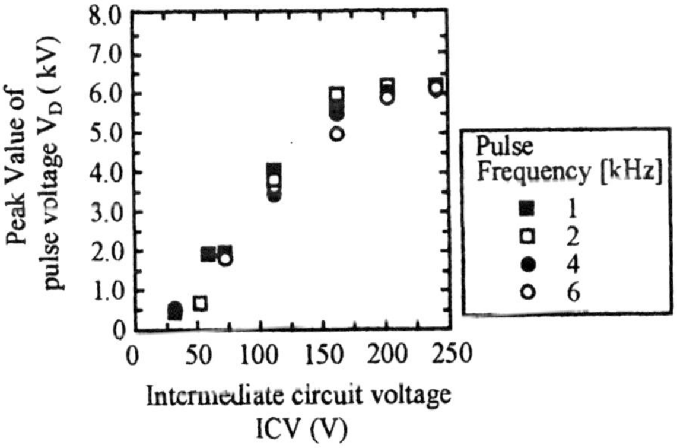

Figure 4 Intermediate circuit voltage (ICV) versus peak value of pulse voltage (CH_4 flow rate: 100 ml/min)

The peak value of the pulse voltage is adjusted by controlling the Intermediate circuit voltage (ICV) (Fig. 2). Figure 4 shows the relation of the ICV to peak value of the pulse voltage at various pulse frequencies. The peak value of the pulse voltage increases with increasing ICV linearly below 140 V ICV, but more slowly above 140 V ICV. No obvious differences in the relation of ICV and peak value of the pulse voltage are observed when pulse frequency is changed. This differs from that when use a pulse voltage having a rise time order of 10 to 20 ns *(28)*.

III. B. Transform efficiency

The transform efficiency of the pulse power supply is an important factor that affects the application of the pulse power supply in industry. Here, the transform efficiency is calculated by measuring powers of P_{in}, P_{out}, and P_D. Figure 5(A) shows each power as a function of ICV. P_{in} and the P_{out} increase with the increase in ICV. As the gap is broken down at an ICV higher than 140 V, P_D occurs above 140 V ICV and increases.

The transform efficiency is then shown in Fig. 5(B). The charge efficiency η_c increases to a level of 40 % as the ICV increases to 230V. The left 60 % is due to the energy consumption including the supplements such as electric fans in the pulse power supply. The discharge efficiency η_d increases with increasing ICV and achieves 80 % at 230 V ICV.

Figure 6 shows the powers and transform efficiency as a function of pulse repetition frequency (*prf*). Each power increases with increasing *prf*. The

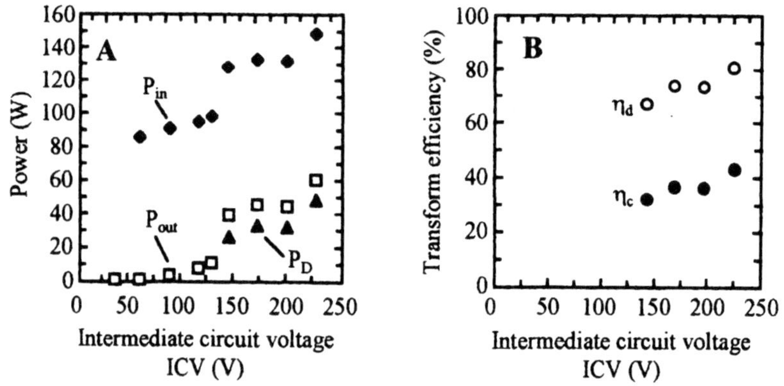

Figure 5 Powers and transform efficiency as a function of intermediate circuit voltage (CH_4 flow rate: 100 ml/min, Pulse repetition frequency: 1 kHz)

discharge efficiency η_d and charge efficiency η_c are improved and peak at 5 to 6 kHz. Those findings suggest that the pulse discharge at a high *prf* and a high pulse voltage promote the transform efficiency of the pulse power supply. It is also possible to improve the transform efficiency by such as matching of the pulse power supply and the load of the discharge gap *(29)*.

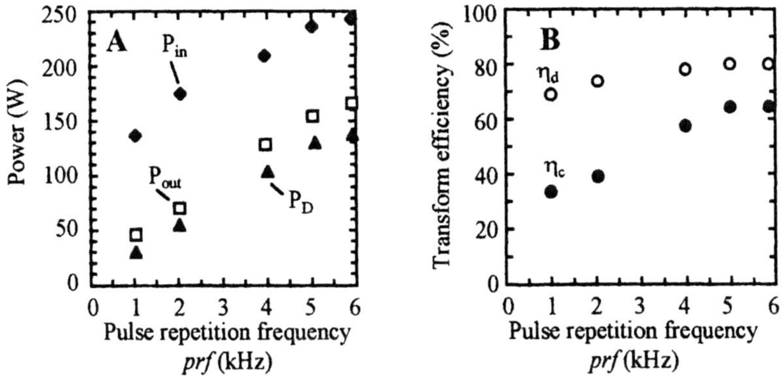

Figure 6 Powers and transform efficiency at various pulse repetition frequencies (CH_4 flow rate: 100 ml/min, DC output V_S: 150 V)

Table.1 Comparison of methane conversion using pulsed plasma discharge

Power supply	CH_4 [ml/min]	f [kHz]	P_{in} [W]	P_{out} [W]	P_D [W]	η_d [%]	CH_4 conv. [%]	Selectivity [%] C_2H_2	C_2H_4	C_2H_6	R [mmol/kJ]
Inv	100	1.0	130	39	27	69	18.3	80.0	2.6	5.4	0.51
Inv	100	5.1	234	152	124	81	54.7	86.8	0.5	2.8	0.32
Inv	300	5.0	245	157	130	82	21.7	77.4	1.4	4.1	0.37
Thy	300	2.0	-	-	-	-	-	62.0	16.0	8.0	1.3
Thy	300	5.0	-	-	-	-	-	82.0	12.0	2.0	1.8

Note: Inv: inverter pulse power supply used in this study

Thy: hydrogen thylatron closing switch used in ref. *(30)*

III. C. Methane conversion

Methane conversion is performed using the pulse power supply and the point-to-point reactor (Table 1). Methane conversion and acetylene selectivity vary from 18.3 to 54.7 % and 77.4 % to 86.8 %, respectively. The maximum value of methane conversion efficiency R is 0.51 mmol/kJ, lower than that using a hydrogen thylatron as a switch that generates a pulse voltage of a rise time 10 to 20 ns. This implies that the methane conversion efficiency can possibly be further improved by such as using a pulse voltage of a shorter rise time.

IV. Conclusions

The transform efficiency of the pulse power supply with an inverter circuit has been evaluated. The discharge efficiency has a peak value of 80 %. Methane conversion efficiency and acetylene selectivity are respectively, 0.51 mmol/kJ and 86.8 % at maximum. The transform and methane conversion efficiencies can be further improved.

Acknowledgements

The New Energy and Industrial Technology Development Organization (NEDO) and The Ministry of Education, Culture, Sports, Science and Technology supported this study. Yao is grateful to NEDO for a fellowship.

References

1. Yao, S. L.; Takemoto, T.; Ouyang, F.; Nakayama, A.; Suzuki, E.; Mizuno, A.; Okumoto, M. *Energy & fuels* **2000**, 14(2), pp.459-463
2. Okazaki, K.; Hirai, S.; Nozaki, T.; Ogawa, K.; Hijikata, K. *Energy* **1997**, 22(2/3), pp.369-374
3. Zhou, L. M.; Xue, B.; Kogelschatz, U.; Eliasson, B. *Plasma chemistry and plasma processing* **1998**, 18(3), pp.375-393
4. Hijikata, K. *Principle of excergy reproduction, abstracts of project research No.256, Supported by the Grant-in Aid of Priority Areas, Ministry of Education*, Japan, 1995; p.4
5. Liu, C.; Marafee, A.; Hill, B.; Xu, G.; Mallinson, R.; Lobban, L. *Industrial & engineering chemistry research* **1996**, 35(10), pp.3295-3301
6. Shepelev, S. S.; Gesser, H. D.; Hunter, N. R. *Plasma chemistry and plasma processing* **1993**, 13(3), pp.479-488

7. Yao, S. L.; Suzuki, E.; Nakayama, A. *Journal of hazardous materials* **2001**, 83(3), pp.237-242
8. Okumoto, M.; Mizuno, A. *Catalysis today* **2001**, 71(1-2), pp. 211-217
9. Prieto, G.; Okumoto, M.; Takashima, K.; Katsura, S.; Mizuno, A. *IEEE Transactions on industry applications* **2001**, 37(5), pp.1464-1467
10. Van Heesch, E. J. M.; Pemen, A. J. M.; Yan, K.; Van Passen, S. V. B.; Ptasinski, K. J. *IEEE Transactions on plasma science* **2000**, 28(5), pp.1571-1575
11. Bhatnagar, R.; Mallinson, R. G. *Methane and alkane conversion chemistry*, Bhasin, M .M.; Slocum, D. W. Eds; Plenum press, NY, **1995**; p.249
12. Liu, C.; Marafee, A.; Mallinson, R.; Lobban, L. *Applied Catalysis A: General* **1997**, 164, pp.21-33
13. Eliasson, B.; Hirth, M.; Kogelschatz, U. *Journal of physics, D: Applied physics* **1987**, 20, pp.1421-1437
14. Gesser, D. H.; Hunter, N. R.; Probawono, D. *Plasma chemistry and plasma processing* **1998**, 18(2), pp.241-246
15. Jogan, K.; Mizuno, A.; Yamamoto, T.; Chang, J-S. *Pol. journal of application chemistry* **1995**, 4, pp.573-583
16. Oda, T.; Takahashi, T.; Nakao, H.; Masuda, S. *IEEE transactions on industry applications* **1993**, 29(4), pp.787-792
17. Ogata, A.; Mizuno, K.; Kushiyama, S.; Yamamoto, T. *Plasma chemistry and plasma processing* **1998**, 18(3), pp.363-373
18. Sekine, Y.; Fujimoto, K. *Natural gas conversion V, Studies in surface science and catalysis*, Parmalina, A. et al. Eds., 1998; 119, pp.277-282
19. Huang, J.; Suib, S. L. *Journal of physical chemistry* **1993**, 97, pp.9403-9407
20. Mizuno, A.; Shimizu, K.; Chakrabarti, A.; Dascalescu, L.; Furuta, S. *IEEE transactions on industry applications* **1995**, 31(5), pp.957-963
21. Masuda, S.; Nakao, H. *IEEE transactions on industry applications* **1990**, 26, pp.374-383
22. Yan, K.; Van Heesch, E. J. M.; Pemen, A. J. M.; Huijbrechts, P. A. H. J. *Plasma chemistry and plasma processing* **2001**, 21(1), pp.107-137
23. Kirlepatric, M.; Fimmer, W. C.; Locke, B. R. *IEEE transactions on industry applications* **2000**, 36(2), pp.500-509
24. Hsiao, M. C.; Merritt, B. T.; Penetrante, B. M.; Vogtlin, G. E.; Wallman, P. H. *Journal of applied physics* **1995**, 78(5), pp.3451-3456
25. Mok, Y. S. *Plasma chemistry and plasma processing* **2000**, 20(3), pp.353-364
26. Yao, S.; Suzuki, E.; Meng, N.; Nakayama, A. *Plasma chemistry and plasma processing, in press*

27. Yao, S.; Nakayama, A.; Suzuki, E. *Catalysis today* **2001**, 71(1/2), pp.219-223
28. Yao, S. L.; Nakayama, A.; Suzuki, E. *AIChE Journal* **2001**, 47(2), p.419
29. Yan, K. *Corona Plasma Generation*, Technische Universtieit Eindhoven, 2001; Chapter 4
30. Yao, S. L.; Suzuki, E.; Nakayama, A. *Plasma chemistry and plasma processing* **2001**, 21(4), pp.651-663

Chapter 22

Factors Affecting the Catalytic Activation of Methane via Microwave Heating

L. Daniel Conde[1] and Steven L. Suib[2,*]

[1]Institute of Materials Science and [2]Department of Chemistry, University of Connecticut, Storrs, CT 06269–3060

For many years Microwave Induced Oligomerization of methane has been extensively investigated. The most important factors that control the oligomerization of methane such as microwave frequency; irradiation power; time the materials are in the reaction zone; and nature of the catalysts were studied here. Oligomers ranging from C_{2s} to C_6 were obtained, where C_{2s} were the most abundant.

Introduction

Natural gas fields which are prospected and developed in conditions somewhat similar to those of oil could have a dominant role in the next decade. Whereas the proven world reserves of oil seem approximately constant, those of natural gas are regularly increasing and could meet the world consumption for 300 years at the present level.[1]

Natural gas is essentially methane (83-97 vol % depending on the origin) and therefore difficult to liquefy and quite chemically unreactive. Methane is thermodynamically stable with respect to its elements. The reactions to make other hydrocarbons, all of which are less stable than methane around 1000°C, have unfavorable free energies of reaction and are strongly limited by equilibrium. They need a considerable energy input, and therefore temperatures above 1000°C are required to transform CH_4 into benzene, acetylene, ethylene and ethane. At 550°C, hydrogen and carbon are more stable.*[2])* Therefore it appears that natural gas, or more precisely methane, is an interesting source of petrochemicals, but its use and transformation are difficult and expensive. Methane is a very valuable fuel from the environmental point of view and the production of energy is so far its main use. However, a more rational use of this resource is that it should also be used to make either petrochemical or gasoline components.

Microwave energy interacts with materials at the molecular level. During microwave heating of a dielectric, internal electric fields are generated within the material. Such fields can produce translations of electrons and ions, and cause rotations of charged species. This movement can be opposed by friction, inertia, and other forces that can lead to attenuation of the electric fields, and to volumetric heating of the material.*[3]* The ability of the microwave electric field to polarize molecules and the ability of these molecules to follow the rapid reversal of the electric field results in the conversion of electromagnetic energy into heat within the irradiated material.

Microwave heating has been investigated for several years in the past decade to catalytically activate methane to produce more valuable products such as higher hydrocarbons and hydrogen.*[4-6]* No other heterogeneous catalytic studies using microwave radiation where either direct or indirect frequency effects, nature of the catalyst, effects of applied power, and irradiation time have been published. In this work we studied the systematic effects of frequency, magnitude of the applied power, nature of the catalyst and irradiation time on the conversion and product distribution of methane oligomerization via microwave heating.

Experimental Section

Microwave Apparatus

The feed was prepared on line using a gas mixing panel. Flow rates to the reactor were controlled with fine metering valves from the panel. High Purity grade methane obtained from Matheson and Ultra High Purity helium purchased from Airgas were used. A mixture of 25% CH_4 in He was fed to the reactor.

Experiments were carried out in a straight 3/8 in quartz tubular reactor, which was mounted inside the center of the cavity. The catalyst bed was held by two pieces of quartz wool inside the reactor. Two different microwave units were used to provide energy to the system in the form of dielectric heat. A Narda mini-Surveyor™ electromagnetic leakage monitor model 8201 was used to detect microwave radiation leaks in the surroundings of the apparatus.

ASTEX Unit

An ASTEX microwave power source model GL139 with a magnetron type GL 130WC was used. The unit was equipped with a three-stub tuner model AX3041 to control the reflected power from the material back to the unit, and an ASTEX single fixed mode applicator model AX7020. This power supply provided up to 1250 W of power at an operating fixed frequency of 2.45 GHz. Power was emitted in pulses with a periodicity of 8.3 msec (120 Hz). Experiments were done at two different power levels (270 W and 500 W) for 3 and 20 min. For comparison, some experiments were done with diluent and some others with pure methane.

Lambda Unit

A Lambda Technologies variable frequency unit, Vari-Wave model LT-502Xb was used for the experiments at different frequencies. Microwave radiation is transmitted from the magnetron to the cavity by means of a high-power coaxial cable. This coaxial line is connected to a high-power coax-to-waveguide (WRD-250) and finally to a broadband microwave cavity launcher. The operational power range for this unit is 0 - 500 W. The incident center frequency can be varied from 2.4 to 7.0 GHz with a bandwidth sweep rate of 100 to 0.1 sec.

The processing cavity is a 12x15x12 inches cavity manufactured from stainless steel. The cavity is access from the front by way of a swing door that provides a positive locking mechanism and enables the unit multiple microwave seals. Unless the door is properly seated, microwave energy cannot be introduced into the processing cavity.

Various frequencies and modes were produced in the microwave cavity. This cavity allows TM_{01n} modes (n = 2 to 9) over a frequency range of 2.4 to 7.0 GHz. The cavity was designed by Lambda Technologies to have different modes for each frequency. In Figure 2 we show how these transverse magnetic modes might be distributed in the cavity and the value of frequency that corresponds to each TM mode. Experiments were done at 2.4 GHz (TM_{012}), at

4.6 GHz (TM_{018}), and at a variable frequency range of 2.4 – 7.0 GHz with a sweep rate of 0.5 sec.

Catalysts

Three different catalysts were tested in the oligomerization of methane. Most of the materials were purchased from different companies and they were used as received. Nickel powder with morphology of branched filaments was used for these experiments. The diameters of the filaments were 2.5 μm as measured by Scanning Electron Microscopy (SEM). Clusters of nickel metal particles were aggregated to form the filaments. Activated carbon MAXSORB produced by Kansai Coke and Chemicals Co was also used. MAXSORB is a very high surface area (2200 m^2/g) activated carbon, produced from petroleum coke and coconut shell char, which results in a low ash content material (less than 2%).*[7]* XRD analysis suggested activated carbon is entirely amorphous. The other catalyst tested was iron powder. As reported by Marun et al.[8] the dielectric properties[9] of these materials do not seem to be an important factor in the oligomerization of methane via microwave heating. Approximately 100 mg of catalyst were placed inside the quartz reactor. The catalyst bed was plugged at both ends with high purity quartz wool. The catalytic quartz reactor was placed inside of the applicator so that the material was positioned in the center the waveguide cavity. In order to purge the atmosphere, helium gas was flowed through the reactor for 30 min prior to starting the microwave absorption.

Product Analysis

Products were trapped using a 4 way-gas sampling valve and analyzed by using an HP5890 Series II Chromatograph equipped with a mass detector. Details about the Gas Chromatographer/Mass Spectrometer (GC/MS) were published elsewhere.*[10]* A carbon balance was used to estimate total conversion. All the compounds multiplied by the number of carbon atoms present in that molecule were summed. We report selectivities (C_{is}) as the selectivities towards the sum of compounds with specific "i" numbers of carbon atoms. Both conversions and selectivities were measured at the end of the irradiation.

The compounds analyzed using gas chromatography were: methane, carbon dioxide, carbon monoxide; C_{2s}, ethylene, acetylene and ethane; C_{3s}, propene, propane, 1,2-propadiene, propyne; C_{4s}, 2-methylpropane, 2-butene, 1-buten-3-yne, 1,2-butadiene, butadiyne, 1-butyne, 2-butyne; C_{5s}, 3-pentene-1-yne; C_{6s}, benzene; C_{7s}, toluene, C_{8s}, ethylbenzene, ethynylbenzene, and ethenylbenzene.

Results and Discussion

The effects of microwave frequency, incident power, irradiation time and nature of catalyst on the oligomerization of methane were studied. In general, C_{2s} were the most abundant products for all the cases. Among C_{2s}, the product distribution regarding abundance was ethane, ethylene, and acetylene, where acetylene was the least abundant. Benzene was also produced with a maximum selectivity of 28%. CO_2 and CO were also present in the product mixture. As we reported*[11]*, there are some absorbed species on the surface of the catalyst such as water; carbon monoxide; carbon dioxide; and molecular oxygen. After reducing the catalysts *in situ* and later performing the reaction, a drop in activity was observed. For reactions with prior reduction, conversions were within the range of 1 to 3% for all the catalysts. Sintering of the catalysts most probably occurred during the reduction process as seen in previous studies[12].

Power Effect

The effect of power on conversion and product distribution was studied using nickel catalyst. Samples were taken after 3 min of irradiation time at a working frequency of 2.4 GHz. As seen in Figure I, conversion increases from 1% to 60% as applied power increases from 380W to 1130W. In some cases, when the same catalyst bed was used at a high irradiation power and for long periods of time, the contrary was observed, showing catalyst deactivation. When the experiments were done, coke deposits were observed on the walls of the quartz reactor. A balance to determine the amount of coke formed during the reaction was performed by GC/MS quantification of methane before and after reaction. In general, the yield towards coke was in the range of 40% to 60% above 1000 W. M. J. F. M. Verhaak *[13]* showed nickel catalysts tend to deactivate in hydrogenation/dehydrogenation reactions due poisoning of active sites by carbonaceous species. Taking into account the product distribution at 1130 W and that the coke yield was 50% at that time, an approximate theoretical hydrogen balance suggested that the amount of hydrogen produced was about 1.5 times the amount of methane originally fed to the reactor. Scanning Electron Microscopy (SEM) photographs *[12]* showed that after high-power reactions, catalysts sinter (specially nickel powder) even at low irradiation times (3min).

As power increases from 378 W to 1130 W selectivity towards CO_2 decreases from 69% to less than 1%. Among C_{2s}, the major product was ethane (24%) for low power reactions. As power increases to 1130 W ethylene became the major product (45%) followed by acetylene (31%).

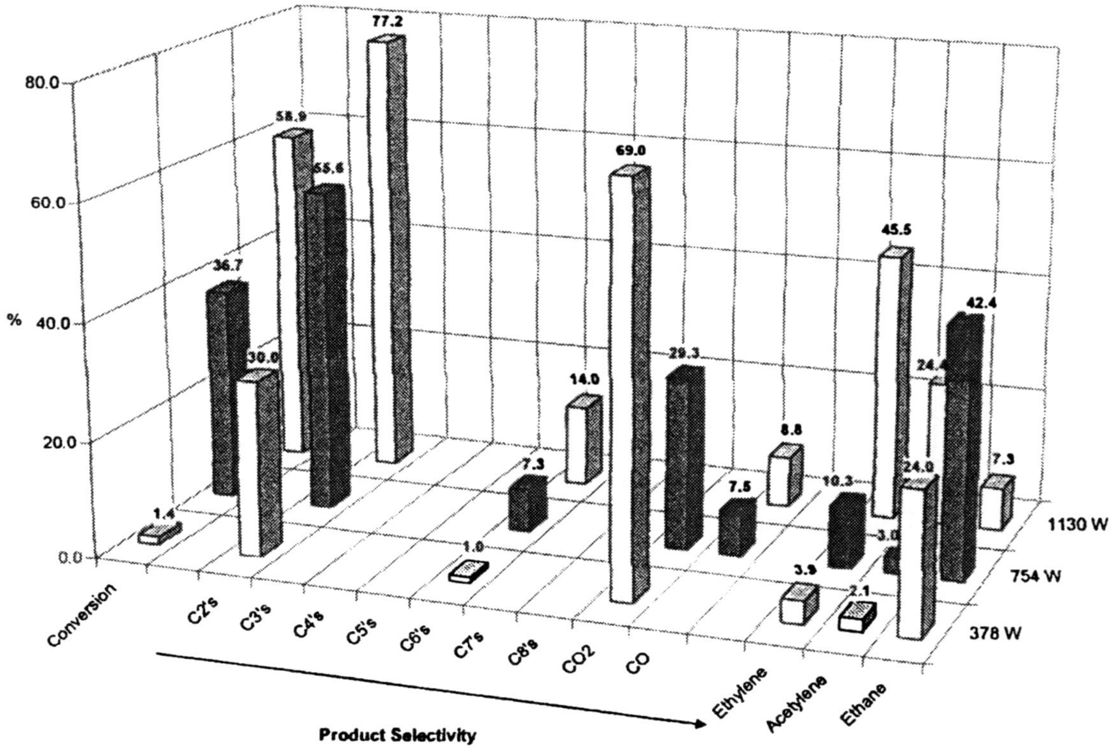

Figure I. Power effect on conversion and selectivity for the reaction using Ni catalyst, 3min of irradiation time, and 2.4 GHz

Time Effect

The effect of microwave irradiation time on the activity and product distribution was investigated with nickel powder at 500 W. The microwave frequency used during the reaction was 2.4 GHz. Figure II shows that conversion of methane increases from 14% to 54% as irradiation time increases from 1 to 30 min. After 40 min of irradiation time the conversion starts decreasing as catalysts deactivate. This loss in activity was also reported by Tanashev*[14]* and attributed to an unusual extra-equilibrium product concentration generated by the microwave radiation. Later, the concentration goes down to its thermodynamic value when time increases.

C_{2s} and C_{3s} presented a maximum selectivity of 55% and 16% respectively at 10 min. Among C_{2s}, ethane was the major product for this set of experiments. After long periods of irradiation time (t > 40 min) and at an intermediate power of 500 W coke is formed on the walls of the reactor. Then, carbonaceous species might also be responsible for the deactivation of the catalysts when time is the variable.

Frequency Effect

The Lambda variable frequency unit was used to explore the effects of microwave frequency on the oligomerization of methane. The results from iron powder catalyst show major differences for different frequencies. All samples were taken after 20 min of irradiation time and a constant power of 500 W was used. Reactions were run at 2.4 GHz and at 4.6 GHz. Variable frequency reactions were also done, where frequency was varied from 2.4 to 7.0 GHz with a sweep rate of 0.5 sec.

Figure III shows an increase in activity as frequency increases. Previous studies*[11]* correlate the microwave radiation frequency with the activity of nickel catalyst. As radiation frequency increases, the particles trying to align with the electromagnetic field are going to move more rapidly, generating more heat than at lower frequency. Therefore, at the same power and irradiation time, the energy density in the material is higher for high frequency radiation. The higher energy density in the material produces different and more energetic transient heating patterns. When the temperature reached by the active sites is too high, the reaction leads to undesirable products such as carbon monoxide and carbon dioxide.

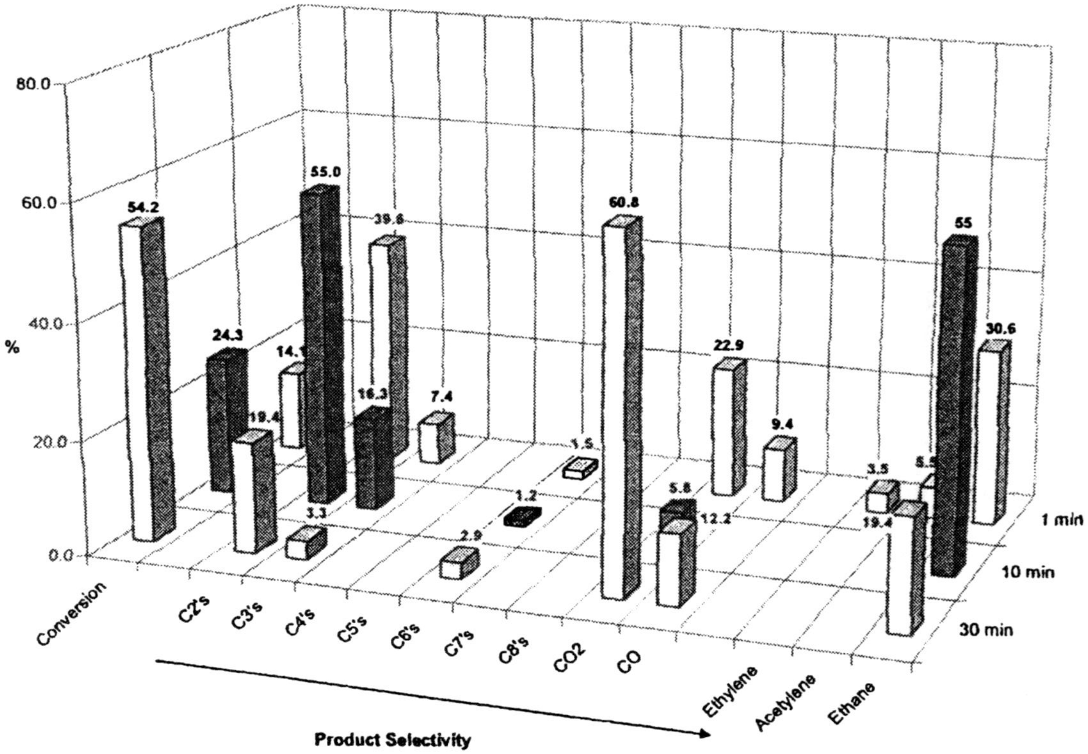

Figure II. Irradiation time effect on conversion and selectivity for the reaction using Ni catalyst, 500 W, and 2.4 GHz

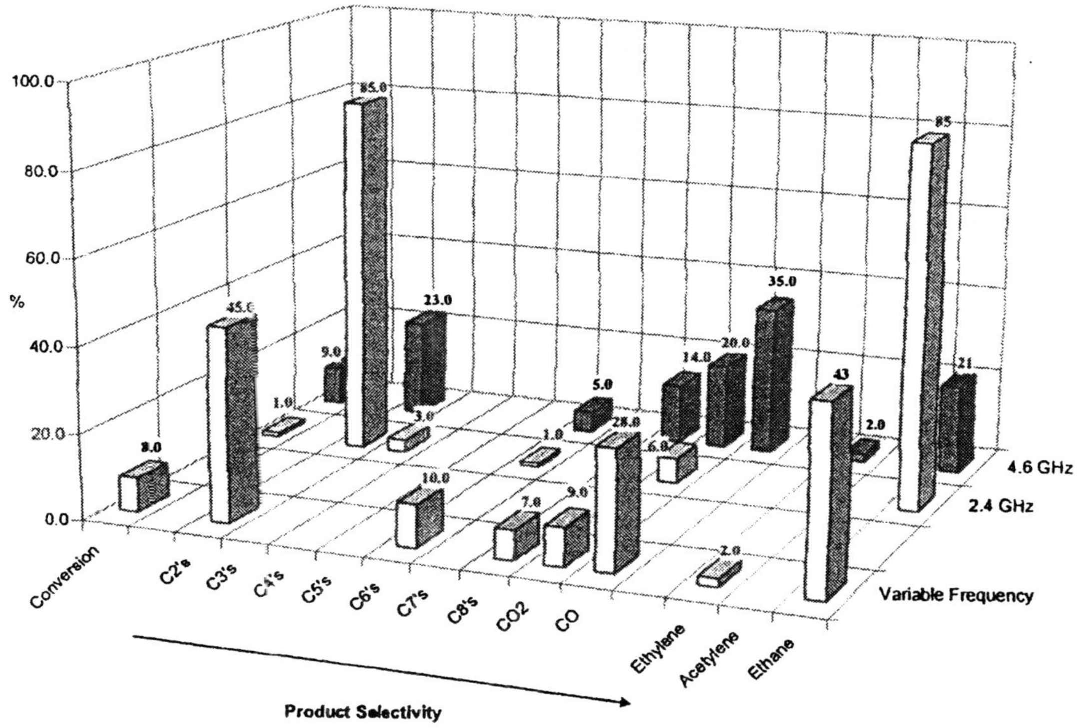

Figure III. Microwave radiation frequency effect on conversion and selectivity for the reaction using Fe catalyst, 20min of irradiation time, and 500 W

When the frequency of the microwave radiation varies over a period of time, the transient heating patterns vary constantly, not allowing the formation of hot spots. Therefore, heating is completely uniform throughout the material. Although the microwave variable frequency radiation can be highly energetic for activating methane, this does not lead to undesirable products. From Figure III it can be seen that conversion is similar (8% and 9%) for high and variable frequency experiments. However, selectivities toward CO_2 and CO are somewhat higher (35% to 28%) for the experiments at 4.6 GHz.

Effect of Catalyst Nature

Three different catalysts were compared. Activated carbon, iron powder and nickel powder were used for the reaction at 500 W. Samples were taken after 20 min of irradiation time. The frequency was set to 2.4 GHz. Figure IV shows the comparison of activity and selectivity for the catalysts. Interestingly, activated carbon is the most active catalyst under our system conditions. Conversion varies from 14% for activated carbon to 5% for nickel powder. Iron powder is the least active catalyst with 1% conversion. The amount of benzene produced is relatively high (28%) when using activated carbon. While ethane is the major C_2 product for both nickel and iron catalyst, ethylene is the major product for activated carbon. Acetylene is only produced when using activated carbon as catalyst (11%). Formation of benzene may be explained by the well-known cyclization reaction. Bamwenda observed similar reactions as reported in a previous publication.*[15]* This is in agreement with the fact that for our system selectivity to benzene increases as selectivity to acetylene also increases.

We have found that activated carbon absorbs microwave radiation more efficiently than nickel powder.*[16]* Consequently, the temperature reached by activated carbon is higher than that reached by nickel.*[10]* This condition of higher temperature on the active sites would favor the conversion of methane on activated carbon.

On the other hand, the surface area of activated carbon is two orders of magnitude larger than the surface area of Ni powder. The data reported here show that conversion and activity of these materials are not proportionally related to the surface area of the catalysts we used. One possible explanation for the lack of direct relationship between surface area and conversion might be the effect of the interaction of microwaves with surface functional groups such as OH groups and other catalytic intermediates like: CH_3, CH_2, CH, and other species. Such functional groups may also influence the adsorption of the reactant species.

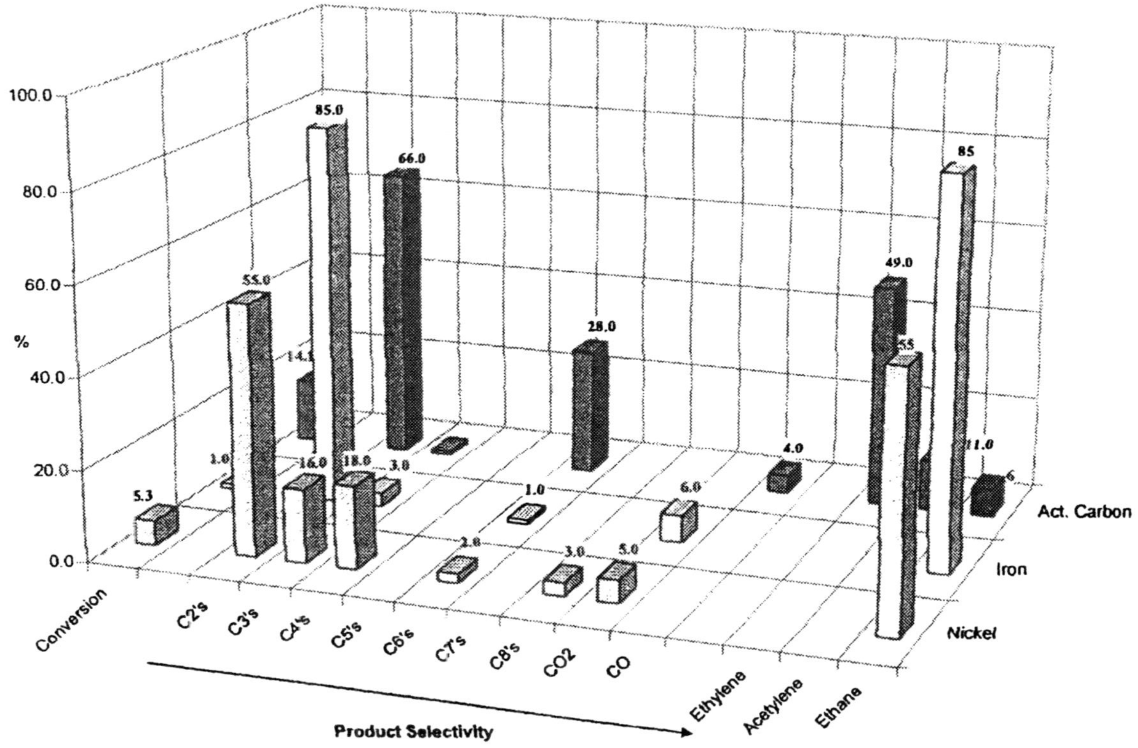

Figure IV. Nature of catalyst effect on conversion and selectivity for the reaction using 500 W, 20min of irradiation time, and 2.4 GHz

Conclusion

The microwave heating oligomerization experiments have shown that nickel, iron powder, and activated carbon can act as selective catalysts for oligomerized products of methane. Oligomers ranging from C_2 to C_6 hydrocarbons have been prepared in good selectivity depending on the nature of the catalyst, microwave frequency, irradiation time, and power levels used in the microwave reactor. Selectivities toward benzene as high as 24% were achieved by using activated carbon as catalysts.

As microwave radiation frequency increased, selectivity towards benzene also increased. The changes in product distribution when using different microwave frequency are most likely due to different transverse modes. Different transverse modes generate different transient heating patterns consequently changing the dielectric properties of the catalyst (loss and dielectric constant). Further optimization of parameters that can influence and maximize the product distribution toward desired products for the oligomerization of methane is in progress and will be reported later.

Acknowledgements

The authors thank the National Science Foundation and the Electrical Power Research Institute under Grant CTS-9413394 of the joint NSF/EPRI Initiative on Microwave Induced Reactions. We thank Dr. Daniel Scola of the Institute of Materials Science, University of Connecticut for support of this research.

References

1. Saint-Just, J.; Basset, J. M.; Bousquet, J.; Martin, G. A. *Recherche.* **1990,** *21*, 733-738.
2. Parkins, N. D. *Chem. Br.* **1990,** *26*, 841-844.
3. Roussy, G.; Pearce, J. A. "Fundation of Industrial Applications of Microwaves and Radiofrequency Fields", Wiley, New York, 1995.
4. Ioffe, M. S.; Pollington, S.D.; Wan, J. K. S. *J. Catal.* **1995,** *151,* 349-354.
5. Pollington, S. D.; Ioffe, M. S.; Westergaard, M.; Wan, J. K. S. *Res. Chem. Intermed.* **1995,** *21*(1), 59-62.
6. Chen, C.; Hong, D.; Dai, S.; Kan, J. *J. Chem. Soc., Faraday Trans.* **1995,** *91*(7), 1179-1184.
7. Mieville, R. L.; Robinson, K. K. "CARBON MOLECULAR SIEVES AND OTHER POROUS CARBONS. Synthesis and Applications", , Mega-Carbon Company: St. Charles, IL., p. 1-23, 1994.

8. Marun, C.; Suib, S.L.; Dery, M.; Harrinson, J. B.; Kablaoui, M. *J. Phys. Chem.* **1996,** *100,* 17866-17872.
9. Thuery, J. "Microwaves: Industrial, Scientific and Medical Applications", Norwood, MA: Artech House, Inc. 670p. 1992.
10. Marun, C.; Conde, L. D.; Suib, S. L. *J. Phys. Chem. A* **1999,** *103*, 4332-4340.
11. Conde, L. D.; Marun, C.; Suib, S. L.; Fathi, Z. *J. Catal.* **2001,** *204*, 324-332.
12. Conde, L. D.; Marun, C.; Suib, S. L. sumitted to *J. Catal.*
13. M. J. F. M. Verhaak, M. J. F. M.; Van Dillen, A. J.; Geus, J. W. *J. Catal.* **1993,** *143*, 187-200.
14. Tanashev, Y. Y.; Fedoseev, V. I.; Aristov, Y. I.; Pushkarev, V. V.; Avdeeva, L. B.; Zaikovskii, V. I.; Parmon, V. N. *Catal. Today.* **1998,** *42*, 333-336.
15. Bamwenda, G.; Moore, E.; Wan, J. K. S. *Res. Chem. Intermediates.* **1992,** *17*, 243-262.
16. Conde, L. D.; Marun, C.; Suib, S. L. *Microwave Induced Oligomerization of Methane over Activated Carbon.* in progress.

Chapter 23

Decomposition of CF_4 by Microwave Heating: A Potential Way to Decrease Greenhouse Gas Emissions

Franz-Joef Spiess[1], Steven L. Suib[1-3], Yuji Hayashi[4], and Hiroshige Matsumoto[5]

[1]U-60 Department of Chemistry, [2]Department of Chemical Engineering, and [3]Institute of Materials Science, University of Connecticut, Storrs, CT 06269–3060
[4]Nagasaki University, 1–14 Bunkyo-machi, Nagasaki, 852–8521, Japan
[5]Fujitsu Laboratories Ltd., 4–1–1 Kamikodanaka, Nakahara-ku, Kawasaki 211–8588, Japan

Carbon tetrafluoride, a potent greenhouse gas, was almost completely destroyed using microwave heating. Several catalysts were explored in this microwave study and activated carbon was the one yielding the highest conversion. The amount of catalyst used was 100 mg having a surface area of 650 m^2/g. The results show that the presence of water is crucial in this reaction and that significant hydrogen production takes place. The maximum conversion achieved is 95 to 100 %. The analysis was carried out using a mass spectrometer and the power levels used were 300 to 1000 Watts. Other major products in this reaction are hydrogen fluoride and carbon dioxide. Experiments to determine the nature of the deactivation of the catalyst were carried, and a reaction scheme is proposed.

Carbon tetrafluoride was found in the air of various European countries and in both hemispheres of the troposphere (0-14 km) in the 1970s (*1,2*). Cicerone's calculations and modeling showed that carbon tetrafluoride is a nearly inert gas in the atmosphere (*3*). The residence time of CF_4 in the atmosphere is calculated to be on the order of 10,000 years, up to 100 times higher than for simple chlorofluorocarbons. Furthermore, CF_4 traps the outgoing planetary infrared radiation in its intense band at about 8 micrometers. These properties make CF_4 a very potent greenhouse gas whose atmospheric concentration is believed to be due to mostly industrial sources, natural sources seems to negligible.

CF_4 is a by-product in the aluminum production and used in the semiconductor industry (*4-8*). In the semiconductor industry, carbon tetrafluoride is one of the gases used in etching (dry etching silicon wafers using radio frequency plasmas) and cleaning (plasma etch cleaning of plasma enhanced CVD reactors after deposition) processes (*4*). The emissions from these processes in the US semiconductor industry, which could be reduced by 98 % using a capture and recycle system or by using plasma or thermal decomposition (as mentioned below), amass to 230 tons per year, about 5-10% of global emissions, compared to 6,800 tons per year by the US aluminum smeltering industry (*4,5*). Carbon tetrafluoride is one of the by-products in the electrolytic smelting of alumina to produce aluminum metal and 1.3 to 3.6 kg CF_4 is produced per ton of aluminum produced (1985) (*6*). According to the 1997 Kyoto protocol greenhouse gas emissions by industrial countries must be reduced by 5.2% over the 2008 to 2012 period, this includes CF_4 (7). Emissions from newer cells are lower, up to 20 times lower in 1999 than in 1993 as reported in a Canadian study (*4,8*). The exceptionally long atmospheric lifetime of carbon tetrafluoride necessitates finding solutions to destroy this potent greenhouse gas.

So far mostly catalytic and plasma discharge methods have been used to decompose CF_4 (9-*17*). Wei et al. reported destruction efficiencies over 99% using a microwave-generated surface wave plasma under optimized input power, pressure, and oxygen admixture (9). Grytsinin et al. showed effective decomposition using a slipping surface discharge (*10*). Tonnis et al. showed the importance of water addition in a Litmas RF abatement reactor (*11*). Xu et al. concluded that water as an additive could give destruction of 99.6% of CF_4, and adds that in the presence of nitrogen NO_x could be formed (*12*). Takita et al. used $AlPO_4$-rare earth phosphate catalysts to decompose CF_4 at conversions of about 50% (*13*). Kanno et al. developed a catalyst, which destroyed CF_4 with a conversion of over 99.9% (*14*). Burdeniuc showed mineralization of chlorofluorocarbons using sodium oxalate and achieved 95 to 100% destruction (15). Similar results were reported by Lee et al.; they showed that by using heated alkali halides carbon tetrafluoride can be effectively destroyed to yield alkali metals (*16*). Several papers like Jacobsohn et al. describe the decomposition as a means to deposit fluorinated amorphous-carbon films (*17*).

Our group performed research on low-temperature AC discharge plasmas involving CF_4 utilizing different reactors at atmospheric pressure compared to the ones mentioned before which were done at lower pressure or in vacuum (*18*). In this research, we first studied the decomposition using AC plasmas with carbon tetrafluoride and nitrogen as the balance gas, but the conversions were too low, so our focus shifted to the use of microwave heating because preliminary data were very promising (*19*). The microwave heating was carried out at medium power (0-1200 W) using water as a supplement and activated carbon as a catalyst. The experiments were performed using an ASTEX microwave unit.

Experimental

A simple setup was used as shown in Figure 1. The feed is a mixture of carbon tetrafluoride, water (introduced by a bubbler), and nitrogen. The content of the feed gas is 4% CF_4, 3% water, and balance of nitrogen at a flow a rate of 15 mL/min. The activated carbon (100 mg) was placed in a quartz tube (3/8 inch outer diameter) and secured with quartz wool. In earlier reactions, the feed gas was 2% CF_4, 3% water, and balance of nitrogen at a flow a rate of 10 mL/min. Sampling was done using a syringe and then injected into a Hewlett Packard Series II gas chromatograph equipped with a Alltech Poropak N 80/100 with a thermal conductivity detector.

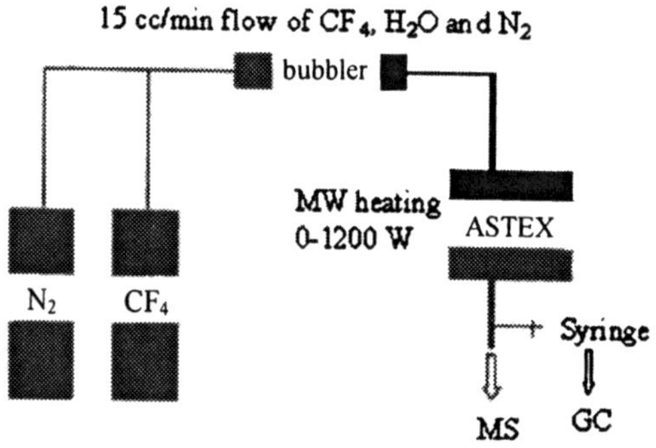

Figure 1. Reaction setup for CF_4 decomposition.

This tube was then placed into the ASTEX microwave cavity (Figure 2). The ASTEX unit consists of an ASTEX microwave power source module GK139 with a magnetron type GL 130WC, a 3 stub tuner model AX3041, and an ASTEX applicator model AX7020. The power was emitted in pulses with a frequency of 120 MHz (every 8.3 ns). All parameters were kept constant. Only the power level was adjusted as necessary.

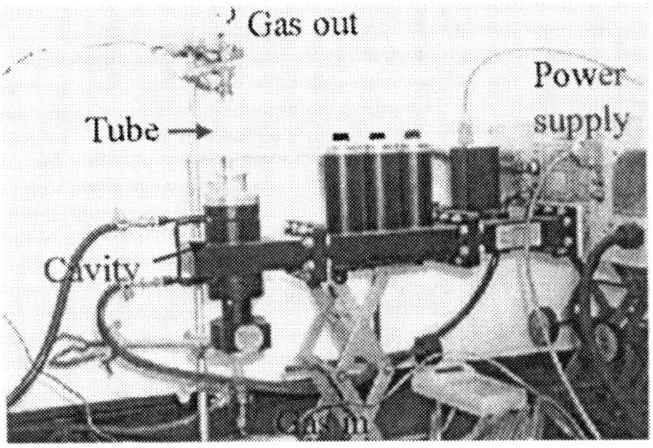

Figure 2.Photograph of the reaction apparatus.

The reaction mixture was continuously monitored with a MKS-UTI PPT quadrupole residual gas analyzer mass spectrometer with a Faraday cup detector and a variable high-pressure sampling manifold. A water scrubber was placed in front of the mass spectrometer to prevent hydrogen fluoride from getting into critical parts of the unit. The reaction system was equilibrated at the beginning and then the power was turned on to the desired level and continued until an equilibrium value was achieved. The catalysts used were either prepared in our research group or used as obtained from commercial vendors. Surface area and pore size measurements were performed using a Micrometrics ASAP 2010 BET system.

CF_4 is a colorless, odorless, and nonflammable gas which at a level of 895,000 ppm/15 minutes can yield to hypoxia with dizziness, disorientation, incoordination, narcosis, nausea, and vomiting. Therefore, these reactions were carried out in a fume hood. The formed hydrogen fluoride is captured in a scrubber to prevent damage to instruments and for safety concerns and will be disposed of properly. The microwave unit is safe to use in the described experiments.

Results and Discussion

Initial studies using discharge plasmas yielded only low conversions (up to 12 %). In these studies, tubular and fan PACT reactors were used to decompose carbon tetrafluoride at high flow rates of about 100 mL/min. Different power supplies were used but the conversion stayed low due to limitations of the power supply or because no uniform plasma could be established in the fan PACT reactors (18). Therefore, more effective ways to decompose CF_4 were sought. MS studies during the discharge plasma experiments suggest that CF_4 alone or CF_4 and water are the most suited systems for the decomposition. Studies using microwave heating exhibited promising results under similar conditions and were therefore pursued.

The first set of reactions was carried out with the setup described above at a constant flow rate of 10 mL/min. This procedure was tried in order to reproduce the conditions of the plasma experiments, but due to limitations of this setup, because the catalyst would not stay in place in the quartz tube, the flow rate was lowered to 10 mL/min. Several catalysts were examined for these types of experiments. The catalysts tested were several different kinds of activated carbon with various surface areas, Ni filament type catalyst, V_2O_5 Supported on Al_2O_3, and a zeolite/activated carbon mixture.

Of these catalysts, only the activated carbon exhibited any change in the CF_4 concentration in these experiments. No reaction takes place without the presence of water vapor.

In the case of activated carbon, several kinds of activated carbon were used. In the earlier studies a high surface area species (3,100 m^2/g) was used. The flow rate in these experiments was 10 cc/min with a feed of 2 % CF_4, 3 % H_2O and balance nitrogen. Figure 3. shows the concentration decrease of CF_4 versus time. Complete destruction of CF_4 is achieved almost initially, but very rapid deactivation also takes place.

Two mixed systems were investigated. A system containing zeolite 5A and the high surface area activated carbon was tested as well as a mixture of the activated carbon and nickel filament type catalyst. These studies were done at the same total flow rate and CF_4 concentration as used before. Both systems show lower conversion than the activated carbon alone. The conversion in the case of the nickel catalyst reaches only about 20%, in the case of the zeolite about 40 to 50%, but the power needed to get a similar conversion is higher as in the case without zeolite. The lower conversions in the case of the zeolite are due to the replacement with the zeolite that can take up a lot more water than does activated carbon. Therefore, more power is needed to release the water adsorbed in the matrix in the case of the zeolite to achieve comparable conversion to the case without the zeolite.

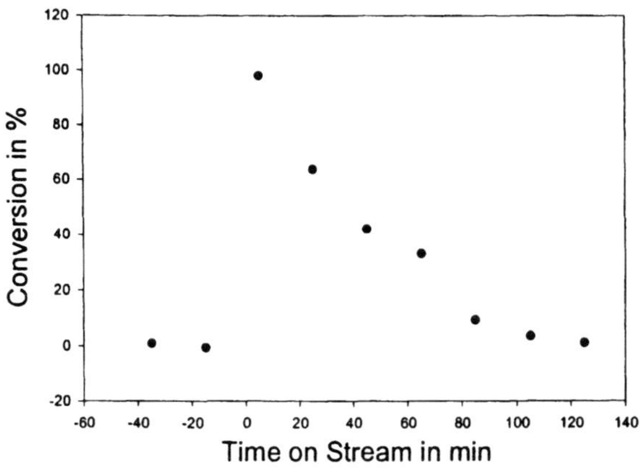

Figure 3. Decomposition of CF_4 versus time using the high surface area activated carbon at 10 mL/min and 2 % CF_4, 3 % H_2O and balance N_2.

In the more recent studies the catalyst was changed to an activated carbon with a lower surface area. Furthermore, the flow rate was increased to 15 mL/min from 10 mL/min, as was the concentration of carbon tetrafluoride from 2 to 4% in the feed gas, the water vapor level remained constant at 3%. The analysis method was changed from the cruder method of syringe sampling and injection into a gas chromatograph to the on-line sampling with a mass spectrometer. In this new setup and under the new conditions, a different activated carbon with a surface area of 2,200 m^2/g was tried as well, but conversions ranged only from 30-40% destruction of carbon tetrafluoride. Only a few reactions were performed with catalyst.

The results presented from here on are the ones for a commercial brand available from Sigma-Aldrich, a lower surface area species. Surface area data will be presented later for this species. A typical decomposition versus time plot is shown in Figure 4 as obtained with the mass spectrometer. The conversion in the case shown below is 74% (based on the disappearance of the CF_4 signal from the mass spectrometer) as is the appearance of one of the major side products, carbon dioxide.

As seen in the time versus conversion plot (Figure 4), the activated carbon deactivates rather quickly. The deactivation of this species is slower and less rapid than in the case of the higher surface species (as seen in Figure 3). In this case, it takes about 100 min for the activated carbon to deactivate, but the catalyst stays relatively active for over 60 min. In other cases for this species similar deactivation times have been observed. Generally, the deactivation tends to be longer if the conversion (and therefore power) is lower, and shorter if the conversion is higher. In the case of the best conversion of 95% with this species, the deactivation time was about 60 min. The deactivation for the high surface area activated carbon is about 60-80 min, but a lot of activity is already lost in the first 30 min of the reaction. Similar observations regarding deactivation times were made using this catalyst type. The deactivation is also generally dependent on power as in the case mentioned above.

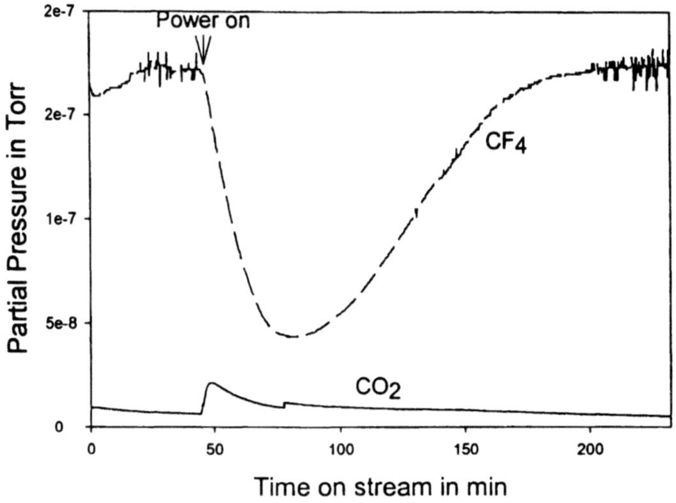

Figure 4. Plot of CF_4 decomposition versus time, top plot is the CF_4 plot (dashed line), bottom plot is the CO_2 plot (solid line), power was turned on to 480 W at 246 min (arrow).

The conversion was studied as a function of power. As shown in Figure 5, the maximum conversion is achieved at medium power levels of 300-400 W (whole range 0-1200 W). The maximum conversion observed is 95% measured as the disappearance of the CF_4 in the MS plot. This is in good agreement with earlier data (data not presented) that showed that the maximum conversion is in the region of 400 to 500 W. The overall conversion in that case was slightly lower than in the data shown in Figure 5. The maximum conversion in that set of data was 75%.

In the mass spectrometric studies, a lot of other species were detected apart from carbon tetrafluoride and carbon dioxide (as shown in Figure 4). The reaction produces significant amounts of hydrogen, about 10-50,000 ppm depending on reaction conditions and power level. Other gases detected were oxygen, methane, nitrogen, due to its presence as a balance gas, methane, and water vapor. Carbon monoxide might be present but its mass coincides with nitrogen, so the signal would be buried under the bigger signal for nitrogen, the predominant gas in the system with a concentration of about 94 % in the feed. Nitrogen oxides might be present as well as discussed by Xu et al. (*12*), but the mass spectrometer was not tuned to that mass, unfortunately.

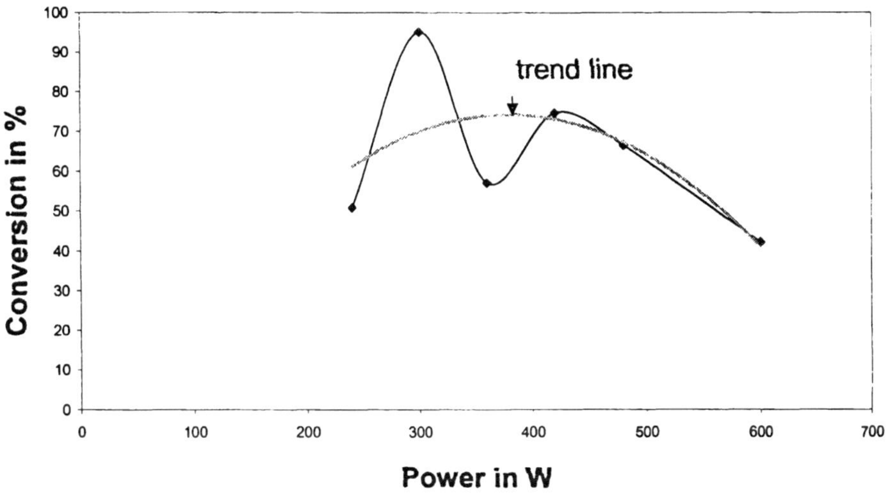

Figure 5. Dependence of the conversion on the power (black) and the corresponding trend line (gray).

The hydrogen level was investigated and its influence on the reaction. Hydrogen evolves as soon as the microwave power is turned on. The hydrogen level in the mass spectrometer peaks shortly thereafter and then levels off again. Figure 6 shows these peak levels at various power settings. The graph clearly shows that in a blank run of the reaction the hydrogen level is higher than in a run with the carbon tetrafluoride present. The activated carbon catalyst is used in both reactions.

The carbon dioxide and oxygen levels were also compared in the same study. The CO_2 level is slightly higher for the blank run compared to normal reaction conditions. The oxygen level normally levels off in the entire course of the reaction, maybe due to absorption in the water scrubber, but in two cases significant oxygen production is observed under normal reaction conditions. In the blank run, the usual pattern is observed, but a short spike in the oxygen concentration is observed when the power is switched on. The water vapor level shows a slight increase when the power is turned, probably due to water coming off the activated carbon where water was adsorbed. Methane is evident in the system as seen in the peaks after the power is turned on at m/e of 15. The main methane peak is at m/e of 16, but this peak coincides with a secondary peak of oxygen at m/e of 16. The line at m/e of 15 is the second strongest line in the methane spectrum at about 90%. Methane may be present in the blank run but this is unknown.

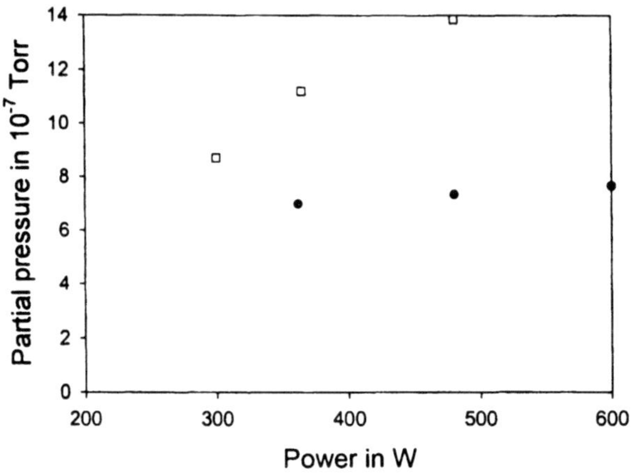

Figure 6. Hydrogen production with (dots) and without CF_4 present (squares), and using the activated carbon catalyst at various power settings.

The extent of the reaction can also be seen, based on how the reaction tube looks after the reaction. The higher the extent of the reaction, the more "decomposed" is the reaction tube due to the glow and HF production as is shown in Figure 7.

The surface area of the activated carbon was determined to 663 m^2/g. The activated carbon consists of meso- and micropores. The surface area below 20 Å is 349 m^2/g, and between 20 and 200 Å, it is 280 m^2/g. The surface area of one activated carbon used in one of the reactions was measured to be 623 m^2/g. This trend shows a slight reduction in surface area and a probable cause will be discussed later.

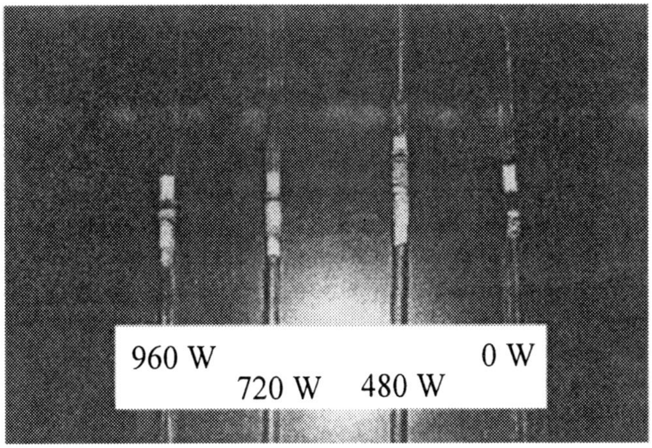

Figure 7. Photograph showing the extent of reaction at various power settings.

The role of the water in this reaction is twofold. Water is the primary energy absorber for the microwave radiation, but also water is the hydrogen and oxygen source needed in this reaction. A schematic of the simplified reaction scheme of the major reactions in the microwave heating cavity is shown in Figure 8.

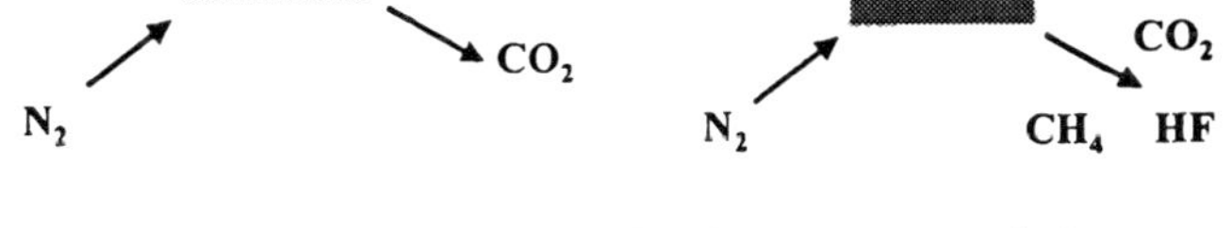

Figure 8. Simplified reaction scheme for the presence and absence of CF_4, activated carbon is presented as AC.

In both cases hydrogen, oxygen, and carbon dioxide are formed. Carbon monoxide and nitrogen oxides might form [reference (*12*) notes that in a system like this they should be expected]. Methane surely formed in the case of CF_4, but might also form without CF_4. The levels maybe different. Hydrogen fluoride is only formed in the presence of CF_4.

When the power of the microwave unit is turned on, water in the feed and on the catalyst (some comes off) is split up in both cases. The formed hydrogen and oxygen radicals or atoms are the active species in this reaction. A reaction mechanism involving this process is shown in Figure 9. In the absence of carbon tetrafluoride, the reaction to hydrogen and oxygen and the reaction to carbon oxides, after the reaction with activated carbon, are predominant. Nitrogen oxides might form as well because the nitrogen signal decreases a little bit when the power is turned on, the formation is also favored under these conditions according to the literature (*12*). Methane may form as well because hydrogen atoms are available to react with carbon of the catalyst. Other products were not observed in the absence of carbon tetrafluoride.

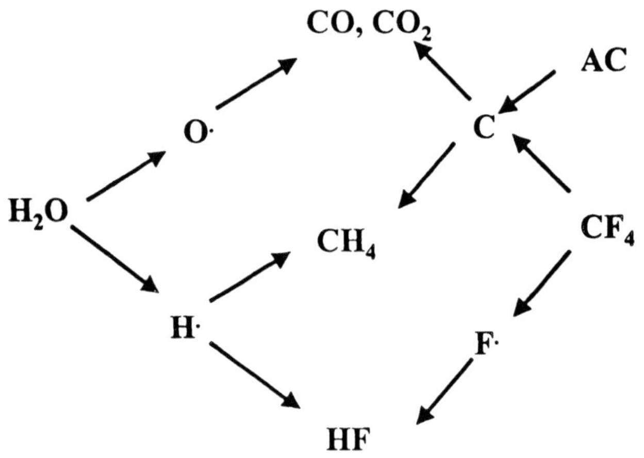

Figure 9. Reaction mechanism for the CF_4.

In the presence of carbon tetrafluoride things are a bit different. The fluorine radicals compete for the hydrogen radicals. This observation is evident due to a decrease in the hydrogen signal in the mass spectrometer. Hydrogen fluoride is formed by the combination of these two radicals. HF is captured in a water scrubber in line before HF could enter the mass spectrometer. Therefore only in two cases HF was observed as a strong signal. Methane is formed by the combination of carbon from the carbon tetrafluoride or from the catalyst (indicated as AC in Figure 9). The oxygen signal seems to be a little bit stronger than in the case of the blank run, favoring the combination of oxygen molecules over carbon dioxide. This explains the slightly lower signal for carbon dioxide, which forms from CF_4 or from the catalyst, in the case CF_4 is present.

Activated carbon plays an important role in the reaction. Water gets adsorbed onto the surface by the means of the surface hydroxyl groups. When the power is turned on, the adsorbed water is released and some additional water is formed by the reaction of hydroxyl groups. Therefore, activated carbon does not have an active site per se; it is more of a mediator.

The reason for deactivation of the catalyst is not completely understood. Initially, hydrogen fluoride formed in the reaction poisons the catalyst, but neither infrared nor EDX analyses showed any fluorine on the catalyst. Therefore, carbon (coke) formed in the decomposition of the carbon tetrafluoride and deposits on the catalyst surface and hinders the access to the active sites. This could be explained by the decrease in surface area, as seen in the BET surface analysis. This coking theory is supported by the literature (20). Thermo gravimetric analysis of this catalyst will be performed in the future and may elucidate the mechanism of coke formation.

Conclusions

The destruction of carbon tetrafluoride by microwave heating represents an efficient and easy means to eliminate this greenhouse gas. Carbon tetrafluoride is almost completely decomposed. The main products are hydrogen fluoride, carbon dioxide, oxygen, and hydrogen. Furthermore, small amounts of methane are formed and are observed. The production of small amounts of carbon monoxide and nitrogen oxides is expected but was not examined. The activated carbon catalyst deactivates after a short period of time of about 100 min. The deactivation time depends in general on the power. Furthermore, water is essential in these reactions. A reaction scheme explaining the observed phenomena was proposed and is justified based on our data.

Acknowledgement

The authors thank JFCC and Planet Japan for support of this research. We also thank Daniel Conde for assistance in the power measurements and reactor setup.

References

1. Gassmann, M. *Naturwissenschaften* **1974**, *61*, 127.
2. Rasmussen, R.; Penkett, S.; Prosser, N. *Nature* **1979**, *277*, 549.
3. Cicerone, R.J. *Science* **1979**, *206*, 59-60.

4. Marinelli, L.; Worth, W. Global Warming: A White paper on the Science, Policies and Control Technologies that Impact the U.S. Semiconductor Industry; Technology Transfer # 9311207A-TR, SEMATECH, 1994.
5. Li, Y.E.D.; Paganessi, J.E.; Rufin, Denis *ACS Symposium Series* **2001,** *766,* 62-75.
6. Weston, R.E. Jr. *Atmos. Environ.* **1996,** *30(16),* 2901-2910.
7. Gosselin, A.; Fradet, C. In *Light Metals 2000, Proceedings of the International Symposium on Light Metals, Ottawa, ON, Canada, Aug. 20-23, 2000;* Canadian Institute of Mining, Metallurgy and Petroleum: Montreal, Canada, 2000; 353-363.
8. Mackay, G.I.; Karecki, D.R.; Pisano, J.T.; Schiff, H.I. In *Light Metals 2000, Proceedings of the International Symposium on Light Metals, Ottawa, ON, Canada, Aug. 20-23, 2000;* Canadian Institute of Mining, Metallurgy and Petroleum: Montreal, Canada, 2000; 339-352.
9. Wei, T. -C.; Fang, Y.S *Abatement of* CF_4, C_2F_6 *and* CHF_3 *in Microwave Generated Surface Wave Plasmas;* 29th IEEE International Conference on Plasma Science, 2002.
10. Grytsinin, S.I.; Korchagina, E.G.; Kossyi, I.A.; Misakyan, M.A.; Silakov, V.P.; Tarasov, N.M.; Temchin, S.M. *Plasma Sources Sci. Technol.* **2001,** *10,* 125-133.
11. Tonnis, E.J.; Graves, D.B.; Vartanian; Beu, L.; Lii, T.; Jewett, R. *J. Vac Sci Technol. A* **2000,** *18,* 213-231.
12. Xu, X.; Rauf, S.; Kushner, M.J. *J. Vac Sci Technol. A* **2000,** *18,* 393.
13. Takita, Y.; Ninomiya, M.; Miyake, H.; Wakamatsu, H.; Yoshinaga, Y.; Ishihara, T. *Phys. Chem. Chem. Phys.* **1999,** *1,* 4501-4504.
14. Kanno, S.; Ikeda, S.; Yamashita, H.; Azuhata, S.; Irie k.; Tamata, S. *Mater. Res. Soc. Symp. Proc.* **1998,** *497,* 59-64.
15. Burdeniuc, J.; Crabtree, R.H. *Science* **1996,** *271,* 340-341.
16. Lee, M.C.; Choi, W. *Environ. Sci. Technol.* **2002,** *36,* 1367-1371.
17. Jacobsohn, L.G.; Franceschini, D.F.; Maia da Costa, M.E.H.; Freire Jr. F.L. *J. Vac. Sci. Technol. A* **2000,** *18,* 2230-2238.
18. Huang A.; Suib, S. L. *Res. Chem. Interm.* **2001,** *27,* 957-974.
19. Spiess, F.-J.; Suib, S.L.; Hayashi, Y.; Matsumoto, H. unpublished results
20. Wiersma, A.; van de Sandt, E.J.A.X.; Makee, M.; Moulijn, J.A. *Appl. Catal., A* **2001,** *212,* 223-238.

Application of Supercritical CO_2

Chapter 24

Application of High-Pressure Phase Equilibria to the Selective Oxidation of Alcohols over Supported Platinum Catalysts in Supercritical Carbon Dioxide

Roger Gläser[1], Jörg Williardt[1], David Bush[2], Michael J. Lazzaroni[2], and Charles A. Eckert[2]

[1]Institute of Chemical Technology, University of Stuttgart, D–70550 Stuttgart, Germany
[2]School of Chemical Engineering and Specialty Separations Center, Georgia Institute of Technology, 778 Atlantic Drive, Atlanta, GA 30332–0100

At conditions above its critical point, the greenhouse gas carbon dioxide may be used as an environmentally benign, nontoxic medium for organic synthesis. The peculiarities and the prediction of the high-pressure phase behavior of the multicomponent reactant mixture related to the selective oxidation of 2-propanol to acetone by molecular oxygen over a carbon-supported platinum catalyst in high-pressure carbon dioxide at 40°C are reported. Very good agreement of vapor-liquid equilibria predictions based on the Patel-Teja equation of state and experimental data obtained with argon as a replacement for oxygen was achieved. From the phase behavior considerations, reaction conditions were derived that ensure single-phase operation for the catalytic 2-propanol oxidation. Optimized product yield and catalyst productivity were reached by balancing the increased reactant concentration, the oxygen/2-propanol ratio and the catalyst deactivation.

A promising utilization of the greenhouse gas carbon dioxide is its application as an environmentally benign, nontoxic, nonflammable, and abundantly available reaction medium for organic synthesis at conditions above its critical point (*1*). Other greenhouse gases have not found comparable interest as solvents for chemical processing. Besides the facile separation of carbon dioxide from organic mixtures by depressurization, its nontoxic properties make it an attractive solvent for pharmaceutical and food-related processes. Supercritical carbon dioxide exhibits unique solvent properties between those of typical liquids and gases that can be sensitively tuned by pressure and temperature. It should be stressed that the term *supercritical* in its strict sense refers to single components only, and is used here for the multi-component mixture in quotation marks as suggested earlier (*2*).

The use of supercritical carbon dioxide is particularly attractive for heterogeneously catalyzed conversions involving mass transfer of a gaseous reactant like oxygen or hydrogen into a liquid reaction phase. Limitations of mass transport can be effectively eliminated since "supercritical" carbon dioxide is miscible with both permanent gases like hydrogen or oxygen and organic reactants over a broad range of compositions and reaction conditions. To fully capitalize on this advantage, the experimental conditions of the single-phase region for a given reaction system have to be accurately known (*3*). In spite of its utmost importance, the issue of high-pressure phase behavior is often not sufficiently addressed in studies of reactions at "supercritical" conditions.

The advantage of working in a single-phase region at "supercritical" conditions has been successfully applied in several studies of heterogeneously catalyzed hydrogenation and oxidation reactions (*4-8*). The fact that carbon dioxide is nonflammable is a particularly obvious advantage for the replacement of conventional organic solvents for oxidation reactions with oxygen (or air) as the oxidizing agent. Specifically, previous studies of alcohol oxidations over noble-metal catalysts have been carried out in "supercritical" carbon dioxide (*7,8*). Conventionally, this conversion is conducted in aqueous solution of the alcohol and, in many cases, the reaction rate is restricted by oxygen transport from the gas into the liquid reaction phase (*9,10*). Other examples for application of high-pressure phase behavior to improve reactions or reaction/separation scenarios include CO_2-expanded liquids as reaction media (*11*), CO_2-induced melting of reactants (*12*), continuous extraction of products from ionic liquid reaction mixtures with CO_2 (*13*), or the CO_2-induced miscibility of organic reaction mixtures and fluorous-sequestered catalysts (*14*).

In the present study, we report on the peculiarities and the prediction of the high-pressure phase behavior of the multi-component reaction mixture related to the oxidative dehydrogenation of 2-propanol (IPOH) with molecular oxygen to acetone (AC) and water in "supercritical" carbon dioxide. The results of these predictions are then applied to choose reaction conditions for experimentally investigating the conversion over a carbon-supported platinum catalyst in a single, homogeneous phase. The effect of a concentration increase of both 2-propanol and oxygen on the catalyst activity was also studied.

Experimental Section

Phase Behavior Calculations

Calculations of the phase behavior were based on the Patel-Teja equation-of-state (*15*). The equation in pressure-explicit form is

$$p = \frac{RT}{v-b} - \frac{a}{v(v+b)+c(v-b)}$$

where p is pressure, T is temperature, R is the gas constant, and v is the molar volume. The parameters a, b, and c are functions of the critical temperature and pressure and two additional variables F and ζ_c, which are adjusted to fit pure-component vapor pressure and liquid density data. Values are given in Table I. The Patel-Teja is similar to other van der Waal's derived equations of state, such as Peng-Robinson or Soave-Redlich-Kwong, but because of the fit of F and ζ_c, it gives better predictions of the pure component vapor pressures and liquid densities. This is especially important for multi-component systems because the sets of equations to solve include equilibrium balance, mass balance, and volume balance.

For mixture calculations, the standard one-fluid van der Waals mixing rules were used where

$$a_{mix} = \sum\sum x_i x_j a_{ij}$$
$$a_{ij} = (1-k_{ij})\sqrt{a_i a_j}$$
$$b_{mix} = \sum x_i b_i$$
$$c_{mix} = \sum x_u c_i$$

There is only one binary interaction parameter, k_{ij}, per pair of components. It is our experience that a more complicated mixing rule, i.e., one with more parameters, may fit the data better, but does not give better predictions based on binary data alone (*16*). In addition, for two of the binary pairs, only 1 data point was available.

The interaction parameters for carbon dioxide–2-propanol and carbon dioxide–1,4-dioxane were fit to experimental high-pressure VLE data at 40°C. It should be noted that the CO_2–1,4-dioxane system exhibited negative deviations from Raoult's law. The interaction parameters for 2-propanol–oxygen and 1,4-dioxane–oxygen were fit to the gas solubility at 20°C. We assume that the interaction parameter will be the same at 40°C. The interaction parameter for CO_2–oxygen was fit to high pressure VLE data at -10, 0, and 10°C. The interaction parameter did not change for these 3 temperatures, and we assume it will remain constant to 40°C. The 1,4-dioxane–2-propanol interaction parameter was fit to low-pressure VLE data. These parameters, listed in Table II, were used to predict the multi-component equilibria.

Materials

2-Propanol (Fluka, puriss. p.a., over molecular sieve), 1,4-dioxane (Merck, p.a., ACS reagent) and carbon dioxide (Air Products, 99.995 %) were used as received. A carbon-supported platinum catalyst (5 Pt/C, 5 wt.-% Pt, precious metal surface: 14.5 m^2/g, specific surface area: 1000 m^2/g, particle size: 90 wt.-% < 80 μm) available from Chempur was applied in the oxidation reactions.

Phase Behavior Experiments

To validate the predictions, two data points were measured for the ternary system 2-propanol/CO_2/argon. We substituted argon for oxygen because of safety reasons (Argon has similar critical properties and essentially the same Lennard-Jones parameters). A detailed procedure is given by Lazzaroni et al. (*22*). Briefly, known quantities of argon, carbon dioxide, and 2-propanol are

Table I. Critical properties and Patel-Teja parameters

	T_C / K	p_C / bar	F	ζ_C
2-propanol	508.3	4.674	1.2723	0.298
1,4-dioxane	587	5.208	0.8032	0.312
oxygen	154.6	5.046	0.4870	0.327
carbon dioxide	304.1	7.38	0.7077	0.309

Table II. Binary Interaction Parameters

Binary pair		k_{ij}	*Reference*
2-propanol	+ 1,4-dioxane	0.045	(*17*)
2-propanol	+ oxygen	0.06	(*18*)
2-propanol	+ carbon dioxide	0.09	(*19*)
1,4-dioxane	+ oxygen	0.13	(*18*)
1,4-dioxane	+ carbon dioxide	-0.02	(*20*)
oxygen	+ carbon dioxide	0.07	(*21*)

loaded in a variable-volume sapphire tube. The initial volume is such that there is only a single phase. The volume is increased isothermally until the first bubble appears. At that point, we assume that liquid composition is the same as the one synthetically loaded.

Catalytic Experiments

The catalytic conversions were carried out at 40°C in a 100 cm^3 batch reactor equipped with a magnetic stirrer (1000 min^{-1}) and a heating jacket. A schematic representation of the set-up is shown in Figure 1. Prior to the experiments, the catalyst was placed in the reactor and dried in vacuum. After adding carbon dioxide, liquid 1,4-dioxane and 2-propanol were introduced into the reactor by means of an injection loop. 1,4-Dioxane (DO) served as an internal standard for closing the mass balance. The mixture was then heated to reaction temperature and further carbon dioxide was added, if necessary. The reaction was started by adding oxygen to the reaction vessel to the desired reaction pressure. To monitor the progress of the reaction, a small amount of the reaction mixture was continuously removed, expanded to ambient pressure through heated tubing and a needle valve and periodically analyzed by temperature-programmed capillary gas chromatography. *Generally, care must be exercised in carrying out oxidation experiments, especially at increased reactant concentrations.*

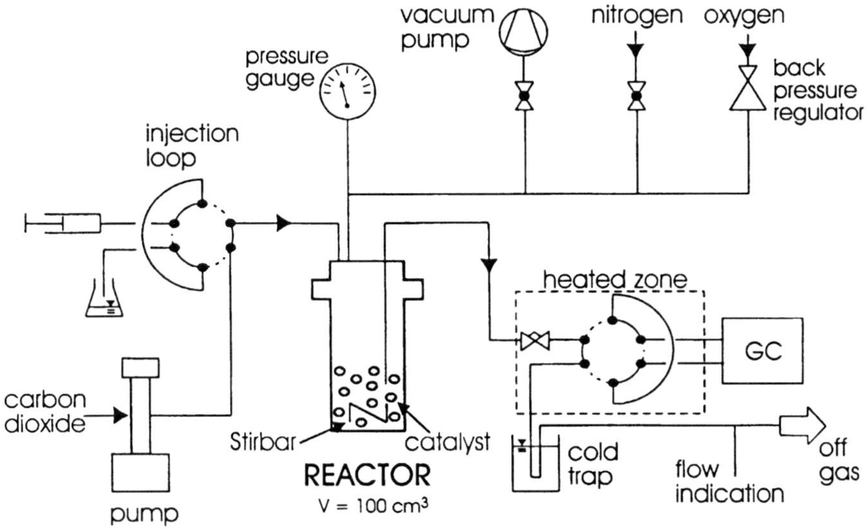

Figure 1. Schematic representation of the experimental apparatus.

Results and Discussion

The VLE curves for the four-component reaction system IPOH/O_2/CO_2/DO, have been calculated at fixed molar ratios of oxygen to carbon dioxide for the temperature of 40°C. This temperature was chosen for the experiments due to its proximity to the critical temperature of pure CO_2 and to demonstrate the feasibility of selective alcohol oxidations at mild conditions. As depicted in Figure 2, the single-phase region for the reaction system is reached at pressures above the equilibrium line indicated for each O_2/CO_2 molar ratio. As expected, a higher pressure is needed to reach the single phase when the oxygen to carbon dioxide ratio is increased.

In Figure 3, the oxygen mole fraction is presented at a constant oxygen to carbon dioxide molar ratio of 5.3 %. The addition of carbon dioxide allows for high concentrations of oxygen in the liquid or the "supercritical" phase without going to high partial pressures of oxygen. It is noteworthy, that, although the reactant concentration were quite low, the presence of 1,4-dioxane which was added to the mixture as an internal analytical standard in a comparable amount as the alcohol, decreases the pressure required for single-phase operation compared to the system containing oxygen, 2-propanol and carbon dioxide only. This is perhaps due to CO_2/1,4-dioxane interactions, which exhibit negative deviations from Raoult's law in the binary mixture.

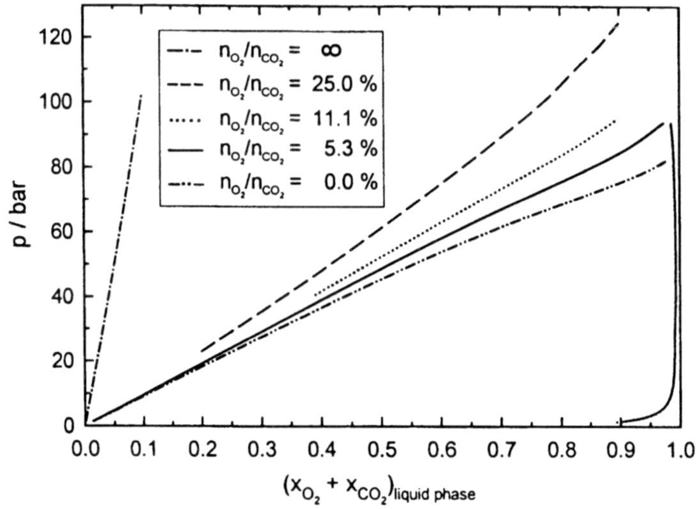

Figure 2. Vapor-liquid equilibrium curves of the IPOH/O_2/CO_2/DO mixture at 40°C for different oxygen/carbon dioxide mole fractions.

In Figure 4, we compare some predictions with experimental data. Since we are only using binary interaction parameters, the lines on Figure 4 are extrapolations with the model, which agree very well with the experimental ternary data. One interesting point is that by increasing the CO_2/IPOH ratio, we require less pressure to obtain high concentrations of argon.

Pertinent results of the catalytic conversion of 2-propanol over 5 Pt/C with oxygen at 150 bar, i.e., sufficiently above the one corresponding to the VLE, are displayed in Figure 5. In all cases, acetone was the only detectable oxidation product indicating the high selectivity of the reaction at the mild reaction temperature. Within the single-phase region, the initial amount of 2-propanol relative to the catalyst mass could be increased above 6 mmol/125 mg, a typical value for the reaction in aqueous solution where catalyst deactivation was observed (*10*). While no catalyst deactivation occurs at a fourfold increase of the IPOH/catalyst ratio, the conversion remains incomplete when eight times the initial amount of 2-propanol was present in the reactor. Nevertheless, a considerably higher amount of reactant could be converted at a higher initial reaction rate r_0 with the same amount of catalyst in the high-pressure reaction medium.

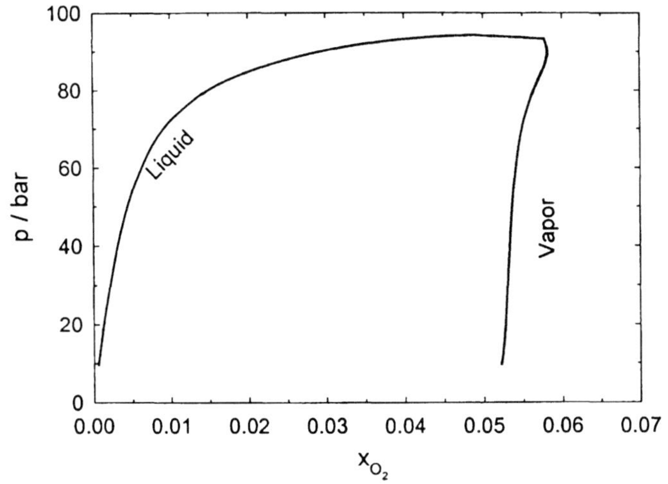

Figure 3. Oxygen mole fraction in the two-phase region for a fixed O_2/CO_2 molar ratio of 5.3 % at 40°C.

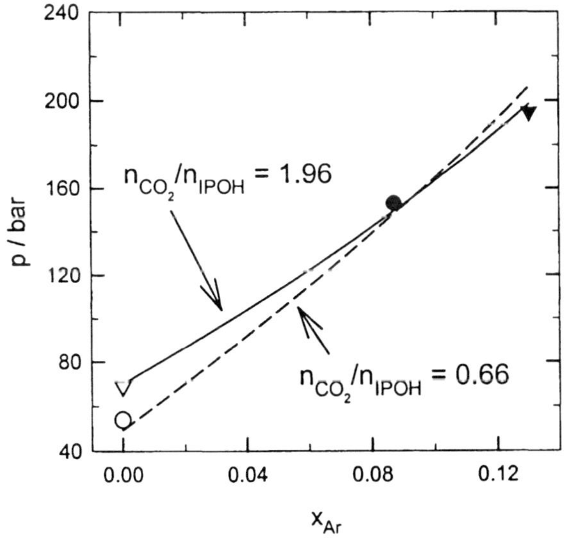

Figure 4. Experimental and predicted mole fraction (solid points – this work, hollow points – interpolated from ref. (19)).

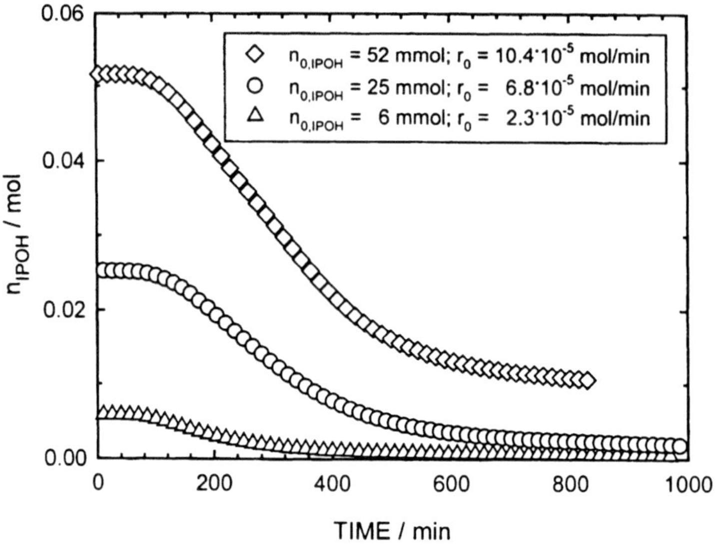

Figure 5. Conversion of increasing amounts of 2-propanol with molecular oxygen over 5 Pt/C in "supercritical" carbon dioxide (T_R = 40°C, x_{O_2} = 2.8 %, $p_{0,CO2}$ = 150 bar, m_{cat} = 125 mg, r_0: initial reaction rate).

To assess the influence of the oxygen mole fraction on the catalytic conversion, increasing amounts of oxygen were added to the same mixture of the catalyst, carbon dioxide, 2-propanol and the standard 1,4-dioxane. It is observed, that the rate of acetone formation increases with the oxygen mole fraction from 0.6 (substoichiometric with respect to 2-propanol) to 2.8 % and, for higher oxygen concentrations, decreases again (Figure 6). This finding is in accordance with results reported by Jenzer et al. (*7*) and can be rationalized by poisoning of the platinum surface due to partial over-oxidation. Although increasing oxygen mole fractions shift the VLE line to higher pressures (Figure 2), the reaction pressure in these experiments may be assumed to be high enough to avoid the occurrence of two separate phases.

Figure 7 shows the catalyst activity for 2-propanol conversion after repeated addition of 2-propanol and oxygen to the same reaction mixture and catalyst. (The reaction pressure has been constant, since during a previous reaction "cycle", the pressure dropped somewhat due to continuous sampling from the vessel.) It can be seen that except with the initial reaction mixture where fresh catalyst has been used, the reaction is incomplete for all later additions of reactants. Additionally, the slower reaction rate (as visible from the increase of

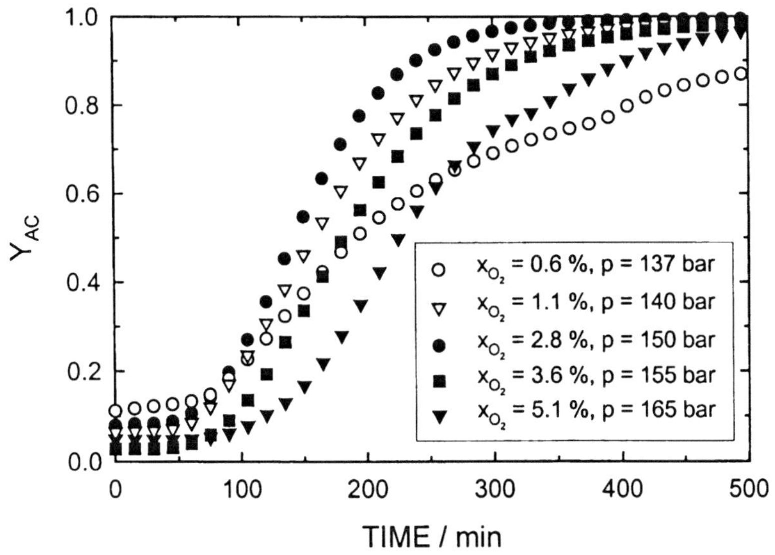

Figure 6. Influence of the oxygen mole fraction on the acetone yield in the 2-propanol conversion with molecular oxygen over 5 Pt/C (T_R = 40°C, p_{0,CO_2} = 135 bar, $n_{0,IPOH}$ = 13 mmol, m_{cat} = 250 mg).

the amount of acetone with time) after each new addition of reactants indicates a successively severe catalyst deactivation. On the other hand, however, the catalyst did remain appreciable activity, although the amount of organic reactants in the reaction mixture continuously increased. As already found for higher initial concentrations of 2-propanol (cf. Figure 5), these results show that higher reactant concentrations can be achieved in the "supercritical" single-phase region, but that the reaction conditions also need to be optimized with respect to the catalyst.

It will be an interesting question of future investigations, whether the selectivity of the catalytic reaction can be influenced or even directed by application of high-pressure phase behavior. Since the oxidation of 2-propanol occurs to acetone exclusively, this reaction is not suited for the corresponding studies. However, the conversion of 1-propanol with oxygen leads to propanal and, via consecutive oxidation, to propanoic acid. As shown previously, propanoic acid is formed in this reaction over a carbon-supported Pt-catalyst in "supercritical" carbon dioxide when the conversion of 1-propanol exceeds ca. 20 % (*23*). Thus, the latter conversion may serve as an appropriate example for future studies on the influence of phase behavior on catalytic selectivity.

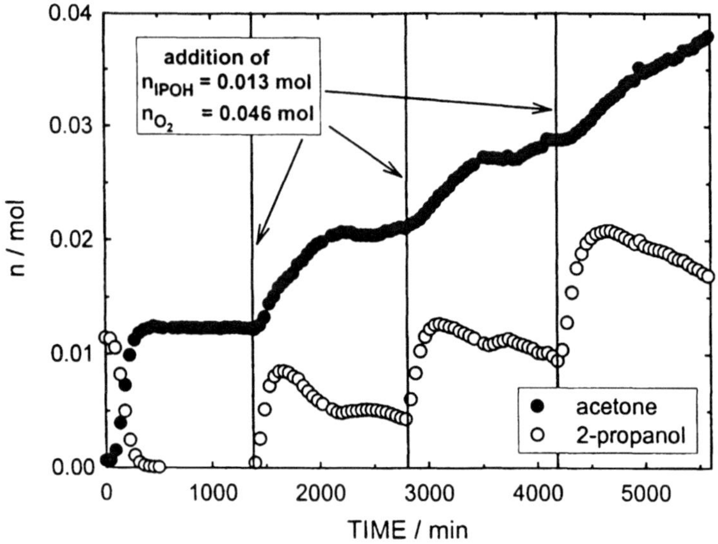

Figure 7. Influence of repeated addition of 2-propanol and oxygen to a reaction mixture on catalyst activity (catalyst: 5 Pt/C, initial reaction conditions as in Figure 6).

Conclusions

A careful consideration of the high-pressure phase behavior is essential for the beneficial utilization of "supercritical" carbon dioxide as a benign solvent for catalytic reactions. This study shows that the phase behavior of the four-component reactant mixture for the selective oxidation of 2-propanol (2-propanol, oxygen, carbon dioxide and 1,4-dioxane as the internal analytical standard), can be successfully predicted by the Patel-Teja equation of state with only binary interaction parameters. This approach is potentially suitable for prediction of other multi-component reaction mixtures consisting of highly asymmetric reactants. Inert components under reaction conditions such as an analytical standard may have a strong influence on the phase behavior and, thus, need to be taken into account in phase-behavior considerations. So far, only the reactant mixture for the oxidative conversion of 2-propanol has been investigated. Future studies will be aimed at including the reaction products and their impact on the phase-behavior as a function of the degree of conversion. These changes are especially important for oxidation reactions, since highly polar and hydrogen-bonding products, such as acids or water, are among the products, and these will have a pronounced effect on the phase behavior.

As shown here for the selective oxidation of 2-propanol with oxygen, conditions for improved rate and catalyst activity can be identified based on phase-behavior predictions. Most importantly, the prediction could be used to find reaction conditions for single-phase operation where all reactants as well as solvent and analytical standard are completely miscible. Although conditions may exist where catalyst activity is limited or declines during the reaction, an optimization of the reaction parameters based on the consideration of the high-pressure phase behavior bears large potential for the improvement of other catalytic oxidations and hydrogenations conventionally carried out in liquid solvents.

Acknowledgements

R.G. thanks the Dr. Leni Schöninger foundation for a habilitation stipend. Thanks are also due to Prof. Dr.-Ing. J. Weitkamp, University of Stuttgart, for generous support. Financial support by the U.S. NSF and the U.S. EPA is gratefully acknowledged.

References

1. *Chemical Synthesis using Supercritical Fluids*; Jessop, P. G.; Leitner, W., Eds.; Wiley-VCH, Weinheim, 1999.
2. Jenzer, G.; Mallat, T.; Baiker, A. *Catal. Lett.* **2001**, *73*, 5-8.
3. Subramaniam, B.; McHugh, M. A. *Ind. Eng. Chem., Proc. Des. Dev.* **1986**, *25*, 1-12.
4. Arunajatesan, V.; Subramaniam, B.; Hutchenson, K. W.; Herkes, F. E. *Chem. Eng. Sci.* **2001**, *56*, 1363-1369.
5. van den Hark, S.; Härröd, M. *Ind. Eng. Chem. Res.* **2001**, *40*, 5052-5057.
6. Tschan, R.; Wandeler, R.; Schneider, M. S.; Schubert, M. M.; Baiker, A. *J. Catal.* **2001**, *204*, 219-229.
7. Jenzer, G.; Sueur, D.; Mallat, T.; Baiker, A. *Chem. Commun.* **2000**, 2247-2248.
8. Steele, A. M.; Zhu, J.; Tsang, S. C. *Cat. Lett.* **2001**, *73*, 9-13.
9. Kluytmans, J. H. J.; Markusse, A. P.; Kuster, B. F. M.; Marin, G. B.; Schouten, J. C. *Catal. Today* **2000**, *57*, 143-155.
10. Nicoletti, J. W.; Whitesides, G. M. *J. Phys. Chem.* **1989**, *93*, 759-767.
11. Brown, R. A.; Pollet, P.; McKoon, E.; Eckert, C. A.; Liotta, C. L.; Jessop, P. G. *J. Am. Chem. Soc.* **2001**, *123*, 1254-1255.

12. Jessop, P.; Wynne, D. C.; DeHaai, S.; Nakawatase, D. *Chem. Commun.* **2000**, 693-694.
13. Bösmann, A.; Franciò, G.; Janssen, E.; Solinas, M.; Leitner, W.; Wasserscheid, P. *Angew. Chemie, Int. Ed.* **2001**, *40*, 2697-2699.
14. West, K. N.; Bush, D.; Hallett, J. P.; Brown, J. S.; Liotta, C. L.; Eckert, C. A. *Proceedings of 2nd International Meeting on High Pressure Chemical Engineering*; Brunner, G., Ed.; Hamburg, **2001**.
15. Patel, N. C.; Teja, A. S. *Chem. Eng. Sci.* **1982**, *37*, 463-473.
16. Scott, L.; Bush, D.; Brantley, N. H.; Eckert, C. A. *Ind. Eng. Chem. Res.* **1998**, *37*, 4512-4519.
17. Rao, K. V.; Ravi, M. V. A.; Prasad, A. R. *Fluid Phase Equilibria* **1998**, *150-151*, 775-787.
18. Lüring, P.; Schumpe, A. *J. Chem. Eng. Data* **1989**, *34*, 250-252.
19. Yaginuma, R.; Nakajima, T.; Tanaka, H.; Kato, M. *J. Chem. Eng. Data* **1997**, *42*, 814-816.
20. Kordikowski, A.; Schenk, A. P.; van Nielen, R. M.; Peter, C. J. *J. Supercrit. Fluids* **1995**, *8*, 205-216.
21. Fredenslund, A.; Sather, G. A. *J. Chem. Eng. Data* **1970**, *15*, 17-22.
22. Lazzaroni, M.; Bush, D.; Hallett, J. P.; Brown, J. S.; Eckert, C. A.; Gläser, R. *manuscript in preparation* **2002**.
23. Gläser, R.; Josl, R.; Williardt, J. *Top. Catal.* **2002**, *in press*.

Methane Combustion

Chapter 25

Catalytic Combustion of Methane over Co–Mn Mixed Oxides

W. B. Li, Y. Lin, and Y. Zhang

Department of Environmental Science and Engineering, Tsinghua University, Beijing 100084, China

A variety of the Co/Mn mixed oxides has been investigated for catalytic oxidation of methane. The results showed that the catalytic activity was increased on the Co/Mn mixed oxide catalysts in the presence of water vapor. The positive effect of water vapor was also observed after the Co/Mn oxide was doped with cerium oxide or supported on Al_2O_3-SiO_2, and the thermal stability of these catalysts was improved by these treatments to Co/Mn mixed oxides.

Introduction

Natural gas has been used as clean fuels in a variety of combustion processes, such as electrical utilities and motor vehicles. Catalytic combustion of methane without flame can not only improve the energy efficiency but also reduce the formation of thermal nitric oxide by lowering combustion temperatures.

Pd based noble metals have been reported as active catalysts for methane combustion. Besides high cost concerns, supported Pd catalysts usually deactivated by water vapor regardless of the type of the supports used, such as

Pd/Al_2O_3 (*1*), Pd/SnO_2 (*2*) and Pd/ZrO_2 (*3,4*). Most of the research works on metal oxides have been focused on perovskite oxides (*5*), only few studies have been devoted to single oxides such as Au, Ag, Fe, and Mn containing metal oxides partly because of their relatively easy deactivation by sintering. On the other hand, considering a great number of literatures on methane oxidation, inhibition by the reaction products, viz., CO_2 and water has not been given much attention (*1*). It is recently found in our laboratory that the reaction activity of methane combustion on the Co/Mn mixed oxides can be improved by the presence of water vapor at low temperatures (*6*).

In this work, a series of Co/Mn Mixed Metal Oxides was prepared by a sol-gel process, and then the Ce modified Co-Mn oxides and the Co-Mn supported catalysts, Co-Mn/Al_2O_3-SiO_2, were prepared by homogeneous precipitation, respectively. The water effect on methane oxidation was investigated in an attempt to improve the thermal stability of these catalysts in the presence of water vapor.

Experimental

Synthesis of Co/Mn oxides

A citrate solution was added to a mixed aqueous solution of $Mn(CH_3COO)_2 \cdot 4H_2O$ and $Co(NO_3)_2 \cdot 6H_2O$ under strongly stirring conditions. 2M aqueous ammonia was slowly added to the above solution until the pH value reached 6.5-7.0. Subsequently, the above transparent solution was water bathed at 95 °C for 10 h and then dried at 120 °C for 12 h to obtain a gel powder, which was further calcined at 450 °C in nitrogen for 1 h and then in oxygen for next 2 h. The available powders were ready for the oxidation activity measurements after being crushed and sieved. The obtained metal oxide samples are denoted as MnOx, CoMn and $CoMn_2$, in which the subscript number stands for the molar ratio of Mn and Co in the samples.

Synthesis of the cerium doped Co/Mn oxides and the Co/Mn oxide supported catalysts

The cerium doped Co/Mn oxides were prepared in the similar procedure above except that urea was used to precipitate a mixed aqueous solution of $Ce(NO_3)_3$, $Mn(CH_3COO)_2$ and $Co(NO_3)_2$ at 120 °C in an autoclave. The resulting sample was then obtained after being washed by distilled water, dried at 120 °C and calcined at 450 °C or 850 °C for 2 h. The powder sample was denoted as Ce-CoMn, in which the molar ratio of Ce:Co:Mn is 0.1:0.9:1.

The Co/Mn supported catalyst was prepared from a mixed aqueous solution of $Mn(CH_3COO)_2$ and $Co(NO_3)_2$ and urea through adding an appropriate amount of Al_2O_3-SiO_2 (BET surface area is ca. 1000 m^2/g) into the autoclave under the same procedure above. The obtained powder sample was denoted CoMn/SiAl, which contained 10 wt. % Co-Mn oxides with the molar ratio of Co to Mn as being 1:1.

Catalytic activity measurements and characterization of the catalysts

Catalytic activities were measured in a fixed bed quartz tubular reactor. A gas mixture of CH_4 (0.5% by volume), O_2 (1.5% by volume) and argon was passed continuously through a 0.1 g catalyst sample bed with a total flow rate of 100 ml/min of argon. The inlet and outlet gas compositions were analyzed after stepwise changes in the reaction temperatures by on-line gas chromatograph (Shimatzu 14B) with a thermal conductivity detector using a 5A Molecular Sieve column and a Porapak Q column.

The XPS data were acquired with a PHI 5300 ESCA spectrometer with an Al Kα 250 W X-ray source at a pass energy of 37.5 eV. Intensities were estimated by calculating the integral of each peak, and fitting the experimental curve to a Gaussian line of variable proportions. All binding energies (BE) were referred to the C1s signal line at 284.9 eV.

Scanning electron microscope (SEM) was taken on the JSM-6301F instrument with pre-coating samples with gold.

Results and Discussion

Methane oxidation activities on the CoMnOx catalysts

Methane conversion was monitored between 300 °C and 800 °C. Figure 1 shows methane conversion as a function of reaction temperature on the Co-Mn mixed oxides. It is observed that Co containing catalysts gave higher methane conversion than the bare Mn catalyst. The reaction activities at low temperatures followed the order of $CoMn_2$>CoMn>MnOx. For example, methane conversion at 500 °C in the absence of water vapor on $CoMn_2$, CoMn and MnOx was found to be 70%, 59% and 47%, respectively. When adding 5% water vapor into the reaction feed, methane conversion on MnOx was decreased in line with the previous report on Pd based catalysts. However, a remarkable increase in methane conversion was seen over CoMn and $CoMn_2$ in presence of water vapor.

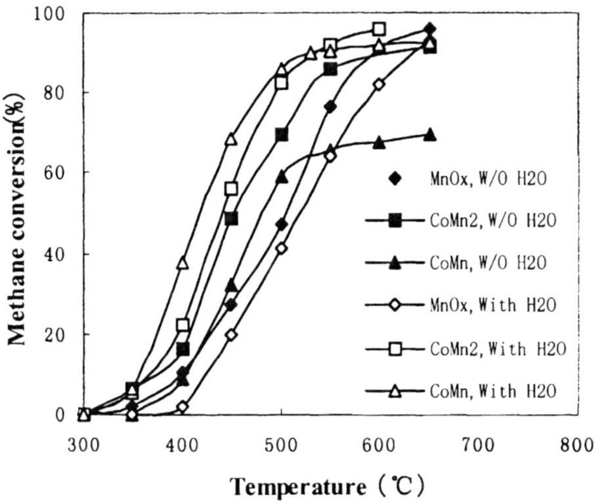

Figure 1. *Methane conversion on Co/Mn mixed metal oxides calcined at 450 ℃ in the absence (solid symbols) and presence (open symbols) of water vapor*

It is shown that methane conversion at 500 °C on $CoMn_2$ was increased from 70% to 83% by the presence of 5% water vapor, and the value on CoMn was increased from 59% to 86% under the similar conditions. It was also noted that such a water enhancement effect became weak at high temperatures. The quantitative data of the XPS spectra on the CoMn sample after pretreatment at 450 °C in air and in the presence of 5% water vapor, respectively, were shown in Table 1. As can be seen, O1s spectra clearly exhibit two strong peaks at 529.63 eV and 531.78 eV, the former peak can be ascribed to lattice oxygen in the form of O^{2-} ion in the oxides, while the latter may relate with adsorbed non-stoichiometric oxygen species in the form of –OH or as O^- [7] associated with surface defect oxides (*8*). Co 2p at 779.98 eV and Mn 2p at 641.74 eV are very close to the values of 780.0 and 641.5 eV for $CoMn_2O_4$. It was obvious that after pretreatment in water vapor the intensity ratio of lower BE O_{1S} peak to higher BE O_{1S} peak became higher, and meanwhile the atomic ratio of Mn to Co became lower, indicating cobalt species could segregate onto the sample surface, which were possibly related with the increase in the activity of methane oxidation upon exposure to water vapor (*6*). The further work to understand the

Table 1. Binding energies (eV) of the CoMn sample

Catalyst pretreatment condition	O_{1s} [a]	O_L/O_S [b]	Mn $2p_{3/2}$	Co $2p_{3/2}$	Mn/Co [c]
at 450℃ in air	529.60 (70.97%) 531.78 (29.03%)	2.44	641.74	779.98	1.59
at 450℃ in 5% H_2O	529.86 (77.74%) 531.89 (22.26%)	3.49	641.79	780.04	0.75

[a] The number in the parenthesis is the molar ratio of the peak intensity

[b] O_L/O_S is the intensity ratio of lower BE O_{1s} peak to higher BE O_{1s} peak.

[c] Mn/Co is the atomic ratio of Mn to Co.

enhancement mechanism may help to develop a type of water-resistant catalysts for methane combustion.

However, when the above Co and Mn containing catalysts were pretreated at 850 °C, the methane oxidation activities were remarkably decreased, e.g., 25% at 750 °C on the Co-Mn catalyst after calcination at 850 °C. The deactivation is believed to be due to the sintering of the active transition metal oxide components. Figure 2 shows the SEM images of the CoMn sample calcined at 450℃ and 850℃ for 3 hours, respectively. As expected, it can be seen that the spherical particles sizing around 60nm were observed on the sample calcined at 450℃; These particles become bigger, i.e., around 450nm upon heating at 850℃, and the aggregation between them was clearly evidenced, which was the possible reason for the deactivation of the Co-Mn catalyst at high temperatures.

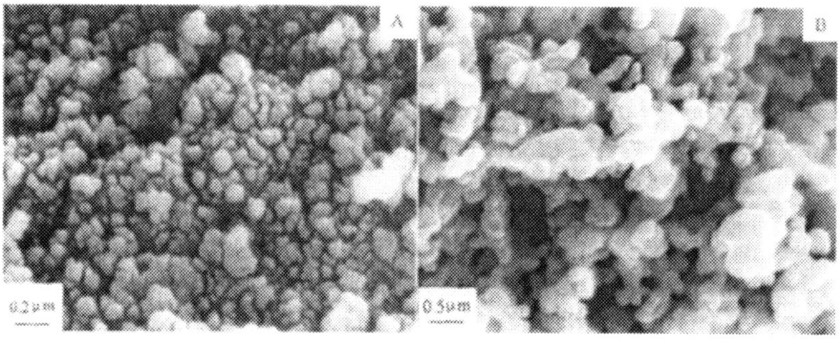

***Figure 2.** SEM images of CoMn calcined at 450°C (left) and 850 (right)*

Methane oxidation activities on the Ce-CoMn catalysts

In order to improve the thermal stability of the Co-Mn mixed oxides, cerium oxide was used to dope the catalyst. Figure 3 shows methane conversion on Ce-CoMn calcined at 850 °C for 2 h. It was observed that a complete oxidation of methane occurred at 720 °C or higher on the cerium-doped catalyst as opposed to <10% of methane conversion on the catalyst without cerium, which indicates addition of cerium components can improve the thermal stability of the Co-Mn catalyst. Meanwhile, the enhancement of the reaction activity was also observed on Ce-CoMn in the presence of 5% water vapor.

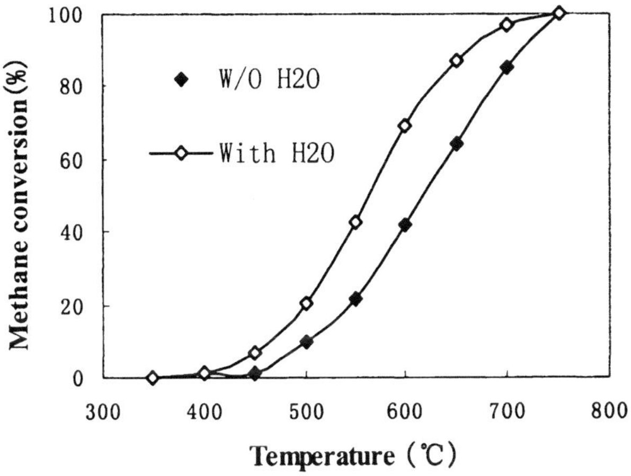

Figure 3. Methane conversion over Ce-CoMn calcined at 850 ℃

Methane oxidation activities on the CoMn/SiAl catalysts

A high surface area Al_2O_3-SiO_2 material was used as a support for the Co-Mn oxide in an attempt to improve the thermal stability. The results were shown in Figure 4. As compared with the results on the CoMn sample in Figure 1, it can be seen that the reaction activities were improved on CoMn/SiAl in the absence or presence of water vapor, which might be related with the better dispersion of Co and Mn species on the support with a high specific surface area. Moreover, after the sample being calcined at 850 °C for 2 h, a similar water enhancement effect was still found on the CoMn/SiAl sample. For example, methane conversion at 700 °C reached ca. 100% in the presence of 5% water vapor as

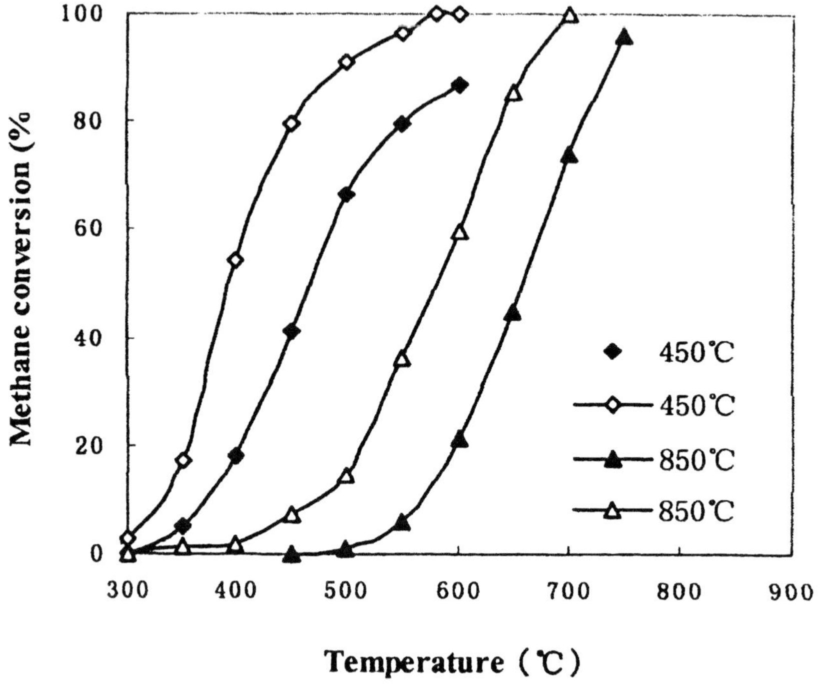

Figure 4. *Methane conversion versus reaction temperature on CoMn/SiAl calcined at 450 ℃ (diamonds) or 850 ℃(triangles) in the absence (solid symbols) or presence (open symbols) of water vapor*

compared to the value of 74% in the absence of water vapor. Clearly, the thermal stability was improved after Co-Mn oxides being supported on an Al_2O_3-SiO_2 support.

The activity of methane oxidation over CoMn/SiAl versus time on stream

Figure 5 shows the changes in the catalyst activity of methane oxidation with time monitored at 500 °C in the presence of 5% water vapor over CoMn/SiAl calcined at 450 °C. It was demonstrated that methane conversion first slightly decreased from 96% to 89% after the reaction running for 1.5 h, and then the conversion was stabilized at ca. 85% for the next 25 h, which was indicating a better thermal stability in the presence of water vapor might be achieved by supporting Co-Mn components onto a support with a high surface area. An attempt to further improve thermal stability of the catalyst is in progress in our laboratory.

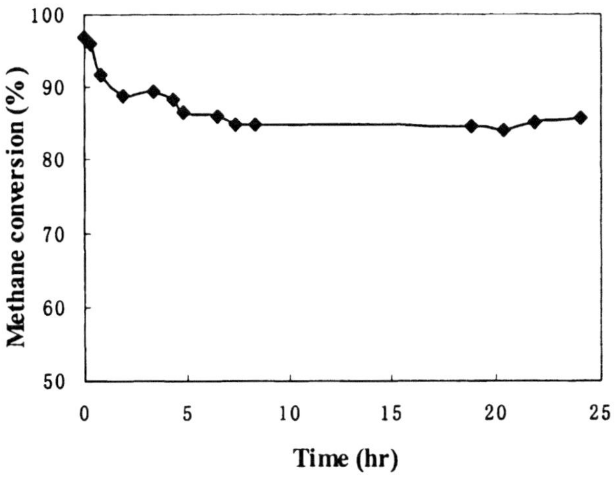

Figure 5. Methane conversion versus reaction time over CoMn/SiAl at 550 ℃ in presence of 5% water vapor.

Conclusion

In summary, a variety of the Co/Mn mixed oxides has been investigated in this presentation for catalytic oxidation of methane. The results showed that the catalytic activity was increased on Co/Mn mixed oxide catalysts by the presence of water vapor. The positive effect of water vapor was also observed after the Co/Mn oxide was doped with cerium oxide or supported on Al_2O_3-SiO_2, and the thermal stability of these catalysts was improved by these treatments to Co/Mn mixed oxides.

Acknowledgment

The work was supported by the Basic Research Foundation of Tsinghua University (NO. JC1999009) and National Natural Science Foundation of China (NO. 29907003).

References

1. Van Giezen, J. C., Van den Berg,F. R., Kleinen,J., Van Dillen, L. A. J., Geus, J. W., *Catal. Today* , **1999**, *47*, 287-293.
2. Sekizawa,K., Widjaja,H., Machida, S., Ozawa, Y., Eguchi,K., *Appl. Catal. A: General*, **2000**, *200*, 211-217.

3. K. Nomura, K. Noro, Y. Nakamura, H. Yoshida, A. Satsuma, and T. Hattori, *Catal. Lett.*, **1999**, *58*, 127-130.
4. K. Fujimoto, F. H. Ribeiro, M. Avalos-Borja, and E. Iglesia, *J. Catal.*, **1998**, *179*, 431-442.
5. T. Seiyama, *Catal. Review*, **1992**, *34*, 281-300.
6. W. B. Li, Y. Lin, *Chem. Lett.* **2002**, *1*, 84-85.
7. M. W. Roberts, *Chem. Soc. Rew.*, **1989**, *18*, 451.
8. L G. Tejuca, A. T. Bell, J. L. G. Fierro and M. A. Pena, *Appl. Surf. Sci.*, **1988**, *31*, 301.

Chapter 26

Supported Pd Catalyst for High-Temperature Methane Combustion: Examining the Combustion Synthesis Preparation Method

M. A. Fraga, M. C. Greca, and L. G. Appel*

Laboratório de Catálise-Instituto Nacional de Tecnologia, Av. Venezuela, 82/sala509, CEP 20081–310 Centro, Rio de Janeiro/RJ, Brazil

The major drawback of the combustion synthesis is that it leads to low surface area materials. The addition of low contents of lanthanides, especially cerium, showed to be effective to enhance the oxides texture. It was also seen that low loading of CeO_2 well dispersed on alumina works as catalyst promoter on methane combustion. Lanthanum does not seem to be a very suitable additive as it reacts with alumina forming $LaAlO_3$.

Catalytic combustion of hydrocarbons, especially methane, has been widely investigated as an alternative technology for power generation (*1,2*). This process is environmentally attractive as it produces less carbon dioxide, a greenhouse gas, per unit of power generated when compared to the combustion of other hydrocarbons. As the combustion reaction takes place on a catalyst surface it reaches much lower temperatures if compared to conventional flame combustion process. As a consequence, low amounts of nitrogen oxides (NO_x) emissions are formed (*2*).

However, as far as the electric power generation involves the use of natural gas-fueled gas turbines, the catalyst characteristics should match the requirements imposed by their operating conditions (*2,3*). High activity at low temperature can be mentioned as a rather important one as it ensures ignition at the compressed air temperature at the catalyst inlet (350 – 410 °C). Moreover, the catalyst should maintain its activity at high temperatures as the adiabatic reaction temperature can reach 1400 °C. Therefore, the use of a thermostable catalyst is compulsory.

Several reports can be found on the performance of different thermally stable systems, which are mostly based on hexaluminates (*4*), perovskites-type oxides of transition metals (*5*) or alumina doped with lanthanides (*1,6,7*) since their stabilizer effect is well known (*8,9*).

Amongst the noble metal-based catalysts, palladium is by far the most studied one (*1,2,6,7,10*). Intensive research has been motivated by the high activities achieved with such systems as well as the high thermal stability of all palladium compounds which are likely to be present during the combustion reaction. Furthermore, the catalyst temperature can be self-controlled due to the PdO/Pd redox process. It has been generally accepted that PdO is the active phase for methane oxidation and, thus, the presence of such sites at low temperatures is desired. After combustion light-off, and a consequent increase in the catalyst temperature due to the adiabatic conditions, such redox process allows the formation of metallic palladium, which is poorly active. Therefore, the rate of the combustion reaction drops steadily until the catalyst temperature allows the reoxidation of Pd to PdO ensuring the combustion re-ignition.

Ceramic oxides obtained by a high temperature process also arise as feasible materials to be used as catalysts supports. Indeed, the preparation of simple ceramic oxides via a route known as combustion synthesis has been gaining reputation as a straightforward process to produce homogeneous, very fine and crystalline powders (*11,12*). The synthesis explores an exothermic, generally very fast and self-sustaining chemical reaction between the desired metal salts and a suitable organic fuel, which is ignited at a temperature much lower than the actual phase formation one. Its key feature is that the heat required to drive the chemical reaction and accomplish the compound synthesis is supplied by the reaction itself and not by an external source. The major drawback of such a route is that it leads to low surface area materials.

In such a context, this work focuses on the preparation of Pd-based catalysts used on the methane combustion whose supports were obtained by combustion synthesis. Alumina was taken as the reference support and the effects brought about by the addition of different lanthanides, namely lanthanum and cerium, in the support texture and the catalytic activity are reported.

Experimental Procedure

Supports Synthesis

To produce a mixed oxide by combustion synthesis, a saturated solution containing the desired cations precursors (nitrates) and a suitable organic fuel (urea) was heated in a wide-mouth vitreous silica basin up to boiling until the mixture ignited and a self-propagating and non-explosive combustion reaction took off. The heat released by the reaction causes the temperature to rise very fast (> 1000 °C) and sustains it, even in the absence of an external source. A large amount of evolved gases results in the formation of a fragile foam that easily crumbles into powder. The basin was then transferred to a muffle furnace preheated at 600 °C and kept for 30 min.

The reactant proportions including the fuel component were calculated based on the desired metal stoichiometry and valency balance in the mixture similarly to propellant chemistry calculations (*13,14*). The supports were prepared in order to obtain oxides with 4 and 12 wt.% of lanthanum or cerium.

Different cerium precursors were used to evaluate the cerium oxidation state influence, namely cerium (III) nitrate ($Ce(NO_3)_3$) and cerium (IV) ammonium nitrate ($(NH_4)_2Ce(NO_3)_6$).

Samples are denoted according to their chemical composition, lanthanide loading and cerium oxidation number, e.g. Al12Ce3 (alumina-ceria mixed oxide containing 12 wt.% Ce prepared from cerium (III) nitrate). Pure Al_2O_3, La_2O_3 and CeO_2 were also prepared by combustion synthesis to be taken as reference.

Catalysts Preparation

The catalysts were prepared by contacting the supports and an aqueous solution of $Pd(NO_3)_2$ in a rotary evaporator for 8 h. The solvent was then slowly evaporated under vacuum at 80 °C; the powder obtained was dried overnight at 110 °C and lastly calcined at 500 °C for 4 h in air. All catalysts were prepared in order to obtain a palladium content of 1 wt.%. These samples will be referred in the same way as their parent supports preceded by Pd, e.g. PdAl12Ce3.

Physical Chemical Characterization

X-ray diffraction (XRD) analyses were carried out just after the powders synthesis in order to identify the oxides phases. The diffractograms were collected in a Rigaku Denki, using Ni-filtered CuK_α radiation. The angular range varied from 15 to 75° with increments of 2°/min. The mean particle size of alumina was calculated from the α-Al_2O_3 (113) reflection using Scherrer equation.

Specific surface areas were determined by the BET method using nitrogen adsorption in a Micromeritics ASAP 2000 equipment.

UV-Vis diffuse reflectance (DRS) spectra in the range of 200 – 800 nm were recorded in a Varian Cary 5 spectrometer using a Harrick diffuse reflectance accessory. Spectra of the catalysts were obtained using pure alumina as a reference. In order to support the species identification, spectra of CeO_2, La_2O_3 and $LaAlO_3$ samples prepared by combustion synthesis were also recorded. Such spectra were collected from a 2% mixture in alumina to overcome specular reflectance.

PdO thermodecomposition was followed by monitoring the oxygen released from the catalysts. Prior to analysis, the samples were treated at 600 °C under helium flow at 50 ml/min for 4 h. The temperature was then raised linearly at a rate of 10 °C/min up to 1000 °C. The oxygen evolution was monitored by a quadrupole mass spectrometer Balzers QMS200.

Catalytic Test

Catalytic tests were performed at atmospheric pressure in a conventional system with fixed bed reactor, monitored by on-line chromatography. Silicon carbide was added for dilution (w_{cat}/w_{dil} = 1/10) in order to avoid temperature gradients in the reactor. A reaction gas mixture containing 2.5% methane and 10% oxygen in a nitrogen balance was fed at 100 ml/min on the catalyst bed (100 mg). Prior to reaction, the catalysts were reduced in situ at 300 °C for 2 h under pure hydrogen flow at 30 ml/min. The activity data were collected just after the catalysts have achieved the steady-state at 500 °C.

Results and Discussion

The synthesized supports were firstly characterized by XRD and their patterns are presented in Figure 1. The lanthanides-containing supports diffractograms exhibited only the reflections of α-Al_2O_3. However, on the high lanthanum load based support (Al12La) a strong disturbance was detected as no α-Al_2O_3 reflection could be observed. No phases related to Ce or La compounds were identified.

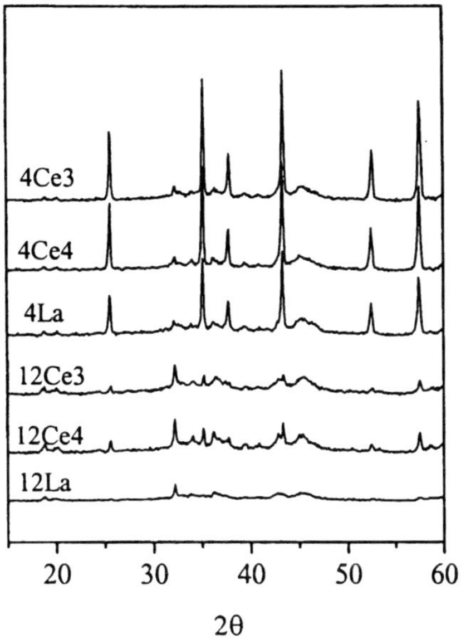

Figure 1 – Diffractograms of the catalysts powders.

Figure 2 displays the UV-Vis spectra of cerium and lanthanum containing samples. The spectra of La-loaded samples (Figure 2a) presented two peaks at 210 and 300 nm. On comparing these spectra with those corresponding to La_2O_3 and $LaAlO_3$ it can be seen that the samples absorbance bands are placed between those values not allowing a faithful distinction between such compounds. The existence of both species, therefore, should not be ruled out.

The spectra of Ce-based supports were identical irrespective the loading or precursor used (Figure 2b). The spectra exhibited basically two peaks centered at 240 and 300 nm, which can be associated with CeO_2. The absorbance at higher wavelength is commonly attributed to three dimensional array of cerium and oxygen atoms while the peak at 240 nm is related to the presence of small crystallites (*15*). Therefore, it can be concluded that the cerium oxide crystallites on the prepared supports are rather small, which is consistent with the previous XRD results that were not able to detect the CeO_2 phase.

Table I collects the BET surface area data of both supports and catalysts as well as the α-Al_2O_3 crystallite average size calculated from diffraction line

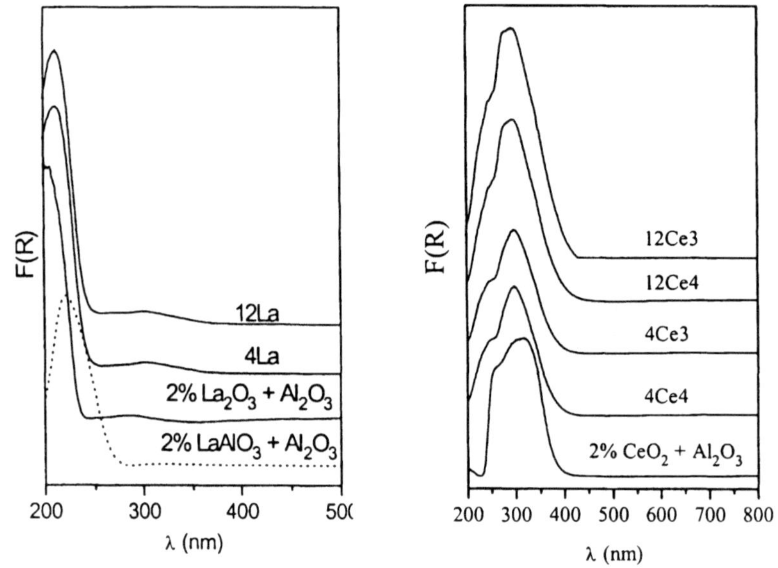

(a) (b)

Figure 2 - UV-Vis spectra of (a) lanthanum and (b) cerium-containing supports.

broadening (2θ = 43.4°). As previously referred, low surface area powders are typically obtained through combustion synthesis method. The values determined by alumina is quite consistent with those reported in the literature (*11,12*).

It can be easily seen that textural changes were induced by adding low contents of lanthanides in the formulation as a dramatic increase in surface area was detected. Such disturbance appears to be more significant when $(NH_4)_2Ce(NO_3)_6$ is used as cerium precursor. The results might suggest that the presence of small CeO_2 crystallites disturbs the crystallization process leading to a decrease in the alumina crystallite size and consequently improving the surface area.

It is seen, on the other hand, that high cerium loading does not interfere the samples textural features, which might be related to the formation of small CeO_2 crystallites agglomerates allowing phase segregation. The occurrence of aggregates of CeO_2 crystallites on alumina has indeed by confirmed by Appel et al in a transmission electron microscopy study (*16*).

As for La-based sample, a completely different behavior was observed; even at high loading the surface area was improved. As no alumina reflections were recorded by XRD it is reasonable to assume the formation of a mixed compound, $LaAlO_3$. Indeed, the formation of such aluminate has been reported in the literature under similar conditions (*17*).

Table I. Surface Area and α-Al_2O_3 Crystallite Size

Sample	*S_{BET} (m^2/g)* *Supports*	*Catalysts*	*d (nm)*
Al	3	3	36
Al4La	23	22	25
Al4Ce3	14	12	28
Al4Ce4	28	24	27
Al12La	20	18	-
Al12Ce3	5	5	37
Al12Ce4	7	6	42

As was mentioned, the PdO/Pd redox process is one interesting feature of methane combustion catalysts. In this work, the PdO decomposition to Pd was monitored by mass spectrometry and the catalysts patterns are displayed in Figure 3. Different thermodecomposition profiles were obtained for all samples and clearly show different degrees of PdO stabilization according to the support used. It should be pointed out, though, that the PdO reduction region falls under 800 °C for almost all systems studied. Temperatures within 800 – 900 °C have been generally reported in the literature over Pd-based systems (*1,6,7*). The main difference should then be firstly connected with the analysis conditions used in this work, besides the intrinsic differences of such catalysts. The PdO thermodecomposition profiles were collected under an inert gas flow, in the absence of oxygen, what provokes an equilibrium shift towards metallic palladium; such a shift is reflected as low decomposition temperatures.

Only one single peak at around 745 °C was observed over PdAl sample, which could be related to the decomposition of bulk PdO. A similar sharp peak was detected for the catalysts prepared over the support with high lanthanum loading (PdAl12La) whereas a rather distinct pattern is seen for the low La-containing sample (PdAl4La). A broad undefined peak is only detected at temperature above 900 °C, which might suggest a strong PdO stabilization effect.

Initially, the distinct behavior observed over the La-loaded samples could highlight some discrepancies. However, these results evidence that different supports were obtained according to the lanthanum content used in the preparation step. A weakly interacting support, probably $LaAlO_3$ as suggested

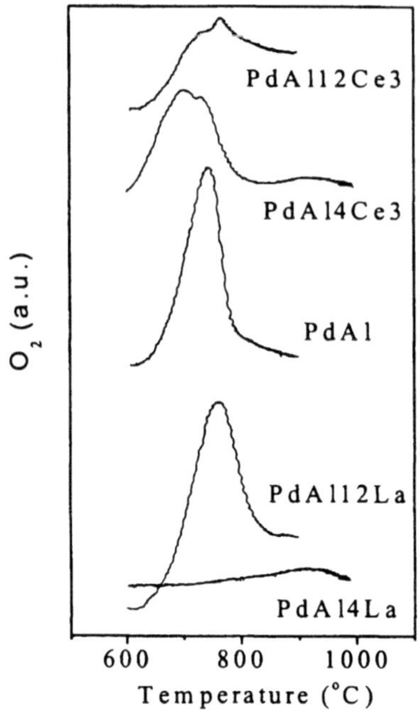

Figure 3 – PdO thermodecomposition profiles.

so far, would be responsible for the PdO decomposition at low temperature (~750 °C) whereas a stabilizing effect is brought about by a La-doped alumina (PdAl4La).

The addition of lanthanides as PdO stabilizer has indeed been reported by some authors (*6,7,18*). As far as lanthanum is concerned, however, the literature data are controversial. No effect was observed on catalyst supported on La-doped alumina (6,*7*), while, in contrast, lanthanum was seen to postpone PdO decomposition and its effect was more pronounced the higher the lanthanum content. No decomposition could even be detected over catalyst supported on La_2O_3 (*18*).

The patterns corresponding to Ce-containing samples are even more interesting. The occurrence of multiple decomposition peaks indicates the presence of more than one PdO species. Taking into account the existence of CeO_2 small crystallites as suggested so far, PdO supported on CeO_2 (PdO/CeO_2)

should be considered. These species would display an easy PdO decomposition due to CeO_2. Some authors have indeed reported such a behavior by comparing Pd/CeO_2 and Pd/Al_2O_3 catalysts (*6,7*). The occurrence of Pd/Al_2O_3 species, on the other hand, would guarantee the higher temperatures decomposition peak.

The slight differences observed by comparing the profiles corresponding to low and high cerium content could probably be bound to their morphological differences.

Catalytic activity data for methane combustion are collected in Table II. It was characterized by the temperatures at which methane conversion achieved 5% and 10% and is given as T_5 and T_{10} respectively. It should be stressed that carbon dioxide and water were the only reaction products detected.

Table II. Methane Combustion over Pd Catalysts

Catalyst	*Catalytic Activity*	
	T_5 (°C)	*T_{10} (°C)*
PdAl	483	-
PdAl4La	441	473
PdAl4Ce3	415	447
PdAl4Ce4	429	460
PdAl12La	470	-
PdAl12Ce3	441	473
PdAl12Ce4	440	472

It can be seen that the catalytic activity markedly depends on the nature of the support. Pd/Al_2O_3 catalyst presents low activity, which could be related to the specific features of such system, especially the rather low surface area, as mentioned above.

The catalytic activity of the La-based systems is dependent on the lanthanum concentration, being favored at low lanthanum loading. As a matter of fact, comparing both catalysts, the difference in the temperature to reach methane conversion of 5% is quite significant, practically 30 °C.

Similar DRS spectra obtained for both samples suggested the presence of the same La species, which would be naturally associated with $LaAlO_3$ based on the XRD results. Nevertheless, the PdO thermodecomposition profiles and catalytic performances do not support such assignment. The activity data evidences a different La species distribution. At low concentration (PdAl4La) the promoting effect on the methane combustion may arise from the La-doped alumina support. On the other hand, the poor performance registered over PdAl12La might be ascribed to the nature of the support, that is, the intrinsic activity of the $LaAlO_3$-supported catalyst.

Cerium revealed to be a better promoter than lanthanum, which is in close agreement with the literature regarding the promoting effects of CeO_2 in noble metal-based combustion catalysts (*19*). The different performance observed amongst the catalysts with low and high contents could be explained by the different distribution of CeO_2 crystallites on the alumina support as previously suggested. High concentration seems to lead to a CeO_2 and Al_2O_3 phase segregation which would allow the existence of different sites, namely Pd/Al_2O_3 and Pd/CeO_2. It is reasonable, thus, that the catalytic activity diminishes due to the intrinsic behavior of Al_2O_3-supported catalyst. As the segregation occurs as a consequence of lanthanide loading, it may be expected that whatever the cerium precursor used, the prepared catalyst will present the same behavior. The results obtained over PdAl12Ce3 and PdAl12Ce4 indeed follow such a trend.

As for PdAl4Ce samples, a model of small CeO_2 crystallites uniformly spread on alumina could be accepted. Such a model is in line with the support textural modifications previously observed (Table I). Nevertheless, differently from what is seen over PdAl12Ce3 sample, the catalytic performance is fairly distinct between samples with low Ce concentration prepared from different salt precursors (PdAl4Ce3 and PdAl4Ce4). Indeed it has been reported that the formation of CeO_2 aggregate crystallites on alumina surface is connected with the salt precursor (*16*). Cerium (IV) ammonium nitrate induces higher heterogeneity for the distribution of cerium on the alumina surface. The formation of small aggregates crystallites of CeO_2 has been observed at concentrations as low as 1.7 wt.% (*16*). Hence, taking into account the difference in CeO_2 dispersion, its promoting effects is softened on PdAl4Ce4 catalyst.

Conclusion

The texture of oxides prepared by combustion synthesis can be effectively enhanced by adding low contents of lanthanides, especially cerium. Lanthanum does not seem to be a very suitable additive as it reacts with alumina forming $LaAlO_3$.

Low loading of cerium oxide well dispersed on alumina revealed to be the most effective promoter for the methane combustion.

Acknowledgements

The authors are indebted to Mr. F. Cano and IME for carrying out the BET measurements. The financial support from CTPETRO/FINEP is gratefully acknowledged.

References

1. Groppi, G., Cristiani, C., Lietti, L., Ramella, C., Valentini, M., Forzatti, P. *Catal. Today* **1999**, *50*, 399-412.
2. Garten, R.L., Dalla Betta, R.A., Schlatter, J.C. In *Handbook of Heterogeneous Catalysis*; Ertl, G., Knözinger, H., Weitkamp, J., Ed.; Wiley-VCH: New York, 1997; Vol. 4, pp. 1668-1677.
3. Beebe, K.W., Cairns, K.D., Pareek, V.K., Nickolas, S.G., Schlatter, J.C., Tsuchiya, T. *Catal. Today* **2000**, 59, 95-115.
4. Sekizawa, K., Machida, M., Eguchi, K., Arai, H. *J. Catal.* **1993**, 142, 655-663.
5. Seiyama, T. *Catal. Rev. – Sci. Eng.* **1992**, 34, 281-300.
6. Kenelly, T., Farrauto, R.J. US Patent 5,216,875, 1993.
7. Kenelly, T., Farrauto, R.J. US Patent 5,214,912, 1993.
8. Burtin, P., Brunelle, J.P., Pijolat, M., Soustelle, M. *Appl. Catal.* **1987**, 34, 225-238.
9. Braun, S., Appel, L.G., Zinner, L.B., Schmal, M. *British Ceramic Transactions* **1999**, 98, 77-85.
10. Chin, Y.H., Resasco, D.E. In *Catalysis*; Spivey, J.J., Ed.; The Royal Society of Chemistry, 1999; Vol. 14, 3-39.
11. Greca, M.C., Moraes, C., Morelli, M.R., Segadães, A.M. *Appl. Catal. A* **1999**, *179*, 87-92.
12. Greca, M.C., Moraes, C., Segadães, A.M. *Appl. Catal. A* **2001**, *216*, 267-276.
13. Fumo, D.A., Morelli, M.R., Segadães, A.M. *Mater. Res. Bull.* **1996**, 31 1243-1255.
14. Jain, S.R., Adiga, K.C., Pai Verneker, V.R. *Combustion and Flame* **1981**, 40, 71-79.
15. Bensalem, A., Muller, J.C., Bozon-Verduraz, F. *J. Chem. Soc. Faraday Trans.* **1992**, *88*, 153-154.
16. Appel, L.G.; Frydman, A.; Perez, C.A.C.; Eon, J.G.; Castner, D.G.; Campbell, C.T.; Schmal, M. *Phys. Stat. Sol. B* **1995**, *192*, 477-491.
17. Kingsley, J.J.; Patil, K.C. *Mater. Lett.* **1988**, *6*, 427-432.
18. Chou, T.Y., Leu, C.H., Yeah, C.T. *Catal. Today* **1995**, *26*, 53-58.
19. Trovarelli, A. *Catal. Rev.- Sci. Eng.* **1996**, *38*, 439-520.

Author Index

Subject Index

B

C

D

E

I

K

L

M

N

O

P

Q

R

S

T

U

V

W

X

Y

Z

Printed in the United Kingdom
by Lightning Source UK Ltd.
133310UK00001BA/18/A

9 780841 238275